I0055211

Failure-Free Integrated Circuit Packages

Systematic Elimination of Failures through Reliability Engineering, Failure Analysis, and Material Improvements

Charles Cohn *Editor*
Agere Systems
Allentown, Pennsylvania

Charles A. Harper *Editor*
Technology Seminars, Inc.
Lutherville, Maryland

McGraw-Hill

New York Chicago San Francisco Lisbon London Madrid
Mexico City Milan New Delhi San Juan Seoul
Singapore Sydney Toronto

The McGraw-Hill Companies

CIP Data is on file with the Library of Congress

Copyright © 2005 by The McGraw-Hill Companies, Inc. All rights reserved. Printed in the United States of America. Except as permitted under the United States Copyright Act of 1976, no part of this publication may be reproduced or distributed in any form or by any means, or stored in a data base or retrieval system, without the prior written permission of the publisher.

1 2 3 4 5 6 7 8 9 0 DOC/DOC 0 1 0 9 8 7 6 5 4

ISBN 0-07-143484-4

The sponsoring editor for this book was Stephen S. Chapman and the production supervisor was Pamela A. Pelton. It was set in Century Schoolbook by J. K. Eckert & Company, Inc. The art director for the cover was Anthony Landi.

Printed and bound by RR Donnelley.

McGraw-Hill books are available at special quantity discounts to use as premiums and sales promotions, or for use in corporate training programs. For more information, please write to the Director of Special Sales, McGraw-Hill Professional, Two Penn Plaza, New York, NY 10121-2298. Or contact your local bookstore.

This book is printed on recycled, acid-free paper containing a minimum of 50% recycled, de-inked fiber.

Information contained in this work has been obtained by The McGraw-Hill Companies, Inc. ("McGraw-Hill") from sources believed to be reliable. However, neither McGraw-Hill nor its authors guarantee the accuracy or completeness of any information published herein and neither McGraw-Hill nor its authors shall be responsible for any errors, omissions, or damages arising out of use of this information. This work is published with the understanding that McGraw-Hill and its authors are supplying information but are not attempting to render engineering or other professional services. If such services are required, the assistance of an appropriate professional should be sought.

Contents

Foreword

The authors of this book are employees of Agere Systems, a leader in communications, data trafficking, and storage integrated circuits. Some authors are as new to Agere as Agere is new to the stock market. Some have tenures of over 30 years, riding out the winds of change that transformed portions of the original 100-year-old AT&T, which spun off the manufacturing leg, Lucent Technologies, which subsequently spun off of the microelectronics division, now called Agere Systems. The semiconductor manufacturing division of Lucent changed its name to Agere Systems in December of 2000. By the beginning of 2001, the new company initial public offering was made at the crest of one of the worst economic downturns in the semiconductor industry to date.

The worldwide electronics industry has experienced a structural change within every segment of the supply chain over the last two years. Agere was no exception. Semiconductor industry growth rates plummeted from +37% to –27% from 2000 to 2001.* It has been projected that more than half of the world's silicon fabrication will come out of Asia in the not-so-distant future, primarily fueled by the growth of foundry manufacturers such as United Microelectronics Corp. (UMC) and Taiwan Semiconductor Manufacturing Co. (TSMC). The increasing cost of maintaining captive fabs; the technology shift toward 300 mm, copper, and low-k; and the increasing skill and competence of the foundry players narrowed the integrated device manufacturer (IDM) playing field. The strategy selected for the fledgling company, Agere, coming from an almost entirely vertically integrated parent company, was to move toward a fab-less or "fab-light" operating strategy. Needless to say, many things changed within Agere. Similar

Source: The Electronics Industry Report, Prismark, 2002.

to the rest of the industry, Agere underwent a restructuring to size itself more in accordance with the more modest single-digit semiconductor industry growth rates of 2002. A greater emphasis was placed on outsourcing of services and supplies. The supply chain has become relatively complex, both logistically and technically. The increasing trend toward outsourcing affected and continues to affect every facet of the company, including capital expenditures on analytical equipment and laboratories. But one thing stayed the same: the company's commitment to the quality of its product. This book is all about the quality of that product and striving to be the best semiconductor component supplier on the globe.

While outsourcing of processes and services that are not key differentiators for the company has become more pronounced, there is no reprieve or shared ownership for the quality of the product. For example, it is not uncommon for the Si designers of the original design company to design proprietary segments of the design and drop those segments into standard libraries supplied by a selected foundry. Simultaneous to the Si design activity, the original design company may work with a package design subcontractor to physically layout the package substrate. The Si may be fabricated in Taiwan at a subcontracted manufacturer who may also probe the wafers. Necessarily, the original design company must work with the subcontracted manufacturer to enable functional test at probe. While this is taking place, the package substrates can be fabricated in Japan. From this point, the wafers and the substrates may ship to a separate assembly subcontractor in the Philippines, where the wafers are thinned, diced, and assembled into packages. The assembly subcontractor may perform final product test or the product may return to the original design company for product test. At each step along the supply chain journey, processes must be statistically controlled, raw material quality must be assured, and shipping and handling must take place expediently and without incurring defects in the product. A number of qualified preferred suppliers are capable of controlling their Si fab and assembly processes, and they stand behind the quality of their service and the materials that go into their processes. However, at the end of the line, the quality of the packaged integrated circuit is the responsibility of the original design company. Quality assurance is made possible by (1) having intimate manufacturing knowledge within the product company, even if they are not manufacturing the product, (2) having materials expertise even if the product company does not make the materials, (3) not only logistically but technically managing the supply partners, and (4) having those skilled in the science of failure analysis and reliability engineering internal to the company.

The authors from Agere are pleased to share some of their experiences and knowledge from this exciting field of failure analysis of integrated circuit packaging. As we all know, there are infinite ways for failures to occur. There are only several ways for one to be successful. The technical challenge is to pull the pieces of the puzzle together to predict and discover the reliable combination of materials and processes.

Melissa E. Grupen-Schemansky, Ph.D.

Preface

It is a pleasure to present this new book, "*Failure Free Integrated Circuit Packages,*" as a valuable addition to the McGraw-Hill *Electronic Packaging and Interconnection Series.* While the numerous books in this series cover the complete spectrum of subjects in the electronic packaging and interconnection field, the packaging and interconnection of integrated circuits (ICs) is surely the heart of the system. Further, the complexity and the multitude of integrated circuit packages, and the logistics of their manufacture and assembly, have made failures and failure analysis of these packages the critical limiting factor in achieving high reliability in modern electronic systems. The numerous channels of outsourcing IC assembly only serve to make these problems more critical. This book thoroughly details all of the important aspects of the systematic elimination of failures in integrated circuit packages through the judicious use of reliability engineering, failure analysis, and material improvements. It is a must for most bookshelves in all areas of the electronics industry.

The book would not have been possible without the exceptional technical depth and experience of the chapter authors. The chapters comprehensively cover all of the critical areas of complex package technologies, reliability, failure analysis, and materials for achieving failure-free integrated circuit packages. The chapters have been arranged for best reader convenience, and with an extensive array of practical, easy-to-understand illustrations and examples. These could, in fact, be readily used as in-plant guides for quality control and failure analysis. Sequential chapters offer comprehensive and clear treatments of package technologies, physics and chemistry of failures, strategies for locating failures, failure analysis techniques, examples of failure modes in organic integrated circuit packages, and material selection and considerations. Preceding the comprehensive presentation in these chapters is an introductory chapter by Dr. Melissa E. Grupen-Shemansky, which clearly explains the subject material covered in each chapter and the interrelationship among the chapters. It is highly recommended that the reader review this introductory chapter before reading the individual chapters.

We would like to extend special credit to all the chapter authors for their time and effort to make this book a reality. We are confident that their combined presentations in this book will find broad use and provide great benefits for major segments of the electronics industry.

Lastly, a book of this scope would not have been possible without the help and support of many other people. We would like to acknowledge the tremendous amount of assistance we received from Ahmed Amin for his failure analysis (FA) work and for providing the images for many of the failures shown in this book. The materials section would not have been possible without the input from Chris Horvath. We are indebted to Ed Nease for his advice on the design file work, Jeffery Gilbert for his input on processing, Brian Vaccaro for his help in the finite element analysis (FEA) sections, and Jeff Klemovage for his outstanding contribution in creating many of the line figures. The authors would also like to acknowledge Ebby Thomas for providing shadow moiré data and images. Finally, we are especially thankful to the engineers at Advanced Semiconductor Engineering (ASE) in Kaohsiung, Taiwan, for their FEA work in portions of Chapter 5.

Charles Cohn
Charles A. Harper

About the Contributors

Joze E. Antol. Member of Technical Staff in the Packaging and Interconnect Technology Dept. at Agere Systems, Allentown, PA. She has worked at AT&T Bell Labs/Lucent Technologies/Agere Systems for 20 years. Her current assignment focuses on the effects of probe and packaging on advanced IC technologies, such as 130-nm copper/low-K. Previous assignments at AT&T/Lucent included die/wire bond process engineer in manufacturing and laboratory analyst responsible for ICP emission spectroscopy. She holds an AS degree in Chemical Engineering from Pennsylvania State University.

Frank A. Baiocchi. Distinguished Member of Technical Staff at Agere Systems, Allentown, PA. Dr. Baiocchi has been with AT&T/Lucent Technologies/Agere Systems for 21 years and is currently an analyst in the Product Analysis Laboratory, involved in failure site analysis on a wide variety of analog and digital technologies. He has had responsibility for plasma etch process development in support of a CMOS manufacturing line and developed an LDMOS wafer fabrication process for an RF power transistor product. His experience includes materials characterization for silicon and III–V based semiconductor devices, plasma process engineering, and silicon process technology. He has published more than 40 technical articles on analysis of thin film materials using accelerator-based techniques such as Rutherford backscattering. Dr. Baiocchi holds a BS degree in Chemistry from DePaul University and a Ph.D. degree in Physical Chemistry from Harvard University.

Anthony J. Bucha. Senior Member of Technical Staff at Agere Systems, Allentown, PA. Mr. Bucha has been with Western Electric/AT&T Technologies/Lucent Technologies and Agere Systems for 24 years and is currently a Senior Analyst in the Product Analysis Lab. His responsibilities include product analysis for the reliability, qualification, and yield improvement programs as well as administration of the labs safety program and ISO 9001/2000 standards and documentation implementation. Previous work included optoelectronic device analysis, wafer yield improvement, and process characterization. He has written and co-authored several papers on product analysis and deprocessing techniques, some which were presented at various conferences including the International Symposium for Testing and Failure Analysis. Mr. Bucha was an active member of SEMATECH International and the International Symposium for Testing and Failure Analysis Society.

JAMES T. CARGO. Technical Manager at Agere Systems, Allentown, PA. Mr. Cargo has been with AT&T Bell Labs/Lucent Technologies/Agere Systems for 19 years. He started his career in AT&T Bell Lab's Integrated Circuits Technology R&D Laboratory. Over the years, he has worked in several capacities including ultra trace analytical chemistry, co-founding AT&T Analytical Services, focused ion beam work, and failure analysis as they pertain to integrated circuits, optoelectronic components, and MEMS devices. He has been actively involved in the Sematech Product Analysis Forum (PAF) and Sematech Assembly Analysis Forum (AAF). In addition, he has authored or coauthored more than 30 research papers and has 3 patents. Currently, Mr. Cargo manages Agere System's Product Analysis Lab, which is responsible for integrated circuit/package analysis associated with customer returns, engineering studies, qualification, reliability, yield improvement, design modifications, and latch-up evaluation. Mr. Cargo holds a BS degree in Chemistry from Syracuse University, an MS degree in Chemistry, and an MBA from Lehigh University

CHARLES COHN. Distinguished Member of Technical Staff at Agere Systems, Allentown, PA. Mr. Cohn has been with AT&T Bell Labs/Lucent Technologies/Agere Systems for 21 years and is currently a substrate technologist in the Packaging and Interconnect Technology Department. He is the lead resource on PCB technology, supporting the development of advanced organic PBGA substrates for wire bonded and flip chip IC interconnections. He maintains close relationships with a broad spectrum of organic substrate suppliers worldwide to continually assess their latest technologies and manufacturing capabilities. Prior to joining AT&T, Mr. Cohn was an engineering specialist at the Singer/Kearfott company involved in electronic packaging and thermal design of airborne electronic modules and inertial navigation systems for the military. He has authored chapters in several McGraw-Hill Electronic Packaging Series handbooks, and presented numerous papers on electronic packaging. He was awarded 11

U.S. patents on IC packaging construction and thermal enhancements. Mr. Cohn holds BS, MS, and ME degrees in Mechanical Engineering from Columbia University.

JOHN M. DELUCCA. Member of Technical Staff at Agere Systems, Allentown, PA. Dr. DeLucca has been with Lucent Technologies/Agere Systems for four years and is currently a product and failure analyst in the Product Analysis Laboratory. He is the lead resource for transmission electron microscopy issues supporting customer returns, developmental work, and intellectual property-related issues. Prior to joining the product analysis group, Dr. DeLucca worked in several capacities, including supply line management, reliability and qualification, and, more recently, customer technical support for optoelectronic products. Dr. DeLucca graduated in 2000 with a Ph.D. in Materials Engineering from the Pennsylvania State University, from which he also holds an MS degree in Materials. He received a BSE in Materials Science and Engineering from the University of Pennsylvania in 1994. He has written and presented several papers on processing and characterization of wide-bandgap semiconductors and was awarded a Materials Research Society Graduate Student award in 1998.

BARRY J. DUTT. Distinguished Member of Technical Staff at Agere Systems, Allentown, PA. Mr. Dutt has been with AT&T Bell Labs/Lucent Technologies/Agere Systems for 23 years. For the past 14 years, he has worked in failure analysis and is currently a senior analyst in the Product Analysis Laboratory. He routinely performs product analysis on qualification, reliability, and manufacturing failures as well as customer returns and engineering evaluations related to design and process development. Previous analysis activities have also included optoelectronic and MEMS devices. While at Bell Laboratories, he was a member of the Digital Signal Processor design team, working in the areas of circuit simulation debug and circuit design. He has co-authored several papers and presented at the International Symposium for Testing and Failure Analysis.

JASON P. GOODELLE. Distinguished Member of Technical Staff at Agere Systems, Allentown, PA, in the Packaging and Interconnect Technology Dept. In his six years with Agere Systems, he has been responsible for developing several advanced package assembly solutions (including flip chip and multichip modules) both internally and through cooperative development efforts at various subcontractors. Dr. Goodelle obtained a BS in physics from Allegheny College and a Ph.D. in Materials Science and Engineering from Lehigh University, specializing in mechanical and physical properties of bulk and thin film polymers. Dr. Goodelle has authored or co-authored several papers in the area of packaging solutions and materials for IC packaging.

KULTARANSINGH N. HOOGHAN. Senior Member of Technical Staff at Agere Systems, Allentown, PA. Mr. Hooghan has been with Lucent Technologies/Agere Systems for seven years as a Failure Analyst in the Product Analysis Dept. His primary responsibility is carrying out Physical Failure Analyses using a Focused Ion Beam (FIB) system. He has three U.S. patents pending on processes for FIB systems. Mr. Hooghan published numerous papers and authored a chapter in a book titled *Introduction to Focused Ion Beams: Instrumentation, Theory, Techniques, and Practice.* He also presented seminars related to FIB Systems. He holds a Masters Degree in Physics from the Bombay University, India, and a Masters Degree in Engineering Technology from the University of North Texas.

JOHN W. OSENBACH. Consulting Member of Technical Staff at Agere Systems, Allentown, PA. Dr. Osenbach has been with AT&T Bell Labs/Lucent Technologies/Agere Systems for 22 years. The first half was spent in the development and reliability of silicon technology processes and materials, primarily dielectrics and metals for application on high-voltage, high-speed CMOS devices. The next 10 years were devoted to metals, dielectrics, and package development for single-mode laser-based products. He helped develop a device and package technology that provided, for the first time, a reliable nonhermetic single-mode laser and photodiode package technology. Currently, Dr. Osenbach is focused on package technology development for silicon integrated circuits. He has authored and co-authored more than 70 papers and was awarded 33 U.S. patents. He organized and chaired a number of conferences on silicon technology and optoelectronic packaging and taught conference courses on reliability of optoelectronic devices. He is a past associate editor of the *Journal of the Electrochemical Society.* In 1999, he was made a Bell Laboratories Fellow for his work on the development of materials and processes needed for the manufacture of reliable nonhermetic optoelectronic and SIC packages. He holds a BS, MS, and Ph.D. in Ceramic Science and Engineering from Pennsylvania State University.

ALBERT C. SEIER. Member of Technical Staff at Agere Systems, Allentown, PA. Mr. Seier has been with Lucent/Agere for six years and is currently a product analyst focusing on Agere worldwide electrical failure analysis needs with emphasis on electrical hardware. In addition, he interfaces with quality managers and lab analysts to ensure timely closure on customer returns. Managing laboratory case data entry is also among his responsibilities. Prior to joining Agere, Mr.

Seier worked in research and development of electrical connectors for Thomas and Betts Corporation, where he sat on the ANSI C119.4 standards committee. He co-authored a paper for IPFA2003 entitled "Overview of Cu/low-K Technology Failure Analysis and Reliability Issues." Mr. Seier holds a BSEE from Lafayette College.

BRIAN A. SENSENIG. Senior Manager, Agere Systems Singapore Pte, Ltd., Singapore. Mr. Sensenig has been with Lucent Technologies/Agere Systems for five years and is currently a Senior Manager in the Global Procurement Organization headquartered in Singapore. He is responsible for the External Contract Manufacturing Assembly and Test services for Agere. Prior relocating to Singapore in September 2003, Mr. Sensenig worked in Agere's headquarters in Allentown, PA, as the lead procurement resource providing purchasing counsel and support to Agere's Packaging and Interconnect Team for new package technologies. He maintains close relationships with a broad spectrum of organic and SATS suppliers worldwide to continually assess their latest technologies and manufacturing capabilities. Mr. Sensenig holds an MBA from DeSales University and is an Approved Purchasing Practitioner (A.P.P.) and Certified Purchasing Manager (C.P.M.) from the Institute for Supply Management (ISM).

JOSEPH W. SERPIELLO. Distinguished Member of Technical Staff at Agere Systems, Allentown, PA. Mr. Serpiello has been affiliated with Bell Labs, AT&T Microelectronics, Lucent Technologies, and Agere Systems over the past 33 years, including 3 years with Solid State Scientific as a Senior Design Engineer in the Memory Products Group involved in high-performance CMOS SRAM design. For the past 18 years, he has worked in failure analysis and is currently a Senior Analyst in the Product Analysis Laboratory at Agere. He routinely performs failure analysis on qualification, reliability, and manufacturing failures as well as customer returns and engineering evaluations related to design and process development. While at Bell Labs, he was a member of the Logic and Memory Devices Group, involved in the areas of automated testing, parameter extraction, device modeling, alpha-radiation measurements of materials, circuit simulation, and SRAM design. He was a former member of the SEMATECH Product Analysis Forum (PAF) and has co-authored several papers that were presented at the International Symposium for Testing and Failure Analysis (ISTFA) and the International Reliability Physics Symposium (IRPS). Mr. Serpiello holds degrees in electrical engineering from Temple University and Lafayette College (BSEE).

MELISSA E. GRUPEN-SHEMANSKY. Director of Packaging and Interconnection Technology Department at Agere Systems, Allentown, PA.

Dr. Grupen-Shemansky has been with Agere Systems for five years, managing high-volume process technology development teams at the Agere factories in Thailand and Singapore, design teams in five locations worldwide, and a core packaging and interconnect technology team in the United States. Prior to joining Agere, she had been with Motorola for ten years in the Semiconductor Products Sector R&D Labs and Manufacturing. Dr. Grupen-Shemansky's research responsibilities involved the development of selective polycrystalline silicon CVD techniques, GaAs RIE, wafer thinning, and wafer bonding. In 1992, she became a fab section manager within the Compound Semiconductor-1 GaAs device manufacturing line. Subsequently, her technical interests shifted to the development of IC and MEMS packaging, which called for collaborations with Motorola-Europe and Motorola-Asia. Dr. Grupen-Shemansky received combined BS and MS degrees in Chemical Engineering from the Pennsylvania State University and her Ph.D. from Arizona State University. Her academic research activities were interdisciplinary combining biomedical, electrical, and chemical engineering skills. She received various technical awards throughout her career, has been issued seven U.S. patents, and has over 30 publications in several technical fields.

RICHARD L. SHOOK. Consulting Member of Technical Staff at Agere Systems, Allentown, PA, in the Packaging and Interconnect Technology Department. Dr. Shook has been with AT&T Bell Labs/Lucent Technologies/Agere Systems for 20 years, where his responsibilities include IC package development and reliability assessment as well as moisture/reflow sensitivity issues. Recent research activities involved Pb-free assembly effects associated with package body warpage and Sn-whisker growth. He has contributed to the U.S. industry standardization of moisture/reflow testing through active participation on JEDEC and IPC committees. Some of Dr. Shook's technical contributions have been the development of the accelerated 60°C/60% RH test criteria incorporated in the joint IPC/JEDEC moisture sensitivity specification J-STD-020 and the development of moisture derating and baking procedures adopted in the joint handling specification J-STD-033. He has published extensively on the topic of moisture/reflow sensitivity and has received an outstanding paper award at the 1999 ECTC conference. He also received JEDEC and IPC technical recognition awards for his work on this topic. Dr. Shook has a Ph.D. in Metallurgical Engineering from the Ohio State University.

RONALD J. WEACHOCK. Senior Member of Technical Staff in the Product Realization Department at Agere Systems, Allentown, PA. Mr. Weachock has been with Western Electric/AT&T Bell Labs/Lucent

Technologies/Agere Systems for 20 years. He has worked extensively as a Failure Analysis Engineer for Qualification and Reliability of application specific integrated circuits (ASICs). He is currently involved in semiconductor root cause failure analysis in the field of advanced packaging and interconnect technologies, including standard packages, ball grid arrays (BGAs), advanced single-chip thermally enhanced packaging, and multichip modules (MCMs). He specializes in defect analysis of high-pin-count flip chip package/interconnects and defect analysis for reliability assessment of advanced electronic package interconnects systems for new product. He has coauthored "Assessment of Flip Chip Interconnect Integrity Using Scanning Acoustic Microscopy" and contributed to several technical papers in the area of electronic-packaging-stress-related-issues. He received his BSEET from Pennsylvania State University.

HUIXIAN WU. Member of Technical Staff at Agere Systems, Allentown, PA. Ms. Wu has been with Agere Systems for four years, working on advanced FA development and integrated circuit/package product analysis associated with technology development, customer returns, qualification, reliability, yield improvement, and design modifications. Ms. Wu received her BS degree both in Material Science and Engineering and Computer Science and Engineering with highest honor in 1996, an MS degree in Semiconductor Materials in 1998 from the Zhejiang University in China, and another MS degree in Reliability Engineering from the University of Maryland in 2000. She is a now a Ph.D. candidate in the Department of Electrical and Computer Engineering at Lehigh University. Ms. Wu has authored or coauthored 18 research papers.

WEIDONG XIE. Senior Member of Technical Staff at Agere Systems, Allentown, PA, in the Packaging and Interconnection Technology Dept. Dr. Xie joined Agere in 2001 and has been involved in new packaging technology developments with emphasis in finite element analysis, thermomechanical modeling, interfacial delamination, solder joint fatigue, and packaging reliability qualification. He has more than nine years experience in structural durability and damage tolerance evaluation and experiment. Dr. Xie holds a Bachelor's Degree in engineering from the Harbin Shipbuilding Engineering Institute, Harbin, China, and a Master's Degree in Solids Mechanics from the Northwestern Polytechnic University, Xian, China. He also holds an MS in Aerospace Engineering and a Ph.D. in Mechanical Engineering from Georgia Institute of Technology.

Chapter

1

Introduction

Melissa E. Grupen-Shemansky

1.1 Overview

The microelectronics package can be thought of as a system of materials serving to both protect an integrated circuit (IC) from the elements of its operating environment and to interconnect (electrically, electromagnetically, optically, and so forth) an IC to the operating system. Often, the package links finer geometries to coarser geometries as the nanometer realm of the semiconductor communicates with the millimeter and larger world of the application system. A package can range from a complex multilayer substrate containing hundreds of components to a single chip directly attached to the system board. Throughout the range of methods used to interconnect an IC to the rest of the system circuit, the interdisciplinary technology of packaging is applied.

The purpose of this book is to introduce the reader to the various packaging technologies in the IC industry and share from the extensive reliability engineering experience of the authors from Agere Systems. In systematically working our way through the multitude of failures that can occur in the engineering of a robust IC package, we make use of material, mechanical, electrical, and chemical areas of expertise. All engineering disciplines are in use in this area of semiconductors, which comes together in the intriguing art of analytical discovery, a form of forensic science. All failures leave evidence. Tracking that evidence back to the root cause is a fascinating and complex challenge. Eliminating the source of the failure is the only acceptable result. Field function depends on the accuracy of this investigative science and the statistical predictability of the failure phenomena, once known. To familiarize our varied readers to packages and packaging technology jargon used in the semiconductor industry, an industry packaging portfolio, of sorts, is presented in Chapter 2. Here, also, the popular interconnect technologies of wirebond and flip chip will be reviewed, as well as some emerging technologies. In Chapter 3, the reader is provided with an overview of reliability refresher. For a more extensive study of the statistics of reliability, references are cited at the end of Chapter 3. We begin to describe the failure mechanisms in the complex composite structure of the IC package in Chapter 4. This is not a comprehensive dissertation of all failure mechanisms, as that would require volumes and is still in the state of discovery. However, the most frequently encountered suspects, such as moisture ingress and thermomechanical mismatches, are examined. This leads us into the investigative science of locating failures in Chapter 5. Some may argue that this is an art, due to the creative detective work often required. However, one does not have to be a master sleuth to follow some of the basic steps toward identifying root causes. These basic steps can get you 80 percent of the way, and, in some of the simpler cases, may solve your problem. Chapter 6 is an overview of the analytical techniques found most useful in packaging failure analysis. The device examples used in this book are electronic in function, so the discovery process begins with a map of the electrical signal using time domain reflectometry (TDR). Largely destructive techniques then are employed to image, open, bend, and stress the samples in search of the cause or origin of the failure. A variety of those failures are reviewed in Chapter 7.

The purpose of this investigative work is to identify and eliminate the cause of failures during the development stages of a device. To be realistic about it, no engineering process is perfect, at least to date; field failures do occur. While it is difficult to acquire commercial product field failure statistics, military product field failures occur because of package-related causes 8 to 14 percent of the time.[1] Many military IC products are packaged hermetically using temperature-resistant ceramic materials. In commercial applications, the trend is toward plastic materials for light weight, low cost, and convenience. Plastics are more moisture and chemical absorbent than ceramics. They have lower glass transition temperatures (T_g), leading to nonlinear material properties at higher temperatures. Therefore, one may expect that a larger percentage of field failures in consumer electronics would be attributable to the plastic package.

Integrated circuit (IC) applications are wide and varied. Today, we find ICs in products ranging from coffee makers to satellites. Accompanying those applications are different environmental operating conditions. For example, military avionics operate under higher than normal background ionizing radiation levels and may require up to 30 years of reliable service. Undersea applications require reliable operation under hydrostatic pressure. Automotive under-hood applications may operate at 150°C ambient conditions, where a sensor in the exhaust manifold must withstand 800°C. Consumer hand-held applications may have relatively short lifetime requirements, but they have to be robust enough to survive accidental drop impacts or bends. Biomedical applications, such as implantable devices, have relatively stable and small temperature fluctuations, but the materials need to be capable of hermetically sealing the active components while maintaining biocompatibility. Ethernet local area network applications now demand greater than 2.5 Gbps (gigabits per second) transmission rates using low-cost conventional plastic materials, which requires packaging engineers to densely pack hundreds of signal traces within millimeters without compromising signal integrity requirements, such as 100-ps or less signal rise time and nanosecond skews. On which applications you focus and which performance factor presents the greatest challenge clearly depend upon the product focus of your company. Integrated circuits are found in all these application spaces and must maintain reliable operation and communication to the outside world under some of the most extreme conditions. Reliable operation in any of these market segments depends on competent and comprehensive failure analysis during the technology and product development phases. Although the emphasis of this book is on electronic products that serve the infrastructure and client divisions of the storage and communications markets, the failure analysis steps and methodologies are generally applicable to other electronic applications. All examples of specific failure modes presented in this book are of organic, nonhermetic packages. The application spaces considered challenge the IC performance with data rates increasing from 2.5 to 6 Gbps this year, and a device energy dissipation as high as 260 W. Application specific devices in this market require more than 2500 electrical interconnections to the system board, frequently requiring twice that in connections to the chip. Increased integration, whether it is on the chip (system on a chip, SoC) or in the package (system in a package, SiP), is pushing body sizes above 45 mm, thus introducing additional concerns with warpage, board-level reliability, and socket testing. The nonhermetic nature of organic material, in general, and the mismatch of the thermal expansion properties of semiconductors and plastics make this segment of failure investigation equally, if not more, fascinating.

All this investigative work does not come without cost. Furthermore, the economic demands in the highly cost-competitive electronics industry place a greater emphasis on the use of standard, low-cost materials and processes. Unlike wafer fab, where the cost of technical advances can be counterbalanced by the cost savings of decreasing real estate (smaller die sizes, more die per wafer), packaging real estate is constrained on the system side, where smaller dimensions increases printed wiring board (PWB) costs by increasing layer counts or decreasing pitches and line geometries. Caught in the middle of increasing chip I/O, shrinking chip dimensions, and increasing off-chip signal speeds,[2] IC packages must accommodate greater interconnect densities, provide the cost effective fanout required on the PWB side, and preserve signal integrity, often with a narrower electrical budget. Getting more performance out of low-cost organic packaging materials requires intrinsic understanding of these compound structures. Chapter 3, Section 3.7, gives some insight into cost and cost avoidance in reliability engineering. Arguably, the cost of reliability engineering is a fraction of the price one pays if the devices fail in the field during operation. It is difficult to put a price tag on the ramification of an IC failing in a pacemaker.

In the pursuit of better reliability, under increasingly extreme operating conditions, the development of new semiconductor packaging materials never ceases. In 2002 to 2003, the semiconductor packaging materials market grew by more than 6 percent and is expected to continue that growth rate over the next 5 years.[3] The newer materials, generally associated with the more advanced package types such as underfill for flip chip or thermal interface materials for more effective thermal dissipation, are growing in market size by approximately 30 to 60 percent. One of the global initiatives that has triggered a rapidly motivated advanced material development is Pb-free. The Restriction of Hazardous Substances (RoHS) directive states that "Member States shall ensure that, from 1 July 2006, new electrical and electronic equipment put on the market does not contain lead, mercury, cadmium, hexavalent chromium, polybrominated biphenyls, and polybrominated diphenyl ethers." The current state of electronic packaging materials is largely in compliance with the new directive already, with the exception of Pb (elemental lead). Anticipating that these restrictions will become laws within the European Union (EU) by 2006, all electronic products made in or shipped to the EU will necessarily contain no intentionally added Pb. In semiconductor packaging and board assembly, Pb is commonly used in solders by which the components and integrated circuits (ICs) are attached to the system board. In moving away from Pb, the substitution of Pb-free solder requires increases in process temperature from approximately 215°C to as high as 260°C. Nearly all package technologies are affected, spanning every electronics market segment. This initiative has not only driven the material change necessary to replace Pb-containing solders, it has had a significant effect on packaging substrates, mold compounds, and underfills. The change in process temperatures exacerbates the issue of moisture absorption. Thus, moisture-induced failures, such as delamination or cracking, become more prevalent. In addition, thermal mismatches of the different materials from silicon to plastic cause more severe bending, which can pose problems at board assembly—if it doesn't destroy the device first. Due to the critical impact of materials and material properties, Chapter 8 is devoted to the emerging material sets. Advancements in this area are the true enablers of failure-free integrated circuit packages under more extreme operating conditions and ever increasing performance demands.

1.2 More about the IC Package

Semiconductor products are composed of a complex set of materials, starting with the semiconductor material—most typically silicon. The delicate microcircuitry on the Si needs to be protected from moisture, heat, electromagnetic interference, ionizing radiation, static discharge, and corrosive elements. Furthermore, it serves little purpose if the electronic functionality is not effectively interconnected to the system or user interface. Integrated circuit (IC) packaging performs this function. Integrated circuit packaging materials are composed of shielding to protect the sensitive electrical system from electromagnetic interference, thermal components to more efficiently dissipate thermal energy, encasement for protection from the elements, interconnection components for communication with the outside world, and, depending on the product, passive components. Shielding constitutes plastic or foam gaskets, aluminum lids, and conductive meshes or tapes, to name a few examples. Examples of thermal components are thermal greases, Cu slugs or heat spreaders, and thermal vias in substrates. Encasement for protection from the elements can be commonly referred to as overmolding, underfill, or lids. Interconnection components in electronic systems are conductors such as leadframes, circuited substrates, solder balls, bump interconnects, or printed circuit boards. This set of packaging materials is composed of metals, plastics, ceramics, epoxies, glass, and silicon. Upon assembly, the materials may be bonded, laminated, or otherwise deposited in liquid, solid, or gaseous state. The packaging of an integrated circuit, yielding a functional component, is a challenging interdisciplinary field applied to a wide variety of products that demand outstanding capabilities in reliability and robustness.

IC packages come in all shapes and sizes to meet the performance, cost, and/or form factor requirements of the application. The smallest of packages is called a *chip scale package* or CSP. By industry standard, it is no more than 1.2 times the size of the IC inside. An example of a true CSP (same size as the silicon IC) is essentially fabricated as an extension of the IC fab technology and is called *wafer-level packaging*. Figure 1.1 displays several examples of wafer level CSPs (WL-CSP) with BGA solder ball interconnects geometrically distributed across the die. Additional dielectric and metal layers are deposited at the end of the normal fab process for the device, distributing the interconnects in a full area array. The additional dielectric layers provide stress relief for the top of the silicon and enable an effective increase in solder ball height. The wafer is then sawed into individual chips which are, at this point, "packaged." The packaged chips can then be solder-attached to a board with or without board-level underfill. Board-level reliability is a function of solder joint height. For fully collapsible solder joint interconnections, pitch is limited by the diameter of the solder joint, which is close to the height of the spherically-shaped interconnect. Therefore, pitch and the area of the chip limit the total number of board connections or input/output (I/O) of this packaging technology to typically less than 100 I/Os. The multilayer dielectric and metal stack-up must serve to protect the IC surface from mechanical stresses imparted by the board while maintaining the integrity of the electrical connection of the solder balls. Because thermomechanical stresses become more severe the farther one gets from the center point, these simple package constructs are typically used with die sizes less than 10 mm. Highly caustic or corrosive applications environments can be detrimental to this package type without the full encapsulation or hermetic lids of other technologies. However, the small form factor makes it ideal for consumer hand-held product applications where light weight and small size are premium attributes.

Figure 1.1 Wafer-level chip scale package. *[(a) Courtesy of K&S and (b) courtesy of Motorola.*[1]*]*

Typically, ICs are mounted on a substrate or interposer before being placed on a board. Figures 1.2 and 1.3 are schematic examples of wirebonded and flip chip packages, respectively. They are shown here with 2- and 4-layer substrates but may range from 1 to 6 layers for organic laminate substrates and 4 to ~12 layers for organic buildup substrates. As can be seen, the wirebond packaged product sits face up on the substrate. The plastic encapsulate, called *mold compound,* serves to protect the front side of the IC and provides mechanical robustness to the component. Electrical connection to the substrate is typically made through Au wires, but sometimes they are composed of Al, Ag, or Cu. The wires are bonded on the semiconductor chip via wedge bonding or ball bonding on a peripheral metal land called a *bond pad*. Today, products have as many as four peripheral rings of bond pads for wirebonding with pitches as low as 45 μm in production; however, 60-μm pitch is more common. The electrical connection continues through the wirebond to a series of transmission lines, metal planes, and vias in the substrate until terminated at the solder ball. During attachment to a system board, these solder balls are heated to liquidus, thus joining to an appropriate metal land or bond pad on the board. This type of attachment process, sometime referred to as *surface mount technology (SMT),* is self-aligning, meaning the surface tension of the liquidus solder will perfectly center the packaged component onto the attachment pads on the board. The location and alignment is dependent on the board attachment pad alignment and tolerances.

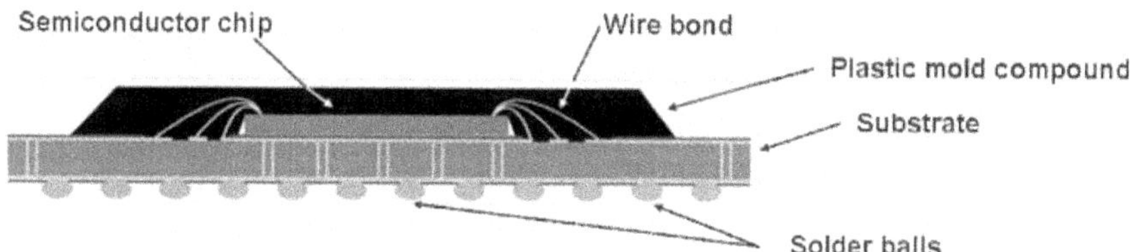

Figure 1.2 Typical wirebonded plastic ball grid array (PBGA) package.

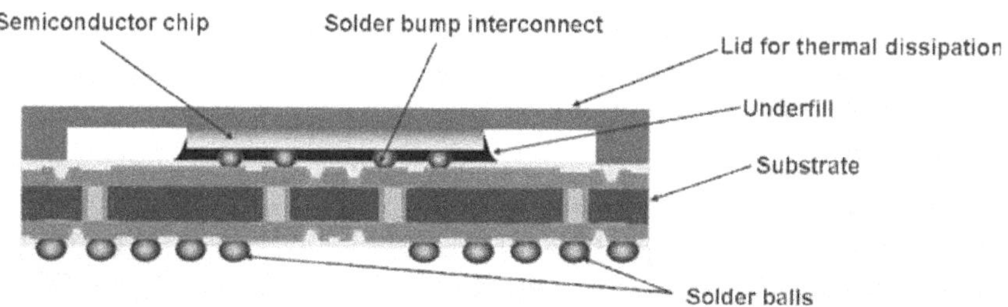

Figure 1.3 Typical flip chip plastic ball grid array (FCBGA) package.

Taking the SMT knowledge further, flip chip packaging was derived. The self-aligning feature of the BGA package attachment to the board is exploited now at the chip level. Solder bumps applied to the IC chip perfectly align the chip to the substrate when temperatures exceed the melting point of the solder. Again, positioning depends on the substrate bond pads and tolerances. In this case, the solder bump is composed of an under-bump metal stack with a spherically shaped solder ball or bump on top. The solder ball can be deposited using a stencil print process and solder paste or by the electrochemical process of electroplating. In practice, bump interconnects can be composed of other metals and geometries, some designed to go entirely liquidus during chip attach and some designed to remain solid. Those interconnect schemes designed to remain solid during the attachment process (thermal-compression bond for example) do not benefit from the self-aligning feature of SMT.

As a result of the flip chip attachment process, the semiconductor device is mounted face down. Underfill is used to protect the circuited side of the IC and provide mechanical integrity to the chip and bump interconnects. The bump interconnects serve two purposes in this configuration. In contrast to a wirebond, the bump interconnect is a mechanical connection to the substrate. Stresses imparted to the chip from the substrate, and vice versa, transmit through the solder bumps. In addition, and similar to the wirebond, the bump interconnect provides the electrical interconnection from the semiconductor device to the substrate. As with wirebond packages, the substrate contains transmission lines, metal planes, and vias terminating the electrical connection of the component at the solder ball. However, substrates used in flip chip packages are often more sophisticated than wirebond packages. The bump interconnect at the chip enables more connections to be made across the surface of the die than in a strictly peripheral wirebond design. Wirebond packages contain as many as 1200 chip interconnections in production today. Flip chip packages enable five times (or more) as many chip interconnections. As a result, the flip chip substrate designs are much denser and contain more layers.

1.3 Multiple Technologies/Multiple Failure Mechanisms

Figure 1.4 gives the reader some idea of the package portfolio breadth necessary to service an infrastructure through client product portfolio. A client product portfolio may consist of products for wireless or wireline connections that include ICs for mobile phones, hard disk drives, modems, and so on. Products may also include application specific ICs

High-volume wirebonded packages

TQFPs
• 7–24-mm package sizes
• 0.4–0.8-mm lead pitch
• U to 208 leads
• 1.2–1.6-mm thickness

FSBGAs
• 7–15-mm package sizes
• Up to 360 balls
• 0.5–1-mm ball pitch
• 1.1–1.74-mm height

Advanced wirebond PBGA

• Double and triple stagger
• Up to 1200 wires
• Critical signals up to 12.6 Gbps
• Cu and Al fab technologies

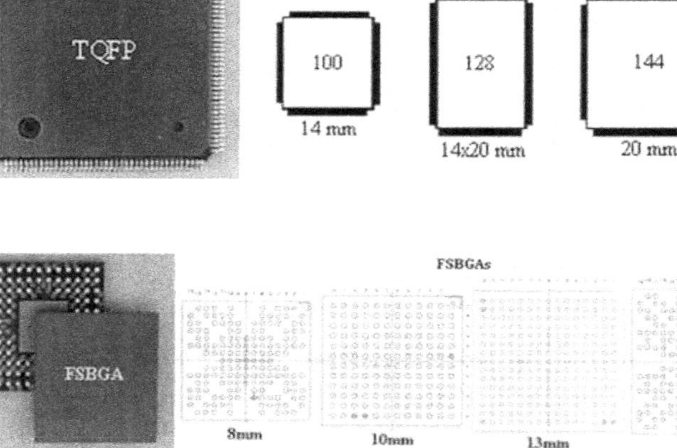

Figure 1.4 Overview of a typical semiconductor package portfolio. *(Courtesy of Agere Systems.)*

Flip chip plastic BGA

- Up to 2500 BGA interconnects
- Excellent thermal performance
- Highest speed signals
- Cu and Al fab technologies

Multicomponent packaging (MCP)
- Leadframe and BGAs
- Wirebond and/or flip chip
- Chip-to-chip interconnection
- Stacked chips

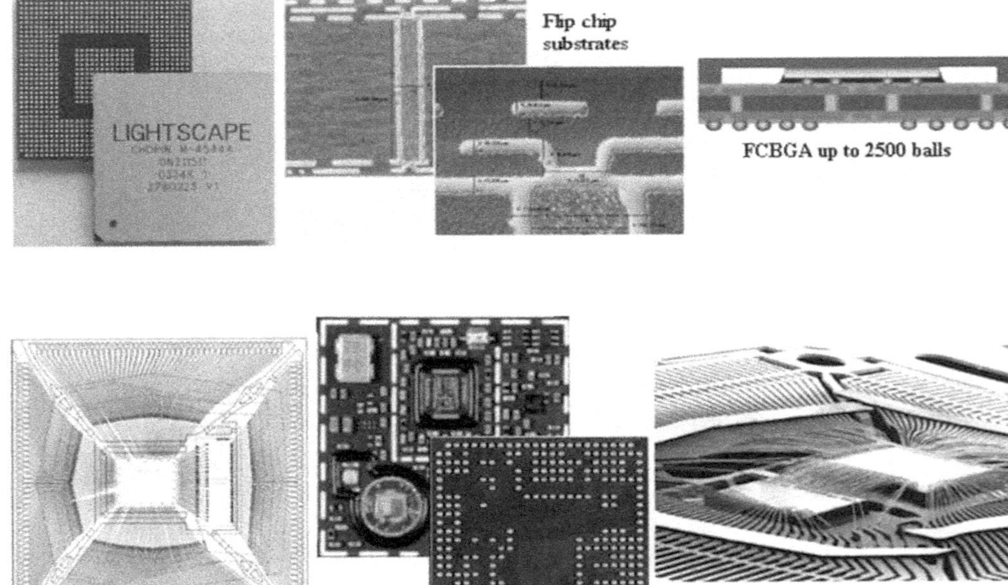

Figure 1.4 *continued.*

(ASICs), network processors, integrated switches, timing devices, and so forth used by the local area networks and metro networks to link the client to the region and ultimately to the rest of the world.

Industry wide, there is continued use of leadframe packages such as thin quad flat packs (TQFPs) and small two-layer matrix BGAs [called fully molded and separated BGAs (FSBGAs) in Fig.1.4] for power amplifiers, memory, and read channel SoCs (systems on a chip), to name a few. Wirebonded PBGAs, using four- to six-layer organic substrates, are used for ASICs, mappers, and microprocessors, for example. Also in this package space, stacked chips are becoming popular, starting out with stacking memory but evolving to logic on logic and more than two ICs. Advances in double, triple, and quadruple wirebonding are extending the previously peripheral wirebond I/O technology to a higher number of interconnections and toward the center of the chip. However, full area array interconnection to the face of the chip is currently realized only with flip chip. As the number of bond pads approaches 1200, flip chip becomes necessary (at least until area array wirebonding is developed). Present-day organic substrates used in flip chip volume production enable approximately 2500 interconnections to the system board (or BGA balls). The number of electrical signal traces that can be connected from the chip to the board is predominately limited by the number of layers in the substrate. In ceramic technologies, more layers are available for routing in the substrate, thus increasing the number of device I/Os possible to well above 2500. The majority of BGA packaged products sold in the industry today are in the 100 to 1000 ball count range, making wirebond and plastic flip chip BGA solutions prevalent. The final category in Fig. 1.4 is multicomponent packaging, sometimes referred to as system-in-a-package, or SiP. One example of a SiP product is a two-chip TQFP storage product with chip-to-chip interconnections that improves the signal-to-noise ratio, as compared to individually packaged chips communicating via a PWB. Another example is a wireless fidelity (WiFi) module that contains 4 die, surface mounted and wirebonded, 1 WL-CSP power amplifier, and 126 additional passive components protected by an RF shield or lid. WiFi enables a wireless connection to the internet anywhere within range of a base station and is several times faster than a cable connection. The final product appears, for all intents and purposes, to be a lidded PBGA, about 1 in (2.5 cm) on a side, that is mounted to the customer's board using a conventional SMT process. This SiP product looks and feels like a conventionally packaged single chip component, but it contains all the major RF components needed to enable an 802.11 wireless LAN (local area network) communication feature in laptop or hand-held devices without requiring the device manufacturer to have in-depth radio expertise. A SiP product approach can provide advantages to a monolithic solution, including faster time to market and lower cost. Compared to individually packaged components, SiP products have demonstrated better performance and have resulted in significant savings in board real estate.

In the package portfolio, shown in Fig. 1.4, die sizes range from 3 to 20 mm per side and package body sizes range from 5 to 50 mm per side. The total number of I/Os within this portfolio range from fewer than 5 to 2500. The performance specifications are equally varied as the physical specs. In the client systems market, packaging trends are toward thinner and smaller packages. A similar trend in client systems IC technology is increasing the power densities; therefore, one will see enhanced thermal dissipation capabilities in the packages used. Because packages in this space typically have fewer total I/Os, they can make use of leadframe technology (a single layer of metal) or two-layer substrates. In infrastructure applications, the trend is toward more complex system on a chip (SoC) or complementary system in a package (SiP). SoC trends demand an increasingly higher density and performance packaging technology. A SiP can be used in cases where a SoC is too costly or technically impractical to implement on a single chip, but the form factor and performance advantages of a SoC are desired. The SiP may accommodate different semiconductor materials (Si and GaAs, for example), various active and passive devices (RF transmit and receive IC, capacitors, resistors, and antennae, for example), or stacked chips. From leadframes to multilayer flip chip ball grid arrays and systems in a package, each package family has different material sets, different mechanical stress distribution, and different possible modes of failure. Within a package family (for example, PBGAs), convergence onto a common materials platform is desirable but can be a daunting task. Hundreds of molding compounds are commercially available today. The requirements for low alpha-particle emission or no lead result in multiple material sets, even within a package family, for at least some finite transition period.

Due to the various package geometries, materials, and application environments, failure mechanisms are varied. Through reliability investigations performed during the development stages of a product or technology, attempts are made to force such failures so that the mechanisms can be understood and later prevented. In the early stages of technology development, we find that it is not too difficult a task to force failures. Devices are subjected to JEDEC-specified accelerated testing which involves exposure of packaged ICs to rapid temperature fluctuations, slower temperature fluctuations, moisture and humidity, electrical bias at higher temperatures, and high-temperature storage conditions. During reliability evaluations and accelerated testing, failure mechanisms are analyzed and their impact on product lifetime assessed. This empirical learning is applied to parametric simulations to (1) avoid failure during the needed operating life and (2) predict lifetimes of similar products. Computer-aided reliability engineering is only as accurate as the assumptions that go into it and the accuracy and applicability of the empirical data set to the operating conditions of the IC application. The failure data must be collected in a statistically significant fashion and derived from accelerated testing that forces early occurrences of failure mechanisms, identical to those in end-of-life operation of the component. To achieve such, it is necessary to have a comprehensive understanding of temperature dependent, nonlinear material properties, physics, thermodynamics, kinetics, and failure mechanics.

Within the technology development phases of some of the packages seen in Fig. 1.4, numerous failure phenomena were observed, and root causes were diagnosed. For example, as plastic encapsulated packages came into high-volume production in the early 1980s, moisture-induced delamination, or *popcorning,* was discovered. Of course, delamination in general can be observed when joining two chemically dissimilar materials forming an interface, be it moisture-induced, contamination-induced, or stress-induced peeling. Stress-induced failures are the result of thermomechanical mismatch of materials. The mismatch is unavoidable in a Si-to-plastic system but can be managed. Mechanical failures can also be induced by improper handling such as overly aggressive test socket clamping that causes microfractures at contact points on the substrate. These microfractures may not be visible after testing at optical inspection, but they may propagate to fracture a copper trace in lower levels of the substrate after being subjected to as few as 100 temperature cycles of –55°C to +125°C. When mechanically related failures occur late in the process, or late in the reliability testing, the mechanical root causes can be as elusive as stress-induced failures. Process-induced failures include saw-shard-induced microfractures to the chip or package, mold compound residue stress concentrations, underfill particle settling that causes CTE (coefficient of thermal expansion) mismatch, and overetched conductors, to name just a few.

As the reader continues through this book, many displays of fractures and failures will be provided. Looking at these pictures can make even the most seasoned failure analysis (FA) veteran pessimistic that a robust, reliable device function can be achieved. But it can be done by systematically working through the failures and understanding their causes. For example, a process-related failure seen during the development of an RF module or SiP proved to be a circuit failure due to excessive leakage currents after storage at higher temperature with humidity. The leakage was traced to a filter in the SiP where two connectors were shorting due to flux residue with a high ionic content. Determining the source of the failure and developing a theory for the mechanism then defined the course of action to correct it, namely using fluxes with lower ionic content and increasing the power amplifier stand-off on the substrate. The component stand-off had two compounding effects. First, capillary forces were effectively drawing flux under the component during assembly; second, the narrow gap prevented flux cleaning agents from flushing the flux residue out. Eliminating the gap under this component was not an option. Rather, the stand-off was increased, thereby reducing the capillary forces drawing the contaminating agent in and enabling cleaning agents that were sprayed on to move in and out of the clearance. In addition, the failure analysis revealed an "ionic bridge" that developed during reliability testing, shorting the input and output terminals of the filter. Chemical analysis of the flux used, and comparison to other commercially available fluxes, revealed that the one used had a higher ionic species content than some others. To further improve process tolerances, an effective assembly process was developed around the use of a low-ionic-content flux. This complex SiP, which contained more than 130 components within a PBGA package, subsequently passed component level qualification and went on to production at an average of 96 percent final test yield.

An example of a mechanical failure occurred on a high-speed PBGA of seemingly normal configuration. The design was in compliance with all known design rules. The only unique aspect was that there were high-speed signal traces in one corner that required additional metal shielding in that region of the substrate. In this case, the observation was made during assembly, before qualification testing began, that unusual discolorations or indentations were made on the organic substrate during the molding operation for this device. Reliability engineering analysis on units that displayed this discoloration indicated that these were deformations and cracks in the solder mask material. Some of these cracks seemed to extend further into the substrate, penetrating the interlayer dielectrics of the substrate, but their impact on performance was uncertain. Upon temperature cycling, FA clearly revealed the cracks propagating into the substrate inner layers and crossing, and in some cases severing, copper interconnect traces. The anomalies seen in the solder mask after molding geometrically corresponded to the mold-clamping tool that contacts the substrate outside the periphery of the die. Process development teams working in concert with design teams derived a working theory and solution. It was theorized that the uneven distribution of metal regions in this substrate, although conforming to metal balancing design rules, was causing the molding tool to apply uneven pressure to the surface of the substrate. Two solutions were proven. Either removing the metal from the clamp area or putting metal into the clamp area in an equally distributed fashion solved the problem.

In some cases, the failure analysis results and investigative process are concise, where the root cause is quickly identified and resolved. In most cases, the complex interaction of materials and stress mechanics causes failures to be seemingly random as second- and third-order parametric effects become significant. In these cases, statistical design of an experiment and analysis is essential. In one such case, solder bump interconnect cracks resulted in electrical opens for a large die (greater than 12 mm) that was flip chip packaged. Failure analysis revealed underfill material in the cracks, thus pointing to the fact that the cracks occurred after the solder had solidified. This evidence narrowed the choices of when the crack occurred during the flip chip assembly process. Attention was then paid to the thermal exposures to which the device was subjected between die placement and underfill cure. After examining the process tolerances for the various thermal excursion steps through a series of controlled statistical DoEs (design of experiments), a thermomechanical finite element simulation was employed. Elasto-plastic models of the bump interconnects predicted trends such as thin die, larger bumps (requiring larger bump pitch in this case), lower substrate modulus, and/or lower substrate CTE that would decrease the strain on the interconnect structure. Ultimately, a lower-modulus, lower-CTE substrate technology was used. This failure discovery and resolution process, working through identification of the problem, process tolerance verification, thermomechanical simulation, empirical verification of the models, sourcing of new substrate technologies and materials, and qualification of the new technologies implemented, took almost 12 months and employed skilled investigators from every engineering discipline. Failure analysis pointed out the defects along the way. Reliability engineering eliminated those defects.

Some of the examples stated here, and many more, will be explored in the chapters that follow. The techniques used in reliability engineering enable the product or technology development engineer to induce failures. Failure analysis methods and analytical tools enable the investigative scientist to hone in on the origin of the defect, systematically putting the first clues together as to the root cause. Statistical design of experiments and theoretical postulation/modeling built evidence around the working hypothesis, which ultimately resulted in successful qualification. This book will walk you through the diverse world of failures in integrated circuits with numerous examples and techniques. The reader will learn how to interpret results and correlate those results in root cause analysis. All examples will be related to organic packaging, following the trend of commercial applications toward plastics because of their light weight, low cost, and convenience. Finally, the reader will gain some insight into the emerging material sets in the constant pursuit of smaller, lighter, and faster microelectronic products.

1.4 References

1. *Microcircuit Manufacturing Control Handbook, A Guide to Failure Analysis and Process Control*, vol. 1, 2nd ed., ICE.
2. Grupen-Shemansky, M., and DiBerardino, M, "When the Package Means as Much as the Chip," EDN, July 10, 2003, p. 51–56.
3. Prismark Partners LLG, "The Electronics Industry Report—2004," November 2003.
4. Wetz, L., White, J., and Keser, B., "Improvement in WL-CSP Reliability by Wafer Thinning," 53rd Electronic Components and Technology Conference, p. 853, 2003.

Chapter

2

Fundamentals of IC Package Technologies

Charles Cohn

2.1 The IC Package

Since the early 1960s, when integrated circuits (ICs) were first introduced, the electronics industry has grown very rapidly to the point at which, in 2002 alone, approximately 80 billion[1] ICs were sold for a variety of system controls.

ICs are typically sensitive to the environment and therefore require special consideration when we are selecting the proper enclosures to protect them from the harsh elements in a given application. The IC may either be directly attached to a PWB and then "glob topped" with an epoxy, or it may be housed in an IC package, which assures protection to the IC from mechanical stress and environmental stress, such as humidity and pollution. In addition, the package provides mechanical interfacing for testing, burn-in, and electrical interconnection to the next level of packaging. The package must meet all device performance requirements, such as electrical (inductance, capacitance, crosstalk), thermal (power dissipation, junction temperature), quality, reliability, and cost objectives. Until recently, the packaging and interconnecting technologies have not been the limiting factors in the performance of most silicon devices in high-volume production. But, as circuit and system demands are increasing, more attention is needed in selecting package technologies, materials, and designs that will meet these challenges. Some package designs may need to provide special design enhancements to accommodate such demands as higher power dissipation, signal terminations, power and ground distribution, and matched impedances.

Within a given package, there is a distribution of traces that connect the IC to the external leads. Large packages (in particular, those with rectangular shapes and side-brazed leads, such as DIPs) require very long traces to the corner leads of the package. Array-type packages, typically square, also contain a distribution of traces, with the longest trace going to the outermost rows of leads in the corner of the package. This distribution of trace lengths within the package, in series with similar distributions of trace lengths on the PWB, could result in undesirable signal delays, missed line driver requirements, or crosstalk between critical signal leads. Hence, the demands on packaging are for smaller physical-size packages with higher-density circuitry, thus minimizing the variation of trace lengths within the respective distributions and reducing their undesirable effects on performance. In addition, lead spacing or pitch and the method of attachment to the next level of interconnection also are important factors in the performance of the device. Figure 2.1 compares the package areas for some typical surface mount (SM) and through-hole (TH) mount packages.

Improved quality, reliability, and cost are other system demands receiving much attention today. System manufacturers are driving down their costs and consequently are putting great pressure on the quality

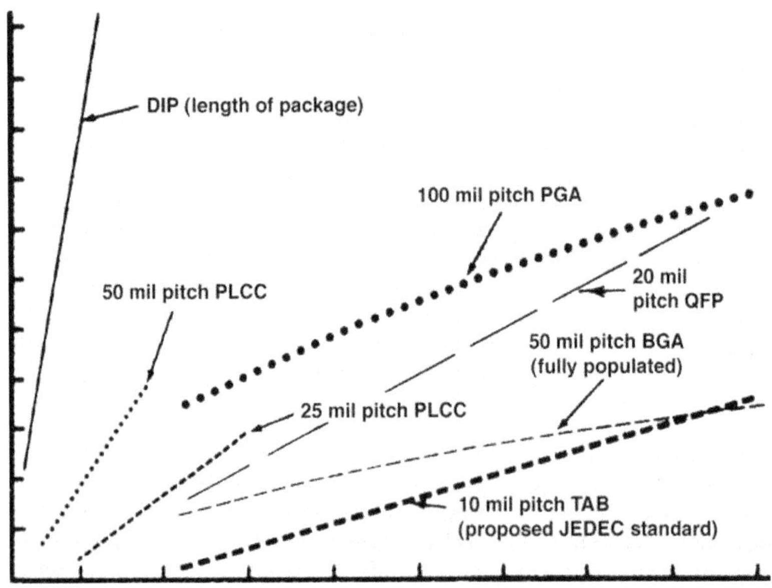

Figure 2.1 Sizes of some common TH and SM packages as a function of number of package leads and pitch.

and reliability of the components they purchase. No longer can a generic package be expected to meet all system requirements without considering all possible conditions that the component will encounter in the field.

2.2 Package Families

IC packages come in a variety of lead arrangements and mounting types (Fig. 2.2). The packages are grouped into families defined by the method of mounting to the circuit board [through-hole (TH) or surface-mount (SM)] and by the physical arrangement of the leads on the package (in-line, perimeter, or array). These package families are governed by standard package outlines that control all dimensions, tolerances, and all other information necessary to define, manufacture, and assemble the package. Package family outlines have been standardized in the Electronics Industries Alliance (EIA), the Joint Electron Device Engineering Council (JEDEC), and the Electronic Industries

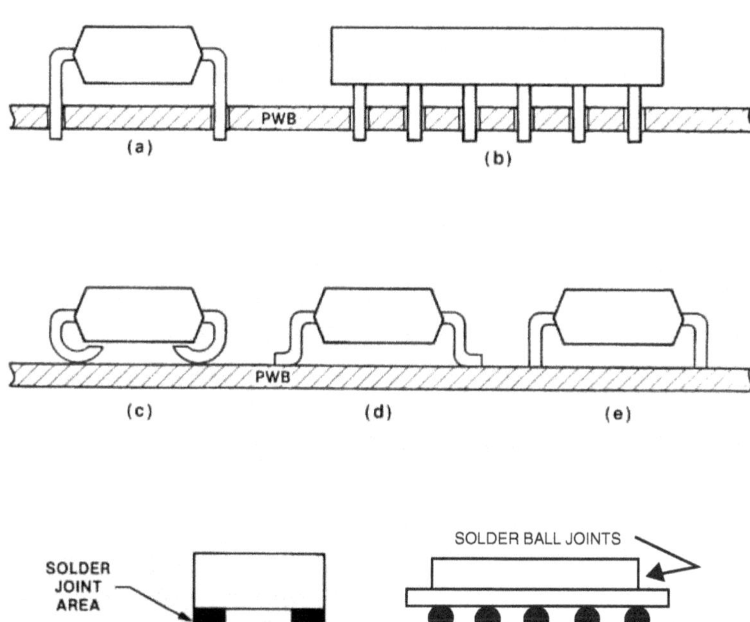

Figure 2.2 Examples of packages, package lead geometries, and mounting methods in current use: (*a*, *b*) typical TH packages, (*c*–*g*) typical SM packages. Lead geometry options are (*a*) DIP, (*b*) PGA, (*c*) SOJ and PLCC, (*d*) SOIC, SOG, QFP, PQFP, and fine-pitch LDCC, (*e*) butt-leaded packages, (*f*) LLCC, and (*g*) BGA.

Association of Japan (EIAJ). JEDEC has been working closely with the EIAJ to reduce proliferation of outlines and to focus similar outlines from each organization into one worldwide standard outline.

Table 2.1 lists the packages available in each of the families, grouped by the mounting method to second-level board assembly (TH versus SM) and by package technologies, as described in Section 2.4. Figures 2.3 and 2.4 illustrate some of the more popular package types used today in the TH and SM families.

Table 2.1 Package Families by Packaging Technology

	Package technology			
	Molded plastic	Laminated plastic	Laminated ceramic	Pressed ceramic
Chip protection	Molded	Molded or encapsulated	Metal or ceramic lid	Ceramic lid
Surface mount	SOIC SOJ PLCC QFP QFN TQFP TSOP TAB	LLCC PBGA FCBGA CSP Matrix BGA	CQFP LLCC LDCC SMPGA CBGA CFCBGA	CERQUAD CQFP
Through-hole mount	DIP SIP ZIP QUIP	PPGA LLCC	DIP SIP PGA	DIP CERDIP

2.2.1 Through-Hole Mounted Packages

For assembly onto a PWB, through-hole (TH) mounted type packages require one plated-through hole (PTH) in the PWB for each package pin or lead, typically 0.035 in (0.90 mm) in diameter on a 0.100-in (2.54-mm) grid. Pin or lead lengths are not critical with TH components, since the assembly process usually calls for the excess length of the pins/leads to be cut off after insertion into the PWB, and then some pins/leads are crimped to provide mechanical backup.

The dual in-line package (DIP) family, introduced in the early 1970s, with dual in-line leads on 0.100-in (2.54-mm) centers (Fig. 2.4), is generally available in three package technologies: molded plastic, pressed ceramic, and laminated ceramic. DIPs can accommodate devices with as few as 8 and as many as 64 I/Os. The family is defined by width, or lead-row spacing, ranging from 0.300 in (7.62 mm) for the

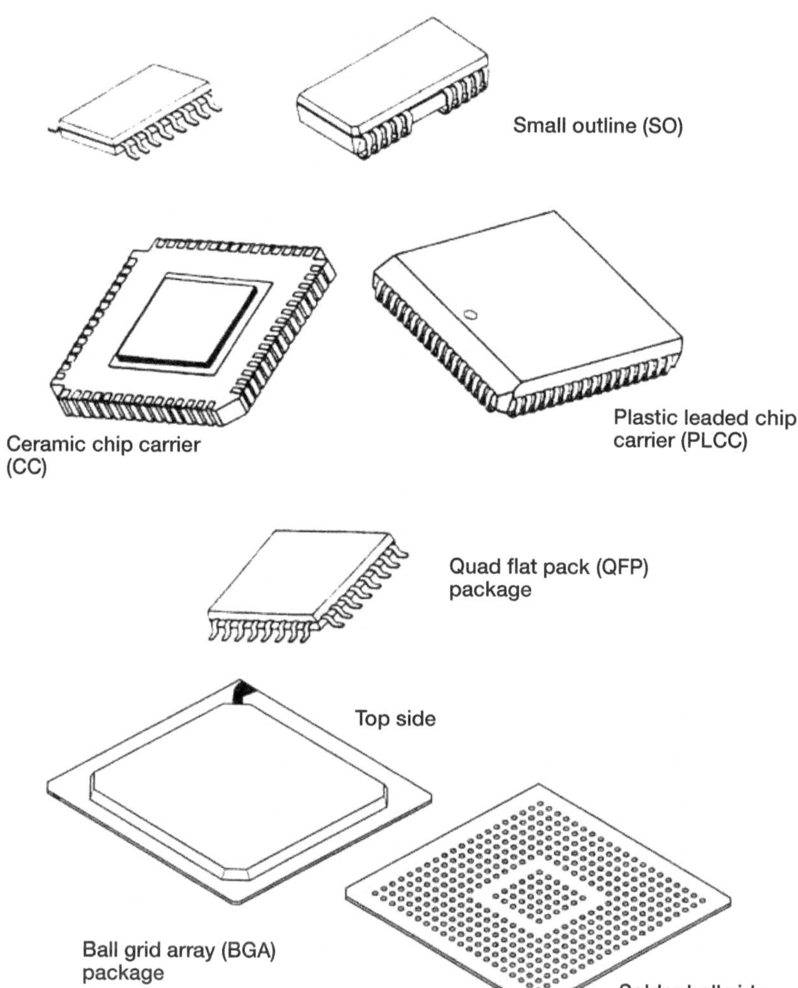

Figure 2.3 IC package families, SM type. Examples of design variations within families are lead geometry (gull wing, J, leadless, or solder balls).

narrowest (referred to in the industry as "skinny" DIPs) to 0.600 in (15.2 mm) for the widest family. Another variation, known as "shrink" DIPs (SDIPs), has the same lead-row spacing as the standard family but with the lead pitch reduced from 0.100 to 0.070 in (2.54 to 1.78 mm) to provide higher pin-count capability for the same size package. Similar DIP families exist in the molded plastic technology, pressed ceramic technology (CERDIP), and in the laminated ceramic technology. In addition, laminated ceramic DIPs come in a wide variety of configurations identified by the method of lead attachment to

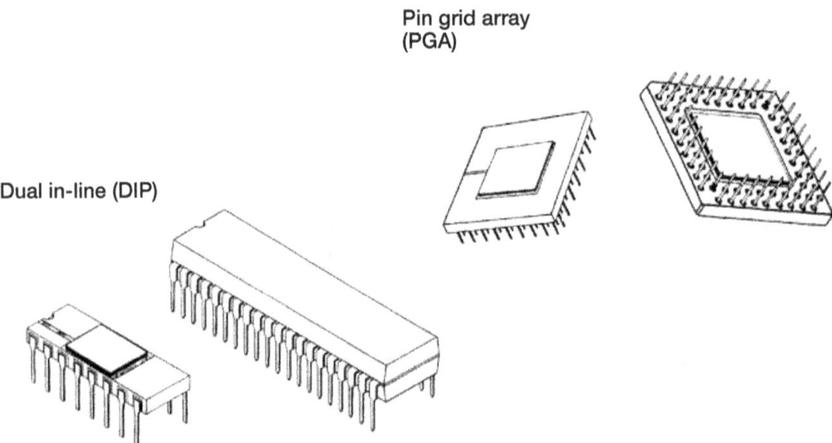

Figure 2.4 IC packaging families, through-hole (TH) type. Design variations exist within families, such as the way leads are attached to or exit the body, as illustrated in the DIP family, and in the direction the cavity faces with respect to the PWB (up or down), as shown in the PGA family.

the ceramic body, which can be side-brazed, top-brazed, or bottom-brazed. Side-brazed DIPs are footprint compatible with the plastic and pressed ceramic DIP families. Top- and bottom-brazed DIPs are generally supplied by a vendor as a flat pack, and the leads are then formed to the specification of the end user, which may or may not be footprint compatible with the side-brazed variety.

DIPs are limited in lead count to 64 leads by the physical size of the package, which approaches 3.0 in (76.2 mm) in length for 64 I/Os. At this level, the IC package becomes very expensive and difficult to handle, and in general, the electrical performance of the device begins to degrade—not to mention that it uses up a lot of expensive PWB real estate.

Other in-line families are the single in-line package (SIP) family, the zigzag in-line package (ZIP) family, and the quad in-line package (QUIP) family. All were introduced to meet special product needs. For example, SIPs were needed for memory chips and memory modules. These package families have leads typically on 0.100 in (2.54 mm) pitch and are available to 40 lead counts for ZIPs and up to 64 for QUIPs. A QUIP has four in-line rows of leads, with the rows generally spaced 0.100 in (2.54 mm) apart, and the leads in each row are staggered with respect to each other to maximize hole-to-hole spacing on the PWB. This package option effectively reduces the package length by one-half over the equivalent DIP of the same pin count.

Area array packages, such as pin grid arrays (PGAs), were developed to extend the pin-count capability for TH-mount-type packages with a significant saving in PWB area. Figure 2.1 compares the physical size of a PGA package, with pins on an area array of 0.100-in pitch to the DIP at an in-line 0.100-in (2.54-mm) pitch. The 0.100-in (2.54-mm) pitch PGA family is available in both laminated plastic and ceramic technologies and can provide pin-count capability above 1000 pins, in both cavity-up and cavity-down configurations.

2.2.2 Surface Mounted Packages

Surface mounted (SM) packages were first introduced in the early 1980s to achieve higher board density and better electrical performance by eliminating the need for through holes. Initially, the industry was reluctant to accept SM packages, but in the 1990s acceptance accelerated to where demand for SM packages exceeded TH packages. Today, the availability of new SM package types has expanded greatly into a very formidable list (Table 2.1).

High-density packaging cannot be achieved with TH-type packages, such as DIPs and PGAs, because they are limited to 0.100-in (2.54-mm) pin/lead pitch. Either perimeter or array-type SM packages with finer lead-pitch capability are needed to satisfy this demand. Perimeter leads for SM packages with a pitch as fine as 0.016 in (0.40 mm) have been achieved, but handling and assembly at these levels has been difficult. Some packages now require more than 1000 I/Os, an unachievable task for perimeter packages. But, given an array-type package like the ball grid array (BGA), with a 0.040-in (1.0-mm) solder ball pitch, the feasibility of meeting the >1000 I/Os requirement becomes realistic.

The demand for SM packages has placed a great deal of pressure on the IC package technology. Some of the most important aspects for SM leaded packages are lead coplanarity, lead finish, lead geometry, lead true position, and thermal effects. To ensure a reliable and high-yield (cost-effective) SM attachment, the perimeter leads on the SM package must be coplanar to within 0.002 to 0.004 in (0.05 to 0.10 mm). Similarly, for SM array packages such as BGAs, the solder balls must be coplanar to within 0.006 to 0.008 in (0.15 to 0.20 mm).

The solder joint is another critical item for SM packages. Because the solder joint provides both electrical connectivity and mechanical attachment of the packaged device, its integrity must be preserved through all the stress encountered in subsequent assembly and test, over the life of the system in its operating environment. Solder embrittlement must be avoided; hence, lead-finish materials on perimeter type packages and the gold plating thickness on BGA solder pads must be specified with care.

SM families, such as the small outline (SO) and PLCC types in molded plastic, CERQUAD in pressed ceramic, and leadless (LLCC) and leaded (LDCC) chip carriers in laminated ceramic, were introduced in the early 1980s. These packages have leads on 0.050-in (1.27-mm) pitch instead of the 0.100-in (2.54-mm) pitch used in the TH families. The SO package is an in-line type defined by body width, or lead-row spacing, similar to the way the DIP family is defined. Both narrow-body and wide-body SOs are gull wing (GW) families. The SO package with J leads (SOJs) has the same body width as the SO; that is, it uses the same plastic molding die but has a J-formed lead for surface mounting.

PLCCs, CERQUADs, LLCCs, and LDCCs are perimeter-type packages with leads on all four sides of the package. The lead geometries vary with the technology used. PLCCs use J-shaped leads designed to provide the highest possible density on the PWB for perimeter-type packages. CERQUAD packages generally come from the vendor as a leaded flat pack, and the device manufacturer, or end user, does the final lead forming (GW or J) prior to assembly to the PWB. The LLCC packages, being made of ceramic, are generally restricted to surface mounting on substrates with similar thermal expansion coefficients. This limitation is overcome by converting an LLCC to an LDCC by means of one of the commercially available soldered clip-lead techniques. LDCCs can be purchased by the device manufacturer or end user either as leaded flat packs with the leads formed to the appropriate shape (GW or J) or as finished packages with the leads in final form ready for SM. Examples of SO, PLCC, and CC packages are shown in Fig. 2.3.

The drive for higher lead counts and higher density has been very evident in the development of fine-pitch SM packages. Since the lead count of a perimeter package is a function of package size and lead pitch, as illustrated in Fig. 2.1, to achieve a high lead count with PLCCs at a 0.050-in (1.27-mm) lead pitch requires very large packages. Difficulties in molding large plastic packages and maintaining coplanarity of all leads necessary to achieve high board assembly yields limits the practical PLCC to a lead-to-lead overall size of 1.699 in (43.2 mm), equivalent to 124 I/Os.

The quad flat pack (QFP), with gull wing leads on all four sides (Fig. 2.5), evolved to fill the needs for high lead counts (376 leads maximum). The QFP family is characterized as fixed body size, variable pitch. That is, a variety of lead counts can be manufactured from a single body size using leadframes of various external lead pitches from 0.040 in (1.0 mm) down to 0.016 in (0.40 mm).

Another family of SM packages, which are shorter, smaller, thinner and lighter, are the TQFPs. Whereas the standard QFP body thick-

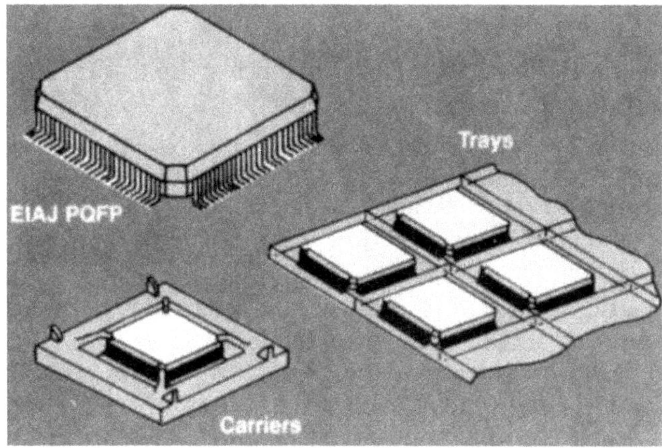

Figure 2.5 EIAJ MQFP packaging and handling options.

ness varies from 2.0 to 3.8 mm, depending on the package outer dimensions, the TQFP body thickness is 1.0 or 1.4 mm. The TQFP packages vary in lead count from the smallest, at 32 I/Os with a body size of 5.0 × 5.0 × 1.0 mm and an 0.80-mm lead pitch, to the largest, with a maximum lead count of 256 I/Os in a body size of 28.0 × 28.0 × 1.40 mm and an 0.40-mm pitch.

A new family of quad flat packs with no leads (QFN) emerged to reduce the cost and enhance the electrical and thermal performance of leadless packages. The near chip scale QFN, also known as the *MicroLeadFrame®* package by Amkor Technology (Fig. 2.7), is a molded leadframe based package with exposed perimeter lands on the bottom to provide electrical contact to the PWB. The die attach paddle can also be exposed at the bottom of the package for thermal enhancement. QFN packages range in size from 3 × 3 mm to 12 × 12 mm with perimeter pads that range from 4 to 100 I/Os on 0.5-, 0.65-, or 0.8-mm LGA pitch.

As the demand for packages with high I/O counts and high performance increases, the industry has turned to BGAs (Fig. 2.3) as an alternative to the fine pitch QFPs, which are difficult to handle. BGAs range from 4 × 4 to 49 × 49 arrays, in sizes of 7.0 mm square to 50.0 mm square.

Reliability of SM packaged devices could be affected directly by the stresses imposed during solder reflow. This is particularly true if the packaged device is inadvertently exposed to high levels of moisture during the period between device assembly and board assembly. A greater insight into IC package reliability is presented in Chapter 3.

2.3 Package Technologies

IC package technologies fall into four basic categories: molded plastic, pressed ceramic, laminated ceramic, and laminated plastic. Another IC package technology, the metal hermetic IC package, is not covered in this chapter because of its limited application in commercial electronics. These packages are used mostly in military applications, where hermeticity and high reliability under the harshest environmental conditions are requirements.

In the molded plastic technology, the package is constructed around the IC assembled on a metal leadframe, in strip form. The IC is first mechanically bonded to the die attach paddle and then electrically interconnected by fine wires from the die bond pads to the corresponding package wirebond fingers. If tape automated bonding (TAB) is used, then the IC pads are directly connected to the bumped leadframe fingers. The final package configuration is formed by plastic molding around the leadframe subassembly. The portion of the leadframe that is external to the package body is subsequently trimmed and formed into specific geometries suitable for either TH or SM attachment. The molded plastic package technology is widely accepted in the industry as reliable and low in cost, but the packages are nonhermetic and can have only a single metallization fanout layer.

The pressed ceramic package contains a leadframe embedded and glass sealed into a pressed ceramic base. The IC is attached and wirebonded to the leadframe and hermetically sealed using a ceramic cap or metal lid. The leads external to the package body are trimmed and formed to the geometries needed for TH or SM attachment.

A laminated ceramic or plastic package consists of a substrate with an integral metallized fanout circuitry and external terminals, which are either leadless, leaded or with solder balls for SM, or pinned for TH attachment. The substrate body may come with a cavity to accept the IC or an in-plane configuration where the die attach area and the bond pads are on the same plane. The substrate material may be ceramic or plastic and of single or multilayer construction, depending on the functional requirements. Both the ceramic and plastic substrates can equally provide high electrical and thermal performance, but the ceramic substrate has hermetic sealing capability, whereas the plastic substrate is nonhermetic and requires some form of epoxy encapsulation for environmental protection of the IC. The electrical interconnections are done using either wirebonding or flip chip techniques.

A comparison of cost, performance, and reliability among the various package technologies shows the multilayer ceramic package to be best suited when the requirements call for high performance, hermeticity, and high reliability, but the trade-off is high cost. A multilayer plastic package, with electrical and thermal enhancements, can provide equal or better performance at a lower cost, trading off hermeticity. For the lower end of consumer electronic products, the package requirements, in order of importance, are low cost, reliability, and performance. These attributes can be satisfied with plastic molded packages and the proper selection of materials and rigid process controls.

2.3.1 Molded Plastic Technology

In the molded plastic technology, the package can be either postmolded or premolded. Of the two, the postmolded package technology is the more cost effective of the two and by far the predominant package technology in use today.

2.3.1.1 Postmolded plastic package. A metal leadframe, usually copper, copper alloy, or Alloy-42 (42 percent Ni, balance Fe), provides the mounting surface (support paddle) for the IC. The leadframe also provides the electrical fanout path from the bonding fingers to the outside leads or pads. The leadframe is usually spot-plated with silver or gold on the paddle and at the tip of the wirebond fingers to provide reliable die attachment and wirebonds. Silver spot plating is preferred for lowest cost and better adhesion to the molding compound. Wirebond interconnects are gold wires, typically 0.80, 0.90, 1.00, or 1.25 mil (0.020, 0.023, 0.025, or 0.032 mm) in diameter, with maximum wire span lengths ranging from 0.10 to 0.25 in (2.5 to 6.3 mm) depending on the wire diameter (i.e., shorter wire span with smaller diameter wire).

After the IC has been wirebonded to the leadframe, thermosetting (cross-linking) epoxy resin is molded around the leadframe-IC subassembly. The external portion of the leadframe is then processed to the final form and lead finish for assembly to the next level. Figure 2.6 il-

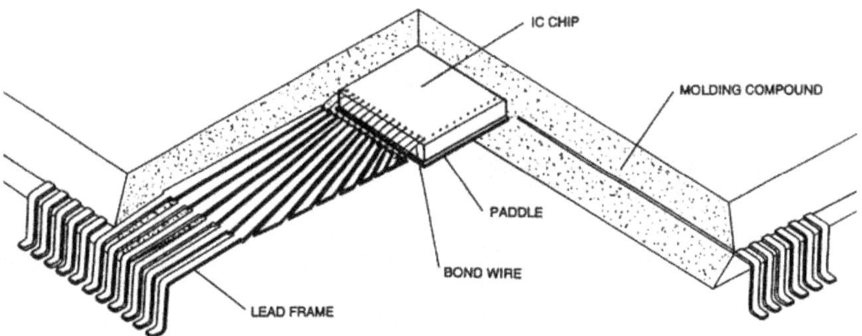

Figure 2.6 Sectional view of a typical QFP illustrating molded plastic (postmolded) technology.

lustrates the construction of a QFP in the postmolded plastic package technology.

The new family of leadless quad flat packs with no leads (QFNs, see Sec. 2.2.2) are fabricated in a batch process to a leadframe matrix. The IC is attached and wirebonded to the leadframe, similarly to the leaded QFPs. One side of the leadframe is molded, leaving the die attach pad and the perimeter pads under the package exposed. A mold-to-leadframe locking mechanism is created by back etching the leadframe between the inner edge of the fingers and the support paddle (Fig. 2.7). By redesigning the leadframe, a flip chip interconnection can be made, with the IC bumped in a single row on the outer perimeter.

Details, such as lead-frame design and format, workholders for die and wirebonders, molding dies, trim and form tools, and handling hardware, will vary according to package types.

2.3.1.2 Premolded plastic package. A premolded package concept, shown in Fig. 2.8, provides a more benign environment for packaging very sensitive ICs requiring low-cost assembly. Delicate wire spans and strain-sensitive features on the die surface are decoupled from the molding process, thus avoiding the stresses associated with postmolded packages.

The package is fabricated using a preplated leadframe, in strip form with multiple sites per strip, and then either transfer molded using a

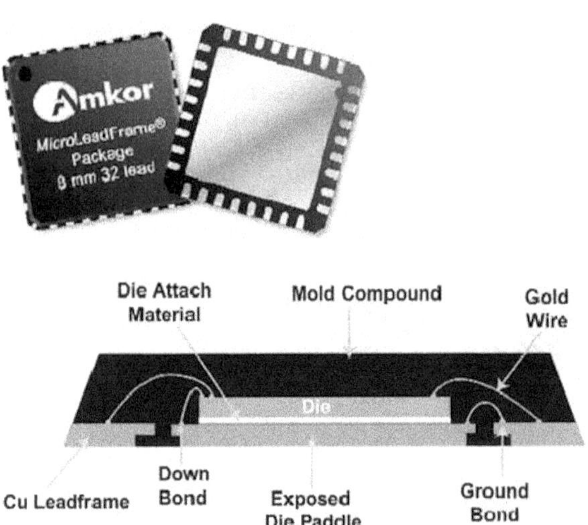

Figure 2.7 Amkor's MicroLeadFrame® package and cross section. *(Courtesy of Amkor Technology.)*

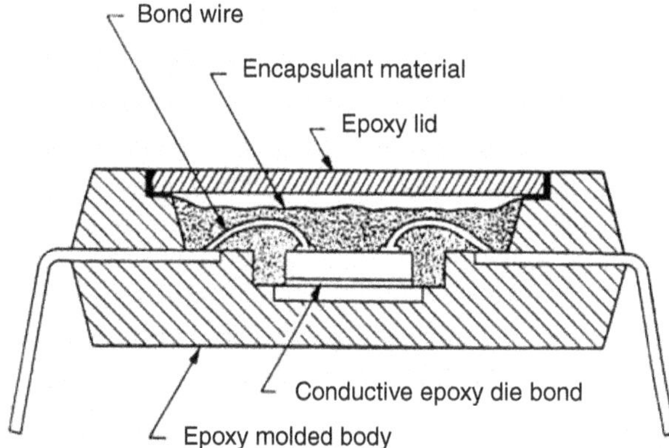

Figure 2.8 Cross-sectional view of a completed premolded plastic DIP, including IC chip.

thermoset epoxy B molding compound or injection molded with a thermoset or a thermoplastic polymer such as polyphenylene sulfide. The leadframe, usually made of a copper alloy for better thermal performance, is spot-plated with either silver or gold on the die-attach paddle area and on wirebond fingers. The external leads may be selectively plated with thin gold to retain solderability for final lead finishing. After molding, the package may need to be deflashed to remove any molding compound flash deposited on the die-bonding areas and then cleaned to remove any residual particulate matter. Die attachment uses a polymer adhesive, either conductive (silver-filled) or nonconductive epoxy, or polyimide paste. The die to leadframe interconnection is by thermosonic or thermocompression ball-wedge wirebonding using 0.80-, 0.90-, 1.00-, or 1.25-mil (0.020-, 0.023-, 0.025- or 0.032-mm) diameter 99.99 percent pure gold wire. This is followed by die encapsulation with a die coating or, preferably, a flow coating to fill the entire cavity, thus environmentally protecting the die and the wedge bond areas in the cavity. Candidate materials include room temperature vulcanized (RTV) silicone rubbers and combinations of silicon gels and cover coats. In some cases, a plastic lid is used instead of encapsulation. Back-end processes include trim and form to the final package configuration, i.e., DIP or plastic leaded chip carrier (PLCC). Final lead finish is then applied by solder dipping. Another option is to use the original thin gold plate as the final finish, making sure that solderability is not degraded in the assembly process. A gross leak tight-lid seal will preclude the subsequent ingress of cleaning solvent (used in PWB assembly) into the die cavity, where it could interact with the flow coat to cause swelling and stress the delicate wirebonds or degrade the interface integrity between die and encapsulant.

2.3.2 Pressed Ceramic (Glass-Sealed Refractory) Technology

Pressed ceramic technology packages are used mainly for economically encapsulating ICs that require hermetic sealing. Hermeticity means that the package, with the assembled IC, must pass both gross and fine-leak tests and also exclude environmental contaminants and moisture for a long period of time. Furthermore, any contaminants present in the package before lid sealing must be removed to an acceptable level during the sealing process. Glass is an effective material for achieving a hermetic seal for high-reliability applications. Leak rate, as defined by MIL-STD-883, is "that quantity of dry air at 25°C in atmospheric cubic centimeters per second flowing through a leak or multiple leak paths from a high-pressure side of one atmosphere (760 mm Hg absolute) to the low-pressure side ≤1 mm Hg absolute." Standard leak rates are expressed in units of atmosphere cubic centimeters per second ($atm \cdot cm^3/s$). Procedures for detecting gross and fine leaks in hermetic packages are detailed in MIL-STD-883. Gross-leak tests involve die penetrants and bubble tests, whereas fine-leak testing procedures use helium and radioactive tracer techniques.

Pressed ceramic piece parts require fewer and simpler process steps, resulting in the lowest-cost piece parts for a hermetic package technology. This technology, when implemented with a leadframe in strip form (that is, multiple sites per strip), can be automated and therefore can be competitive with the plastic technology for low-cost packaging. There are three pressed ceramic options: (1) basic piece parts (bases and caps), (2) partially assembled subassemblies (embedded leadframe and base), and (3) a prefabricated package also known as a window-frame package.

In the first option, the package is constructed around the IC using basic piece parts, i.e., ceramic base, leadframe, and ceramic cap. When done in house, this is the lowest-cost option, because the IC manufacturer controls all of the assembly process steps. Option 2 is a minor variation of option 1, where the IC manufacturer purchases the leadframe already embedded into the base ceramic and ready for die mounting. This option is attractive to manufacturers who do not have, nor wish to invest in, the equipment needed for embedding the leadframe. These options are known in the industry as the CERDIP or CERQUAD packaging technology. The assembly structure is shown in Fig. 2.9*a*.

In both options, the IC is attached to the leadframe using either AuSi eutectic or silver-loaded glass adhesive technologies. Hermetic sealing is done with a glassed ceramic cap or lid. Due to the relatively high temperature used for glass sealing, typically greater than 400°C, an all-aluminum monometallic system is used. This consists of the

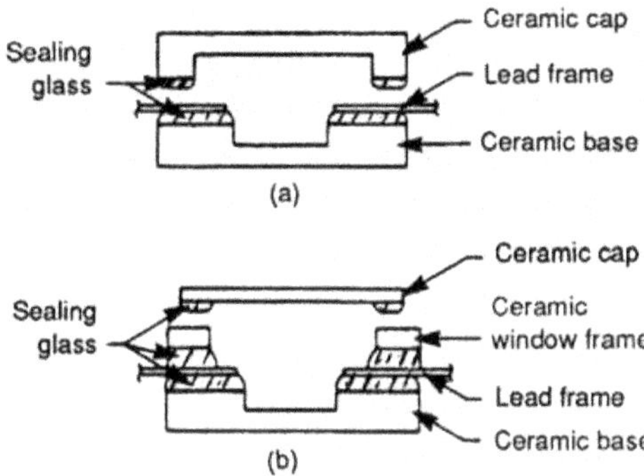

Figure 2.9 Structures of *(a)* CERDIP and *(b)* window frame. *(Courtesy of Kyocera America, Inc.)*

aluminum bond pads on the IC, ultrasonic aluminum wedge-to-wedge wirebonding, and leadframe fingers that have aluminum deposited or clad surface finishes in the bond areas. Lid sealing is done in a controlled nitrogen atmosphere conveyor-type furnace in which the heat is transferred by conduction and convection or by infrared radiation. Back-end processes include leak testing, lead finishing, and lead trim and form. External leads are typically plated with matte tin, which provides a finish that is compatible with the subsequent lead trim and form operation and final assembly to the next-level of packaging. CERDIP leadframes are procured already formed and require only to be finished and trimmed prior to testing. Leadframes for SM types such as the CERQUAD usually come planar and are processed planar through lead finishing, then trimmed and formed into the desired shape for the final package configuration.

The third pressed ceramic option, known in the industry as a window-frame package, is the structure shown in Fig. 2.9*b*. This design concept provides the option of decoupling the IC from the high-temperature glass-sealing procedure by providing a low-temperature solder seal capability for the final hermetic seal. This option is used for temperature-sensitive ICs that cannot go into a glass-sealed package without degrading their performance or reliability. In addition, this option also allows the use of thermosonic or thermocompression gold ball-wedge wirebonding equipment, which extends this package technology to those manufacturers who have the wirebonding capability but not the ultrasonic aluminum wedge-to-wedge bonders needed to achieve the desired all-aluminum monometallic structure for glass seals. Back-end processes are similar to those used with the other options.

Seal glass is an important constituent of the pressed ceramic technology, because it influences both electrical performance and reliability of the packaged device. The dielectric constant of the glass used to embed the leadframe can affect the capacitance between the leads, i.e., the speed of the I/O signals through the leadframe. Seal glass is exposed to the cavity; hence, alpha-particle emission from the glass fillers may affect the soft error reliability of some sensitive devices. The hermeticity of the finished package is influenced by (1) the adhesion of the seal glass to the ceramic, (2) the seal glass to the leadframe, (3) the mechanical strength of the glass, and (4) the dissolution of fillers in the mother glass. Any residual stress developed in the mother glass is caused by the mismatch in the thermal expansion between the ceramic, Alloy-42 leadframe, and the glass. The particle-size distribution of the filler in the glass also influences the mechanical strength or the ability to provide reliable hermetic seals.

The raw material for the pressed ceramic technology is alumina ceramic (90 percent minimum Al_2O_3) in a black-brown characteristic color. Package vendors are also providing other ceramic materials, such as aluminum nitride (AlN), silicon carbide (SiC), and low-thermal-expansion sealing glasses (4.7 ppm/°C), for improving hermeticity yields of pressed ceramic.

2.3.3 Cofired Laminated Ceramic Technology

The cofired laminated ceramic technology is one of the most reliable IC packaging technologies currently available. In addition, the technology is capable of meeting today's demands for higher performance at the device and system levels. This section addresses the technology for single-chip packages. The cofired laminated ceramic technology is also being used for such applications as hybrids and multichip packages (MCPs).

2.3.3.1 High-temperature cofired ceramic technology. A dispersion or slurry of alumina ceramic powder (90 to 92 percent Al_2O_3) in a liquid vehicle (solvent and plasticized resin binder) is first prepared, then cast into thin sheets by passing a leveling or doctor blade over the slurry. After drying, the sheets are in the green-tape stage, ready for cutting to size, punching of via holes (through the dielectric layers for future interconnection), and forming the cavities. Via holes are usually filled with a tungsten paste, and then custom conductor paths are screened onto the surface with the same metal paste. Several of these sheets are press-laminated together in a precisely aligned fixture, and the entire structure is fired at approximately 1500 to 1600°C, in a reducing atmosphere, to form a monolithic sintered body. The cofired laminated refractory ceramic technology is a complex process requiring careful control throughout.

After the laminate has been sintered, it is ready for the finishing operations of lead/pin attachment and metallization plating. For packages that require either leads or pins, nickel is first plated over the exposed tungsten in preparation for lead/pin brazing. The lead/pin material is either Kovar (an Fe-Ni-Co alloy) or Alloy-42, and the brazing material is an Ag-Cu eutectic alloy. After lead/pin brazing, all exposed metal surfaces are electroplated or electroless plated (usually gold over nickel) for bondability and environmental protection. A typical single-chip package could contain up to seven tape layers and four screened dielectric layers. The thickness of a taped layer, with tungsten-filled vias, may range from 0.002 to 0.025 in (0.05 to 0.64 mm). Screened dielectric layers are typically 0.001 to 0.003 in (0.025 to 0.076 mm) in thickness.

The cofired laminated ceramic package technology is very effective for constructing complex packages with signal, ground, power, bonding, and sealing layers. Electrical characteristics, such as line capacitance, propagation delay, impedance, and inductance, can be customized for a particular application by design layout and specification of layer thicknesses.

This technology, however, has three technical areas of weakness: hard-to-control tolerances caused by high shrinkage in processing, a high dielectric constant of 9.5 (which affects signal-line capacitive loading), and a modest thermal conductivity of Al_2O_3. The use of beryllia (BeO), AlN, or SiC instead of the 90 to 92 percent Al_2O_3, will result in a greatly superior thermal performance and a significantly lower dielectric constant.

Die attachment is done with either an AuSi eutectic, AuSn eutectic, or a polyimide system to achieve a high yield and highly reliable mechanical and electrical interconnection. AuSi and AuSn preforms vary from 99.99 percent pure gold, with dopants to provide a reliable electrical back contact, to alloys of a variety of constituents, such as 80Au/20Sn for lower eutectic temperature bonding. Polymer die attachment in hermetic packages has also been used. Filled and unfilled polyimides were found to withstand the hermetic assembly processing temperatures. Electrical interconnections are generally done using thermosonic ball-wedge gold wirebonding with wire diameters typically at 1.00 mil (0.025 mm). Wire spans are usually limited to a maximum of 0.150 in (3.8 mm) to meet shock, vibration, and acceleration requirements. Excessively long wire spans are avoided by going to a two tiered, or in some cases three-tiered, wirebond construction, as similarly shown for a PPGA package in Fig. 2.10. Final hermetic sealing is done with an 80Au/20Sn braze, in a dry nitrogen ambient, either in a belt furnace with a temperature profile peaking at about 350°C or in a controlled-ambient glove box containing a parallel-gap

Figure 2.10 Tiered ball-wedge wirebond configuration in a PPGA package. *(Courtesy of Agere Systems.)*

seam sealer. The seam seal brazing operation confines the heat locally to the seal ring area; thus, the package and IC are not significantly heated by the operation. The environment in the sealed package is predominantly nitrogen, and the moisture content is typically at or below 1000 ppm H_2O by volume, which meets the military standard requirement for high-reliability applications.

Back-end processes are similar to those for pressed ceramic, i.e, leak test, lead trim or lead trim and form, and sometimes lead finish for SM packages. The most popular lead finishes available range from electroplated gold for soldering and socketing to tin plating for soldering applications only. In between, there are several other combinations involving alloys of Au-Pd-Ni, Au-Ni, and electroless gold and nickel. Some of these finishes are not compatible with one or more of the assembly processes and may degrade solderability to the point at which the SM solder reflow quality may be jeopardized. Leads with thick gold or alloys with significant amounts of gold will cause brittle solder joints that cannot be tolerated in SM. Fine-pitch leaded ceramic chip carriers (LDCCs) may require a solder dip coating to achieve high yields and reliability.

Ceramic packages come with brazed leads/pins for SM or TH mount and leadless perimeter or array pads for SM. The package is delivered to the IC assembly manufacturing line with leads/pins, cavity, and bond sites finish-plated to customer specification and ready for all subsequent assembly steps.

Dual in-line package (DIP). A typical DIP construction (Fig. 2.11*a*) consists of three cofired laminated ceramic layers with a chip cavity formed by

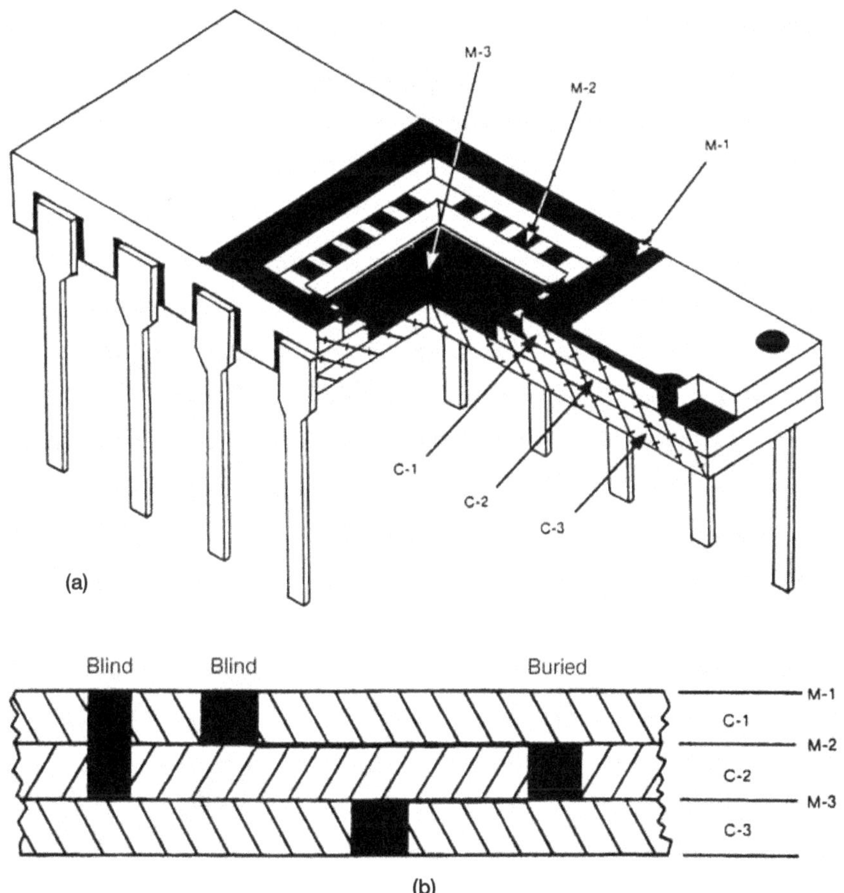

Figure 2.11 *(a)* Isometric sectional view of cofired laminated ceramic technology in DIP configuration and *(b)* cross section illustrating blind and buried vias in structure of part *(a)*.

the middle layer and a wirebond ledge formed by the top layer. There are two buried metallization layers and one top-layer in the construction. The bottom buried layer provides the metallization for the chip cavity base, while the top buried layer provides the signal fanout path and bond pad from the wirebond ledge to the external pins. The different levels can be interconnected by vias, as shown in Fig. 2.11*b*. Top-layer metallization provides the seal ring for hermetic sealing. The seal ring may be tied to a ground plane through a via, which may be located near the notch, as illustrated in Fig. 2.11*a*. The same construction applies to a wide variety of DIP configurations, such as the package with two cavities where the ICs are electrically interconnected to each other (Fig. 2.12).

Ceramic pin grid array package (CPGA). As demand for high I/O pin counts increased, the DIP, due to its size limitations, was no longer able to fulfill the IC packaging needs. As a result, a TH-mount ceramic pin grid array (CPGA) package evolved to accommodate the increase in I/Os at an effective pin count to area package configuration.

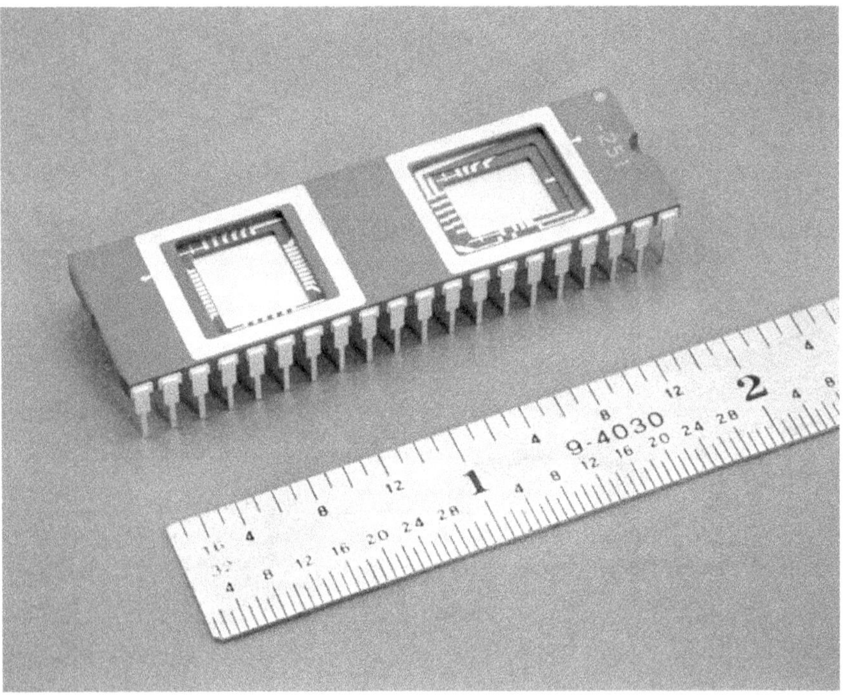

Figure 2.12 Ceramic dual in-line package (DIP) with two electrically interconnected cavities. *(Courtesy of Agere Systems.)*

The CPGA is a ceramic cofired multilayer package (Fig. 2.13) with a cavity facing up or down to accommodate the IC, and pins attached to the underside in an array pattern. The base of the cavity is normally gold plated for eutectic die bonding. Within the cavity are wedge bond pads, located on single or dual ledges, for wirebonding the IC to the package. The CPGA is manufactured of alumina (90 to 92 percent Al_2O_3) or aluminum nitride (AlN) ceramics and follows the same processing steps as for any other cofired laminated ceramic multilayer package (e.g., the DIP).

Kovar or Alloy-42 (a Fe-Ni-Co alloy), 0.018-in (0.46-mm) diameter pins are brazed on to the underside of the package, in an array configuration, at a 0.100-in (2.54-mm) pitch or staggered rows. This results in an effective pitch of 0.050 in (1.27 mm) in-between rows. The four corner pins have small tabs brazed-on at approximately 0.050 in (1.27 mm) from the package base so as to provide a stand-off height when the CPGA is TH mounted to the PWB. The stand-off height provides pin compliance and facilitates cleaning under the package. The finish on the pins is either gold or solder plated. A metallized seal ring

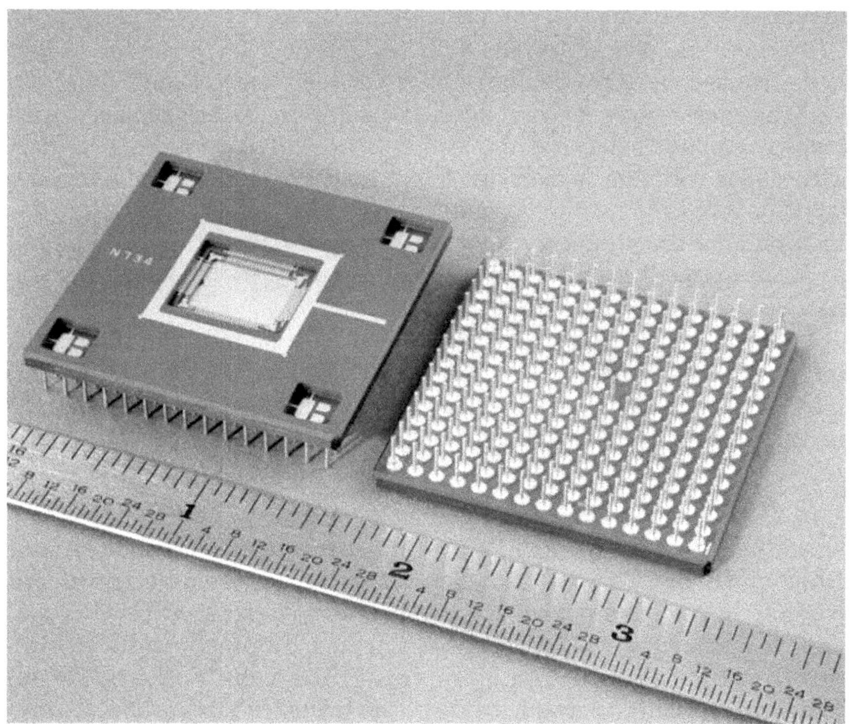

Figure 2.13 Ceramic pin grid array (CPGA) package. *(Courtesy of Agere Systems.)*

around the cavity provides for either seam sealing or eutectic bonding a metal lid that hermetically seals the chip cavity.

Ceramic ball grid array (CBGA) package. With the advent of high-density, high-pin-count. and high-performance requirements for SM packages, the ceramic ball grid array (CBGA) evolved to meet related demands. The CBGA is constructed in a similar fashion as other laminated ceramic packages but, instead of perimeter leads and area array pins, the CBGA contains area array pads for solder ball attachment. The substrate consists of a single or several cofired laminated ceramic layers, with tungsten or molybdenum metallization and with vias interconnecting the circuitry on the top side to the solder pads on the bottom side. The IC is eutectically attached to the ceramic substrate and interconnected to the substrate by wirebonding or by flip chip techniques. The substrate is then lidded or capped to protect the IC, and solder interconnects are attached to the underside to complete the assembly. The solder interconnects consist of high-temperature 90Pb/10Sn solder balls, which are attached to the substrate with eutectic 63Sn/37Pb solder. The solder balls are 0.030 in (0.75 mm) in diameter for grid arrays on 0.050 in (1.27 mm) pitch and 0.025 in (0.63 mm) in diameter for a 0.040-in (1.0-mm) pitch.

To overcome the thermally induced fatigue strain on the solder balls, caused by the CTE mismatch between the ceramic substrate and the glass-epoxy motherboard, the solder balls are replaced with more compliant solder columns; thus, the CBGA becomes a CCGA. The solder columns, which are 0.020 in (0.50 mm) in diameter and 0.050 or 0.087 in (1.27 or 2.2 mm) high, can be directly attached to the underside of the alumina substrate, or 63Sn/37Pb solder can be used as an interface for attachment. The different types of solder interconnects are shown in Fig. 2.14.

The CBGA and CCGA packages offer high reliability and high performance with interconnects exceeding 2000 at 0.040-in (1.00-mm) pitch. For packages with a fewer number of interconnects, the area array pitch can be as small as 0.020 in (0.5 mm).

2.3.3.2 Low-temperature cofired ceramic technology. Multichip packages (MCPs) with functional substrates are gaining in popularity for RF applications. The most common use of the low-temperature cofired ceramic (LTCC) technology is to provide embedded inductors, capacitors, and resistors for functional substrates. The manufacturing processes of the LTCC substrate are exactly the same as those of high-temperature cofired ceramics, except the firing for the LTCC substrate can range from 850 to 1000°C. Most of the Japanese manufacturers use their own proprietary material and green tape for LTCC manufac-

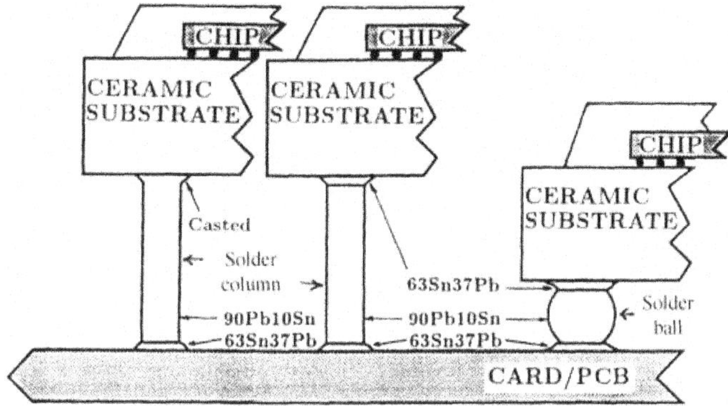

Figure 2.14 IBM's ceramic ball grid array (CBGA), showing different types of solder interconnects and attachments. *(After Lau.[3])*

turing. Several other manufacturers in the U.S. and Europe may use green tapes provided by DuPont. The most common metallizations in use for vias and internal conductors are Cu and Ag. For the external conductors, Cu, Pt/Ag, or Pd/Ag can be used. The LTCC technology faces a number of challenges, such as cost, shrinkage factors, quality, and accuracy of the embedded component.

2.3.4 Laminated Plastic Technology

The laminated plastic packaging technology is a spin-off of the traditional PWB technology wherein bare chips are mounted directly to the substrate in a technique known as chip on board (COB). Early vintage laminated plastic packages used FR-4 epoxy substrates, which exhibited high contamination levels (due to the flame-retardant fillers) and suffered from a low glass transition temperature (140°C), making them incompatible with many standard IC package assembly processes. Substrate materials such as high-temperature epoxies, polyimides, and triazines have replaced the deficiencies of FR-4 substrates. These materials exhibit glass-transition temperatures in the range of 180 to 240°C and have a much lower level of contamination than the FR-4 substrates.

The substrate may have single-sided, double-sided, or multilayer circuitry, depending on the application needs, and can be provided with or without a cavity to house the IC. The exposed copper metallization is generally electro- or electroless plated with nickel and gold to prevent oxidation and provide wirebondable pads. The most popular package types using this technology are the plastic pin grid array (PPGA) for TH mounting and the plastic ball grid array (PBGA) for SM.

2.3.4.1 Plastic pin grid array package (PPGA). The PPGA, illustrated in Fig. 2.15, is a nonhermetic, prefabricated package that houses the IC, providing electrical interconnection to the motherboard and protection from hostile environments. The package consists of a square plastic PWB body with round pins press-fitted and solder reflowed to the underside. The pins can be configured in any desired footprint as long as the pin field is in a straight or staggered array of 0.100-in (2.54-mm) pitch and is restricted to the outside of the die attach and wirebond pad areas. The pin array configuration results in the highest density of pins per package area and highest pin count of any TH type package. PPGAs come in various forms: chip-down facing the motherboard or chip-up mounted to the top of the package, double-sided or multilayer substrate construction, and with or without a chip cavity. In addition, thermally enhanced PPGAs may also contain plated THs as "thermal vias" under the die or a copper slug to spread the heat dissipated by the IC.

The PPGA package is constructed by first fabricating the substrate from a rigid double-sided or multilayer PWB, then patterning the Cu circuitry by either the subtractive or semiadditive process. The substrate contains plated-through holes and, depending on design complexity, may also have buried and blind vias for electrical interconnection.

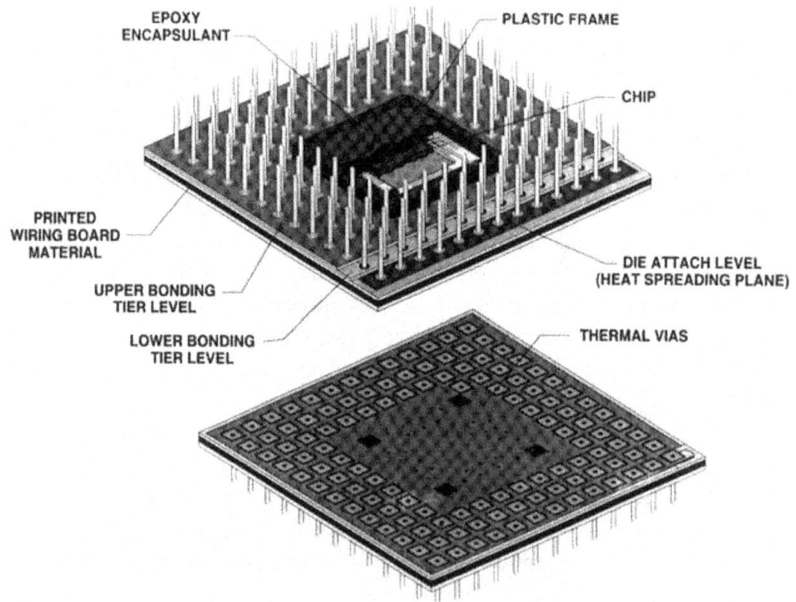

Figure 2.15 Multilayer PPGA with chip cavity facing down. *(After Cohn et al.*[6]*)*

Pins are made of phosphor bronze, typically 0.018 in (0.46 mm) in diameter, instead of the more expensive Kovar or Alloy-42 used with ceramic packages. The solder-plated (60Sn/40Pb or 90Pb/10Sn) pins are inserted through gold or solder plated-through holes in the substrate, and the connection is made by a press fit and solder reflow. A star or knurl is incorporated into the shank of the pin to facilitate a more reliable interconnection.

Solder reflow is accomplished by either dipping the entire length of the pin into a solder bath or laser spot heating the base of the pin. During solder dipping, the solder wicks up into the interface between the plated hole and the shank of the pin and at the same time forms a fillet at the base of the pin. This provides for a strong, reliable contact but results in a variable solder thickness along the length of the pin, which may affect solderability when tested per MIL-STD-883.

Another solder reflow process uses a laser beam to spot-heat the base of individual pins at a rate of approximately 25 pins per second, causing the solder in the hole-to-pin interface to reflow. Sufficient solder must be available at the hole-to-pin interface so that, when reflowed, it will produce a strong interconnection. Since laser reflow heating only affects the base of the pin, the original solder plated thickness along the length of the pin remains unchanged.

The IC is assembled to the PPGA using most of the same assembly processes and techniques used for the plastic molded technology. The IC is first attached using one of the polymer adhesive systems, such as silver-filled epoxies or polyimides, then wirebonded by either the thermosonic gold ball-wedge or the ultrasonic aluminum wedge-wedge process. The IC is then polymer-coated for environmental protection using a "glob-top" liquid epoxy, flow-coated silicone gel, or RTV rubber. When encapsulating an IC that is not in a cavity, it is desirable to use a resin dam to confine the encapsulant to the IC and wirebonded areas. A metal or nonmetal lid may be attached in a chip-up configuration, mainly for physical protection of the IC when gels or RTV rubber systems are used.

The PPGA comes in either a low-cost or an enhanced, high-performance configuration, depending on the application and customer's design objective. The low-cost version, shown in Fig. 2.16, consists of a conventional 0.060-in (1.5-mm) thick, copper-clad laminate processed as a single- or double-sided substrate with no cavity. Pins are press fitted and solder reflowed. The IC is encapsulated with either a "glob-top" liquid epoxy or covered with a silicone gel or RTV and lidded.

A performance-driven device will require a multilayer substrate with power and ground planes and multiple bonding tiers to yield the appropriate interconnection density within the allowable wirebond design rules. Liquid epoxy or a lid is used to encapsulate the IC. Pins

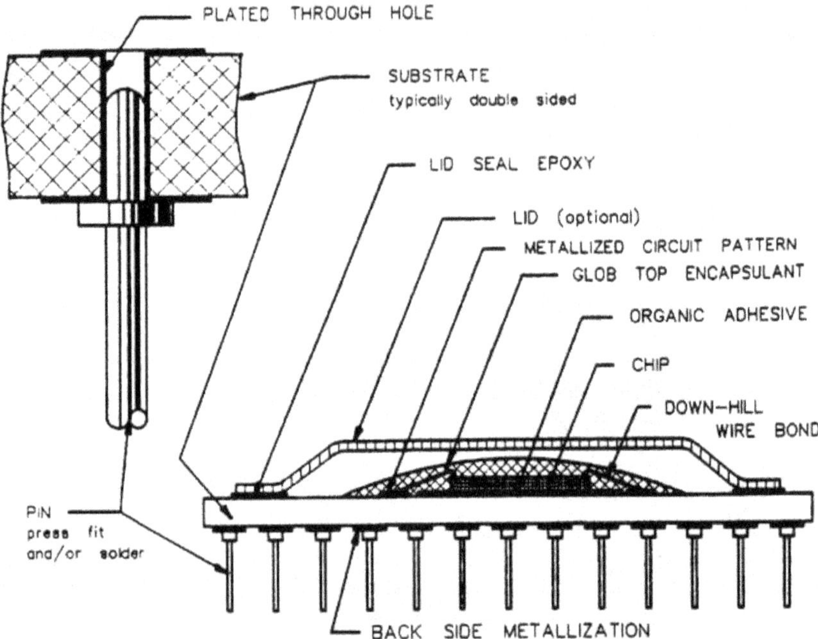

Figure 2.16 Cross section of a low-cost PPGA in laminated plastic technology. The substrate is either single- or double-sided P'WB with no cavity for IC. The chip is environmentally protected by glob-top epoxy encapsulant. A metal or plastic lid is optional for physical protection.

may be inserted to provide either a cavity-up or a cavity-down configuration. A typical cross section through the chip cavity of such a laminated PPGA, showing the multilayer construction, two-tier bond pads and thermal vias, is illustrated in Fig. 2.17. For ICs dissipating approximately ≥3 W, and depending on the environmental conditions and the maximum IC junction temperature (T_j) allowed, it may be necessary to mount the IC directly to a heat-spreading copper slug of a cavity-down package configuration as shown in Fig. 2.18.

The low dielectric constant of PPGA substrate materials (in the range of 4 to 5) and the ability to provide thermal vias under the die, or a heat-spreading copper slug, result in equivalent or better electrical and thermal performance when compared to cofired laminated ceramic CPGAs using Al_2O_3.

2.3.4.2 Plastic ball grid array (PBGA) package. Another IC package that utilizes the laminated plastic technology is the plastic ball grid array (PBGA) package. The PBGA (Fig. 2.19) leverages on the PPGA technology but replaces the pins with solder balls for surface mounting (SM).

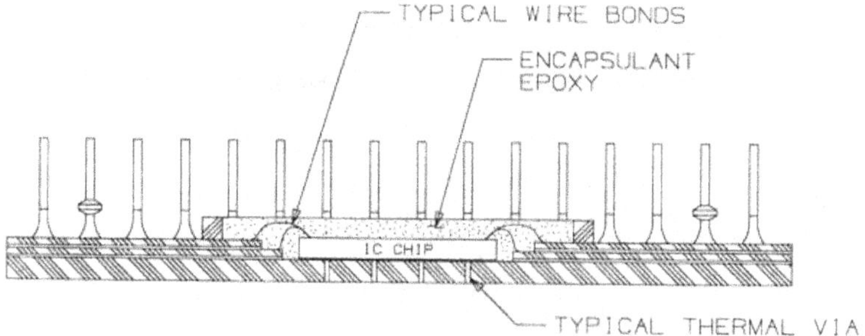

Figure 2.17 Cross section of a multilayer PPGA with chip cavity facing down.

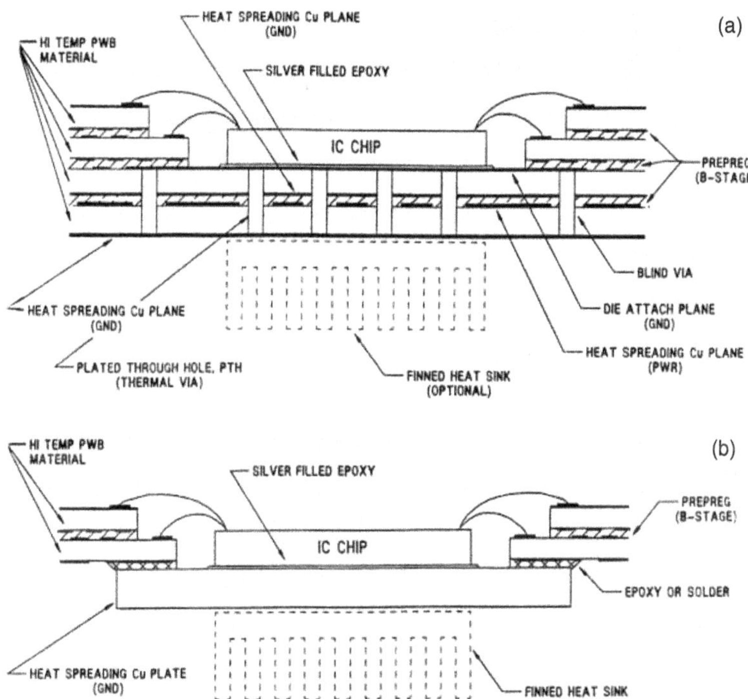

Figure 2.18 *(a)* PPGA package thermal enhancements using heat-spreading Cu planes and thermal vias, and *(b)* PP'GA package thermal enhancements using a heat-spreading Cu plate. *(After Cohn et al.*[6]*)*

PBGAs were first introduced by Motorola in 1989 for low-I/O devices and were then called *overmolded pad array carriers (OMPACs)*. The PBGA has since emerged as the SM package of choice, because it offers many attractive features, such as the following:

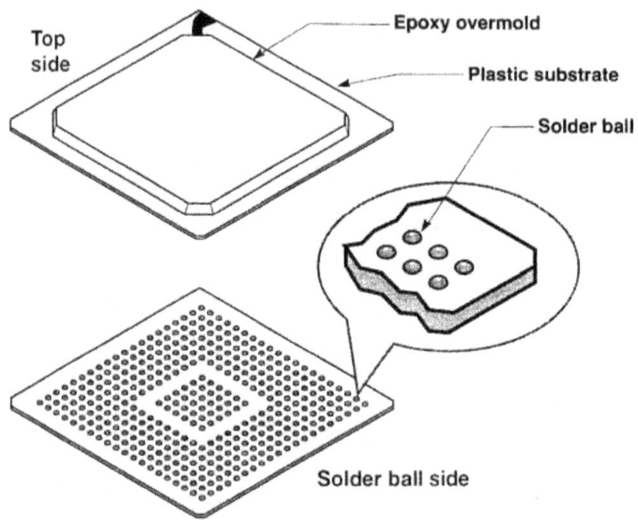

Figure 2.19 Plastic ball grid array (PBGA) package (overmolded type).

- High I/O to package area ratio
- Multilayer substrate capability
- Improved thermal/electrical performance
- Increased interconnect density
- Ball coplanarity less critical during SM, ≈0.006 in (0.15 mm) to 0.008 in (0.20 mm)
- Self-centering during SM solder reflow
- Elimination of fine pitch solder paste printing
- Pad-to-pad shorting diminished
- Low profile
- Expected solder joint defect level <5 PPMJ

A PBGA package consists of an IC mounted and interconnected to a square or rectangular, double-sided or multilayer PWB substrate. Typically, plated-through holes interconnect the top surface signal traces to the respective solder pads on the underside of the substrate. After die attachment and wirebonding, the assembly is overmolded using a transfer or injection molding process. Another encapsulation method is to "glob top" the IC and its bond wires with a liquid dispensed epoxy, which is confined within a resin dam or a chip cavity. Attaching solder balls to the underside of the package then completes the assembly.

The PBGA substrate provides structural support for the IC, encapsulant, and solder balls. It also facilitates electrical interconnection between the IC and motherboard and helps dissipate the power generated by the IC. Some of the issues encountered with substrates are electrical interference to the IC's functionality (electrical noise, current leakage, high trace resistance), affinity for excessive water absorption that may lead to internal delamination (popcorning), warpage affecting coplanarity, and CTE mismatch that may affect solder joint reliability. Proper design and material selection can minimize these issues.

The PBGA substrates are usually fabricated in singulated or strip (multiple sites) form (Fig. 2.20), according to customer's design specifications, and then shipped to the semiconductor manufacturer for assembly. When the substrates are fabricated in strip form, with multiple sites, the vendor identifies by marking ("x-out") those sites that are mechanically or electrically defective. During assembly, the chip pick-and-place machine avoids placing ICs on an x-out site. The substrates may be either selected from standard open tooled artwork configurations, available from vendor stock, or custom designed to fit special applications. General-purpose substrate designs help reduce assembly turnaround time and inventory logistics problems. In addition, open tool designs eliminate tooling charges and take advantage of possible volume discounts.

The most common PBGA substrate material used in the industry today is bismaleimide triazine (BT), although other materials are being used as well, but to a lesser extent. The BGA pads are sized according to the solder ball diameter used and second-level solder joint reliability effects. The pads are arranged in a standard JEDEC area array on a pitch of 0.020, 0.026, 0.030, 0.040, or 0.050 in (0.5, 0.65, 0.8, 1.0, or 1.27 mm). The exposed Cu metallization is electroplated with nickel and gold to prevent oxidation and facilitate wirebonding to the bond pads and solder ball attachment to the solder pads.

The composition of the solder balls is 63Sn/37Pb or 62Sn/36Pb/2Ag or, for lead-free designs, 95.5Sn/4.0Ag/0.5Cu. The solder ball diameter varies from 0.012 to 0.030 in (0.30 to 0.75 mm) and depends on the ball pitch used for the given application. Table 2.2 shows the industry's commonly used sizes for solder ball/pad diameter versus ball pitch.

Where device heat dissipation is critical, the PBGA substrate design may contain plated THs as "thermal vias" under the die to provide a direct path for the heat to flow to the solder balls and dissipate into the motherboard. PBGAs contain many materials with various coefficients of expansion, thermal conductivities, and elastic behaviors (Table 2.3). Maintaining adequate thermomechanical performance, mechanical tolerances (e.g., flatness, coplanarity), moisture sensitivity, and overall

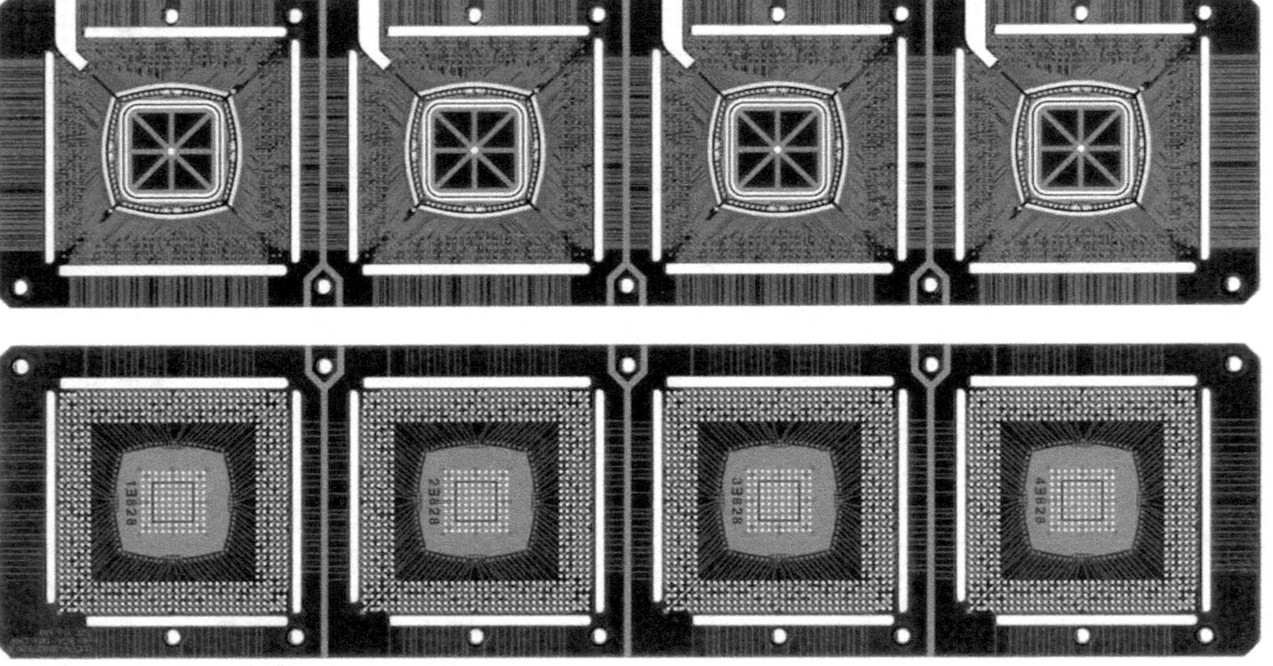

Figure 2.20 Laminate strip with slots outlining the four BGA substrates. *(Courtesy of Agere Systems.)*

Table 2.2 Typical PBGA Solder Ball Pitch vs. Ball/Pad Diameters

		SMD		NSMD
Solder ball pitch, mm	Solder ball dia., mm (mils)*	Cu pad dia., mm†	Solder mask opening dia., mm	Cu pad dia., mm
0.5	0.30 +0.10/–0.05 (12)	0.35	0.25 ± 0.05	0.30
0.65	0.30 +0.10/–0.05 (12)	0.35	0.25 ± 0.05	0.25
0.8	0.40 +0.10/–0.05 (18)	0.42	0.32 ± 0.05	0.32
	0.50 ± 0.10 (20)	0.55	0.45 ± 0.05	0.45
1.0	0.50 ± 0.10 (20)	0.55	0.45 ± 0.05	0.45
	0.63 ± 0.10 (25)	0.55	0.45 ± 0.05	0.45
	0.63 ± 0.10 (25)	0.63	0.53 ± 0.05	0.53
1.27	0.75 ± 0.15 (30)	0.73	0.63 ± 0.05	0.63

*Before solder reflow.
†Based on solder mask positional tolerance (metal edge to solder mask edge) of ±50 μm.

Table 2.3 Typical PBGA materials

Item	Specification
Substrate material	High glass transition temperature (T_g > 175°C) Epoxy/glass laminate—bismaleimide triazine (BT), FR-5, etc.
Metallization	0.5 oz (18 μm) min. Cu on outside layers and 0.5 oz (18 μm) min., 1 oz (35 μm) min. or 2 oz (70 μm) min. Cu on internal layers (depending on power dissipation)
Plating	0.3 μm min. Au over 4.0 to 20.0 μm Ni over Cu
Die attachment	Silver-filled conductive epoxy or nonconductive epoxy
Bonding wire	0.9, 1.0, or 1.20 mils dia. gold wire
Encapsulant	Epoxy overmold or dispensed epoxy
Balls	Eutectic solder, e.g., 62Sn/36Pb/2Ag, 63Sn/37Pb or, for lead-free devices, 95.5Sn/4.0Ag/0.5Cu

device reliability with these materials is a major challenge to the industry.

Figure 2.21 shows the cross section of a typical low-cost, double-sided PBGA substrate with the IC mounted to the top surface, wirebonded, overmolded, and with solder balls attached to the underside. For enhanced power dissipation, the substrate contains "thermal vias" under the die and a center array of thermal solder balls that are nor-

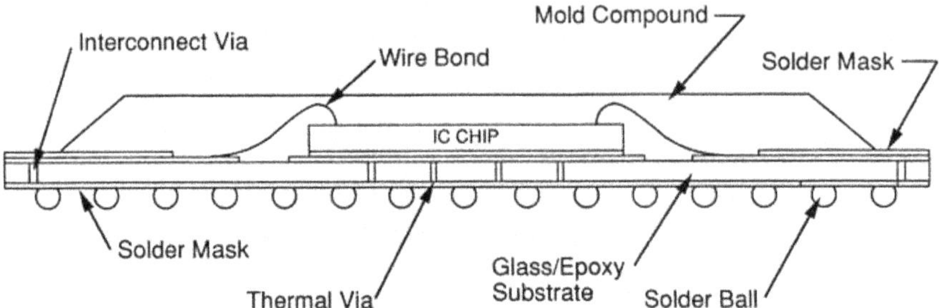

Figure 2.21 Cross section through a PBGA with a double-sided metallization substrate. *(After Cohn et al.[4])*

mally grounded. In a high-performance PBGA configuration, the substrate is multilayered with internal power and ground planes for improved electrical inductance and heat-spreading capability (Fig. 2.22). The internal Cu planes are normally 1 oz (0.035 mm) thick, but, for better heat spreading and lower junction-to-air thermal resistance (approximately 10 percent improvement, depending on substrate size), a 2-oz (0.070-mm) Cu thickness is used. The substrate for a high-power-dissipating device typically contains a chip-down cavity with single or double bonding tiers and a heat-spreading copper slug to which the IC is directly attached (Fig. 2.23).

Wirebonding is the leading IC interconnection process used today for PBGAs, with flip chip bonding being a close second.

Each substrate size (27 × 27 mm, 35 × 35 mm, and others) has a given plastic overmold configuration that does not change with different artwork or ball count. This enables the package designer to take advantage of existing mold cavities.

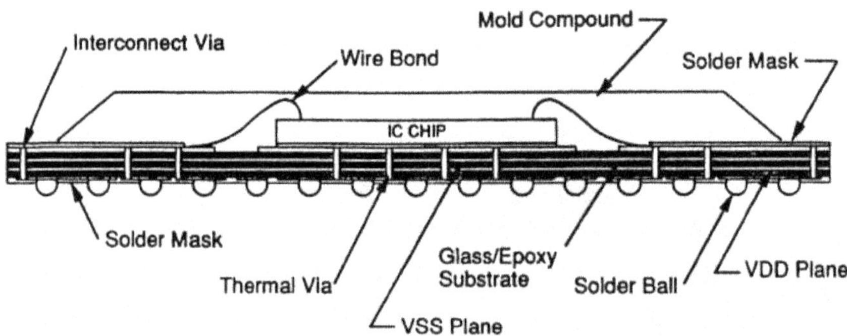

Figure 2.22 Cross section through a PBGA with a multilayer substrate. *(After Cohn et al.[4])*

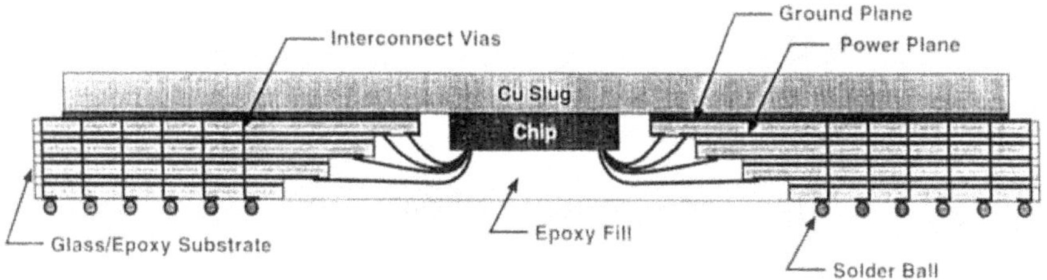

Figure 2.23 Typical cross section through a thermally enhanced PBGA.

A typical PBGA assembly sequence, processed in strip form with wirebond interconnection, consists of the following:

- A conductive (silver filled) epoxy or a nonconductive epoxy is transfer printed onto the die attach area, the IC is placed over the epoxy and the assembly is cured.
- The IC is thermosonically ball-wedge gold wirebonded and visually inspected.
- The strip sites are overmolded or "glob topped" with a liquid dispensed epoxy resin system and cured.
- After overmolding, the sites are code marked, and solder balls are attached to the underside of the individual sites.
- The individual PBGA devices are separated from the strip by routing or punching, then inspected, cleaned, electrically tested, burned-in (when required), electrically retested, placed in bakeable shipping trays, baked/bagged, and shipped to customer.

2.3.4.3 Tape BGA (TBGA). A tape BGA is defined as an array-type package having a copper/polyimide flex tape for a substrate. The tape contains one or two metallization layers interconnected by plated-through holes. Tape substrates have been utilized in matrix BGAs, thermally enhanced cavity-type packages, low I/O flip chip BGAs, and so on.

In a cavity-type TBGA (Fig. 2.24), one side of the tape contains the wedge bond pads and the circuitry to fan out to the solder ball pads. The metallization layer on the other side of the tape can be utilized as a ground plane.

Another example of a TBGA package (Fig. 2.25), developed by IBM, utilizes flip chip (FC) bonding to a copper/polyimide tape containing two metallization layers. The top circuitry layer fans out the I/Os from the IC to solder ball pads on the bottom layer via plated-through holes. The remaining space between the BGA pads is utilized as a

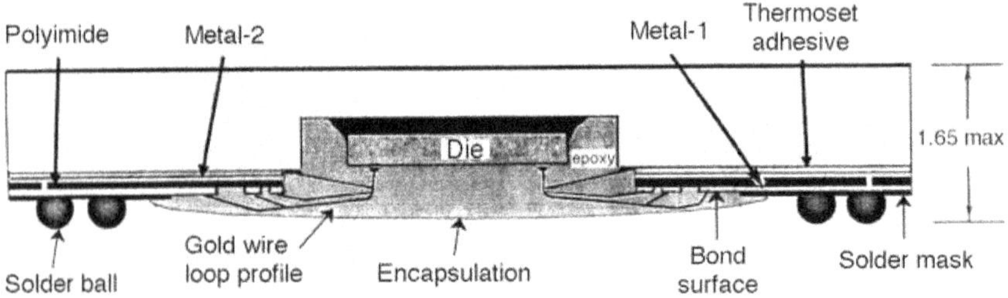

Figure 2.24 Cross section through a thermally enhanced tape BGA. *(After ASAT.*[8]*)*

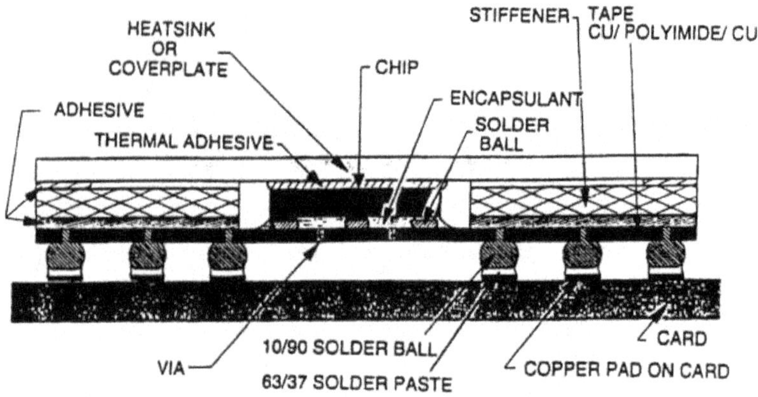

Figure 2.25 Cross section through a flip chip tape BGA. *(After IBM.*[10]*)*

functional ground plane. The circuitry is formed by electroplating Cu onto the polyimide tape, as defined by the resist mask. A stiffener ring is attached to the tape, using an adhesive, to provide rigidity and solder ball coplanarity to the package.

The interconnection between the IC and the tape is accomplished by a modified solder bumped flip chip process developed by IBM, which consists of an IC with 97Sn/3Pb solder bumps in an area array placed on matching footprints on the tape. Bonding to the tape is achieved by a partial reflow of the solder bumps without flux while applying a slight pressure to the bumps.

An epoxy underfill is applied between the die and the substrate tape. The underfill provides moisture protection to the die surface and reduces the thermal stresses induced in the solder joints due to coefficient of expansion mismatches of the bonded materials.

On the BGA side, 0.025-in (0.635-mm) diameter, 10Sn/90Pb solder balls on a 0.050-in (1.27-mm) pitch are attached to the pads on the tape, using a resistance heating process. At the center of each BGA pad is a 0.008-in (0.200-mm) plated-through hole that is used for electrical interconnection and to anchor the solder ball. To enhance the thermal characteristics of the TBGA, a Cu plate is attached to the stiffener ring and to the back of the die to provide heat spreading.

As a result of the relatively short conductor lengths and close proximity of the ground plane, the FC-type TBGA has low signal-line inductances as well as low power and ground inductances.

The capability to route higher-density circuitry on polyimide tape is an advantage of tape over laminate substrates. This advantage is slowly eroding because of advances in laminate technology, which have resulted in laminate design rules that approach those of tape. In addition, the polyimide tape is more expensive than a laminate substrate of the same size, except at very high volumes.

2.3.4.4 Chip scale package (CSP). The CSP, which is slightly larger than the IC itself (max. 1.2 × the area of the die), comes closest to direct chip attachment but without the responsibilities usually imposed on the board assembler for handling, attachment, wirebonding, and encapsulation of known good die (KGD). Using CSPs, the above tasks are shifted back to the semiconductor manufacturer in that the device comes fully tested, burned-in, and ready for SM assembly to the PWB by conventional means, like any other IC package. Several chip scale package configurations were developed to provide a lower-cost, high-I/O, high-density, compact package. One such package was called the Micro-BGA (μBGA®), developed by Tessera Inc. It combines a unique IC interconnection with BGA technology that results in a highly miniaturized package (Fig. 2.26). The μBGA® consists of a copper/polyimide flex tape substrate with an array of pads interconnected to Cu ribbon leads extending from the substrate edge and

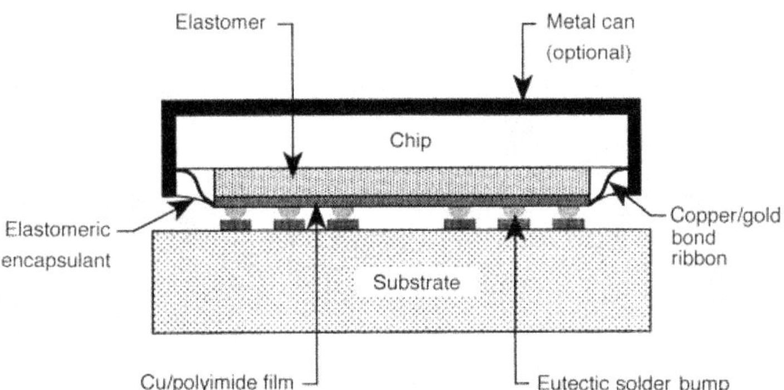

Figure 2.26 Typical cross section through a μBGA®. *(After Tessera.[7])*

joined at their tips to a common bus bar. Both the pads and ribbon leads are Au plated. The ribbon leads are thermosonically bonded to the perimeter pads on the die and separated from the bus bar at the same time. A silicone elastomer layer between the flex tape and the die surface cushions any impact to the IC during socketing or board assembly. The compliant layer also reduces any stresses caused by the CTE mismatch between the silicon and the solder ball joints when surface mounted to a PWB. A molding compound encapsulates the ribbon leads and supports them during package handling.

2.3.4.5 Matrix BGA (MBGA). The matrix BGA (MBGA) (Fig. 2.27), also called fpBGA™ by ASAT and ChipArray® BGA by Amkor, is a conventional PBGA with the overmold extending to the edge of the substrate. The thickness of the overmold, being uniform across the substrate, provides protection to the bond wires, even at the edge. The substrate is designed with the wedge bond pads placed close to the edge and interconnected to the solder ball pads on the BGA side through fan-in circuitry and plated-through holes, mostly located under the die. A solder mask covers the entire surface on both sides of the substrate, with the exception of clearances to expose the wedge bond pads and solder ball pads.

Typically, MBGA substrates range in size from 4.0 × 4.0 mm to 17.0 × 17.0 mm, with ball counts from 24 to 256 I/O, in an area array of 0.5-, 0.65-, 0.8-, or 1.0-mm pitch. The organic substrate is usually made of bismaleimide triazine (BT) or equivalent, with two metallization layers, although additional inner layers of power and ground planes can be incorporated for electrical and thermal enhancement. The substrate consists of multiple sites in an array configuration with tie bars connected to common buss bars to facilitate Ni/Au electroplating. For substrates where the tie bars are not cut after substrate fabri-

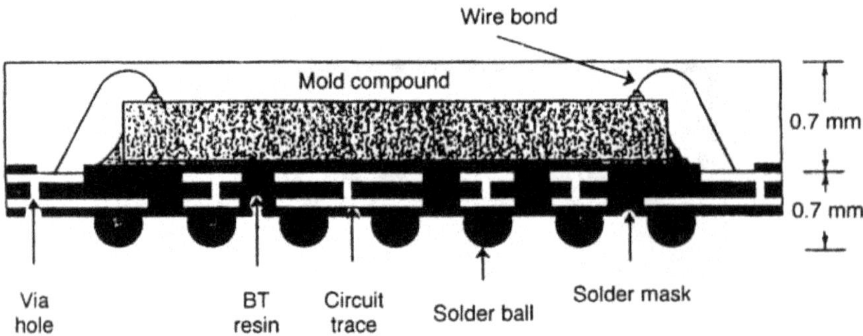

Figure 2.27 Typical cross section through a matrix BGA. *(After ASAT.[8])*

cation, to isolate each site for electrical testing of opens and shorts, the sites are usually checked prior to applying the solder mask by automatic optical inspection (AOI) equipment. The defective sites are identified with markings that will prevent placement of an IC on the affected sites during assembly.

Assembly follows the same process flow as for conventional PBGAs, with the exception of overmolding and singulation. The overmold covers the entire array of sites instead of individual sites. After code marking and ball attachment, the sites are singulated using a wafer saw that cuts through the overmold and substrate.

The overall thickness of an MBGA, including solder balls, is 1.40 mm (Fig. 2.27). The demand for thinner packages has forced the assembly vendors to reduce the overall thickness to 1.25 mm by reducing the thickness of the substrate from 0.36 mm to 0.25 mm.

The MBGA package has an interconnection density (substrate area to solder ball count ratio) that comes close to that of the chip scale package (CSP).

2.3.4.6 Flip chip BGA (FCBGA). Combining a BGA package configuration with flip chip (FC) bonding results in a most effective high-density interconnection (Fig 2.28). The solder bumped chip is placed face down onto matching bonding pads on a multilayer organic BGA substrate, and the assembly is reflowed. Prior to IC assembly, the bonding pads on the substrate may be bare of any solder or solder may be applied to the pads (SOP) and then reflowed and coined. After flip chip attachment, the IC is underfilled with an epoxy resin formulated to relieve the stresses induced by the thermal mismatch between the IC and the substrate and to prevent moisture from getting to the chip

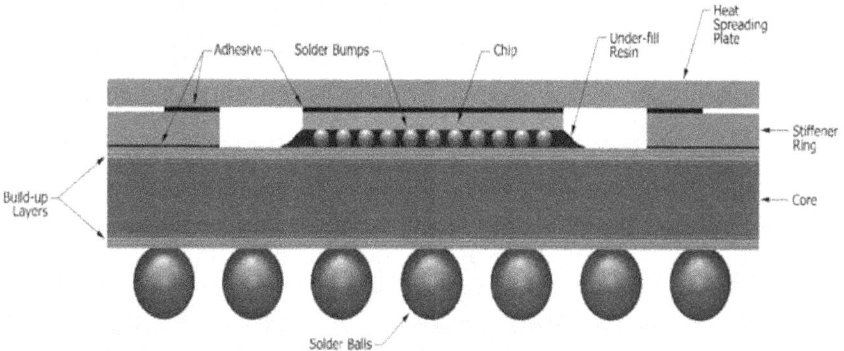

Figure 2.28 Typical cross section of a flip chip (FC) BGA.

surface. The flip chip can be capped or topped with a liquid epoxy, which is retained within the confines of a resin dam. To reduce cost, the back of the flip chip is left exposed, but this leaves the edges of the chip susceptible to chipping during handling. Where the chip power dissipation is of concern, or to improve solder ball coplanarity, a Cu stiffener ring slightly thinner than the chip stand-off is attached to the substrate prior to IC assembly. After IC assembly, an adhesive is applied to the top surface of the stiffener ring, and the back of the chip is covered with a pliable conductive adhesive. A Cu plate with a Ni surface finish is placed over the assembly, making contact first with the adhesive on the back of the die. The plate is pressed on the chip adhesive until a predetermined adhesive thickness is reached. At the same time, the Cu plate also becomes attached to the adhesive on the stiffener ring, completing the lidding process. The purpose of having a slightly higher chip stand-off than the stiffener ring thickness is to maintain a constant adhesive thickness on the back of the chip, regardless of the tolerances in thickness that either the chip or stiffener may have. The heat dissipated by the IC is conducted through the adhesive to the heat spreading Cu plate. A heat sink can be attached to the plate to further lower the case-to-ambient-air thermal resistance. After chip encapsulation, solder balls are attached to the underside of the substrate to complete the FCBGA.

The high-density interconnection on the IC, at the current bump array pitch of 200 μm (expected to be further reduced to 180 μm and 150 μm in the near future), requires equal high-density circuitry on the substrate. To achieve these circuit densities, the substrates have to be fabricated by the more expensive build-up process, with semiadditive circuit patterning. The construction may be of a 2-2-2, 3-2-3, or other combinations, where the numbers represent (number of build-up layers)-(number of core metallization layers)-(number of build-up layers).

Substrate vendors utilizing the build-up process have reported current production capabilities of 23/23 μm lines/spaces and 70/100 μm via/land dia. The exposed metal surface finishes on the substrate are electroless plated with Ni/Au or coated with an organic solderability preservative (OSP).

The intermetallics forming at the interface between the FC/BGA solder and the Ni/Au electroless plated pads may at times cause embrittlement and weaken the solder joint. This has forced the industry to look at alternative pad surface finishes, such as applying the solder directly to the FC and BGA copper pads. Prior to applying the solder, various measures are being considered to protect the exposed copper surfaces from oxidation, such as (a) applying an organic solderability preservative (OSP) or (b) a flash or electroless Au plating, or a light copper etch.

Flip chip bonding is becoming more widely used, in particular for high I/Os and where high electrical performance is required.

2.3.4.7 Stacked die BGA package. The stacked die BGA technology has evolved to meet the demand for smaller packages with more functionality for use in products with minimal board real estate, such as cell phones, videos, and still cameras. In a stacked die package configuration, two or more ICs are stacked on top of each other, with spacers in between, and interconnected to the substrate[9] (Fig. 2.29). The ICs may all be wirebonded to the substrate, or the bottom IC may be FC bonded and the rest wirebonded. For two or more ICs to fit within the same overmold height of a single chip package, each IC is background to approximately 6 mils thick. Assembling stacked die packages requires advanced wafer thinning, multiple passes through die attachment, and wirebonding. The advantages of stacked die BGA packages are as follows:

- Lower assembly cost (because two or more IC packages are reduced to one)
- Increases device functionality while maintaining package size
- Reduced PWB routing density
- Utilization of standard package footprint

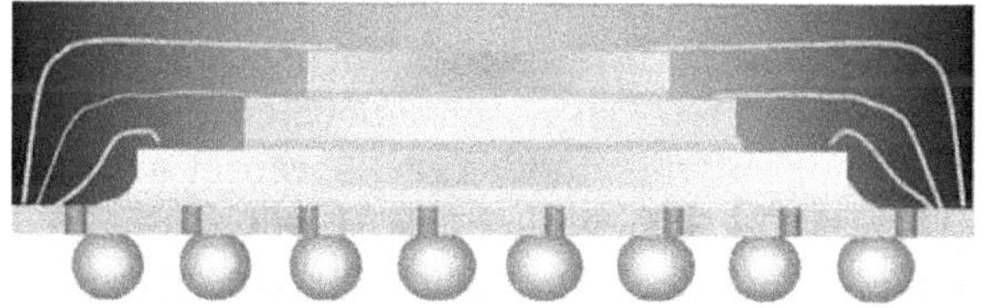

Figure 2.29 Stacked die BGA package. *(Courtesy of Amkor Technology.[9])*

2.4 Comparison of Package Technologies

A comparison of the attributes and trade-offs of the four IC packaging technologies is shown in Table 2.4. Pressed and laminated ceramic technologies have hermetic sealing capabilities that are best suited for environmental protection of the IC. The technologies provide a leak-tight seal, a cavity free of contaminants, and moisture that is controlled to an acceptable level. Nonhermetic molded and laminated plastic technologies can provide equal reliability performance for most applications.

The electrical, thermal, and mechanical performance of a given package technology is greatly influenced by the properties of the materials inherent in that technology. For example, the capacitance between adjacent leads is a property that is influenced by the dielectric constant of the material in direct contact with the leads. Table 2.5 lists the materials and dielectric constants for each package technology. Plastic technologies have the potential for best electrical performance due to their lower dielectric constants. Electrical performance is further optimized by the use of power and ground planes and high-density routing within the package. Both laminated plastic and ceramic technologies have this as a standard design capability.

Thermal performance is influenced by the thermal conductivity of the basic package materials (Table 2.5) and the use of thermal design enhancements (Table 2.4). The thermal conductivity of alumina ceramic materials is an order of magnitude better than that of plastic materials. Materials such as AlN and SiC, with higher thermal conductivity and heat-spreading copper-tungsten slugs, have further improved the thermal performance of ceramic packages. Design enhancements in laminated plastic packages (e.g., thermal vias under the die, center array of thermal solder balls, heat spreading internal planes or copper slugs, or heat spreaders in molded plastic packages) have closed the gap and are competing in thermal performance with laminated ceramics.

Mechanical performance is related to internal stresses developed within the package as a result of the mismatches of the coefficients of thermal expansion (CTE) of the different materials used in the package construction. CTEs of the silicon IC and of the substrate material to which the IC is mounted are listed in Table 2.5 for each technology. The best match is between silicon ICs and ceramic-type packages. Further complications are encountered in technologies where the IC is not decoupled from other materials, such as molding compounds in molded plastic or in laminated plastic technologies. ICs in the pressed ceramic and cofired laminated ceramic technologies are usually decoupled from other materials; however, their assembly processes, which include high-temperature exposure during chip mounting and her-

Table 2.4 Package Attributes and Trade-Offs vs. Package Technologies

Attributes and trade-offs	Molded plastic	Pressed ceramic	Laminated ceramic	Laminated plastic
Hermeticity	No	Yes	Yes	No
Power/ground distribution capability	Doable custom design	Not feasible	Standard	Standard
High-density routing capability	Custom TAB design	Not feasible	Standard	Standard
Power dissipation capability	Fair	Fair	Excellent	Excellent
Thermal design enhancements	Heat spreaders	New materials	New materials	Thermal vias and heat spreaders
Capability for handling bond-pad-limited chips	Yes	Yes	Yes	Yes
Automated assembly	Excellent	Feasible	Good	Excellent
Max. I/O capability	375	196	1000+	1000+
Cost rank order (lowest = 1)	1	2	3	2+

metic sealing, can present problems, particularly for very large die, greater than 0.500 in (12.7 mm). Other material properties, such as Young's modulus and Poisson's ratio, also affect the magnitude of stresses in packages where the ICs are not decoupled. The finite-element analysis is a useful CAD tool for modeling stress conditions in new package designs before committing them to development and manufacturing stages.

The cost of packaging is best addressed by using a rank-order system rather than an absolute cost. There are many factors and assumptions that enter into a cost analysis of packaging, and these are beyond the scope of this chapter. Molded plastic technology packages, such as plastic quad flatpacks (QFPs), which are assembled on leadframes, and matrix ball grid arrays (MBGAs), which are assembled on double sided, multiple-site strip substrates, rank low. In contrast, CPGAs or PPGAs, which have multilayer laminated substrates with cavities for mounting the chip in and come singulated, need to be assembled in "boats" and thus rank higher.

The ability of a package to accommodate bond-pad-limited ICs is an important factor influencing the cost of the IC. For example, bond-

Table 2.5 Physical Properties vs. Package Technologies

Physical properties	Molded plastic	Pressed ceramic	Laminated ceramic	Laminated plastic
Dielectric constant at 1 MHz				
Plastic overmold	5.0			5.0
Ceramic body		9.0	9.0	
BT resin lamina				3.9
Thermal conductivity, W/m K				
Plastic overmold	0.80			0.80
Ceramic body		20.9	20.9	
BT resin lamina				0.26
CTE, ppm/°C				
Chip mount surfaces				
Copper LF	17.2			
Alloy-42 LF	8.0			
Ceramic body		6.0–7.7	6.0–7.7	
BT resin lamina				15.0
Plastic overmold	13.0			13.0
IC chip (silicon)	2.7–3.5	2.7–3.5	2.7–3.5	2.7–3.5

pad-limited ICs in molded plastic and pressed ceramic packages are sometimes limited by the leadframe feature size, which is dictated by its thickness. The use of thinner leadframes, such as in fine-pitch high-I/O packages, has extended the capability for the plastic molded package technology. The laminated ceramic and plastic package technologies accommodate wirebonded-pad-limited high I/O die by using tiered wedge bond sites or a tighter wedge bond pad pitch, thus bringing the wirebonds closer to the IC and minimizing the wire span length. Flip chip interconnection, in laminated ceramic and plastic technologies, has made it possible to further reduce the size of the die, but it requires a more advanced substrate technology that is costlier to fabricate. Thus, the cost savings realized by shrinking the die size are sometimes reduced by the higher packaging cost. Each application must be evaluated thoroughly to determine the bottom-line cost for both the IC and the package.

Table 2.6 compares most of the attributes discussed in this section by rank order, providing a simple format to evaluate trade-offs between the competing package technologies.

2.5 Summary and Future Trends

Four package technologies have emerged over the years, and each has enjoyed its share of success. Demand for higher pin counts will un-

Table 2.6 Attribute Rank Order vs. Package Technologies

Physical properties	Molded plastic	Pressed ceramic	Laminated ceramic	Laminated plastic
Moisture sensitivity	2*	1	1	2
Lead capacitance	1	3	2	1
Power dissipation	3	2	1	†
CTE mismatch, chip-to-chip-mount surface	‡	1	1	2
Automated assembly capability	1	1	2	§
Capability of accommodating bond-pad-limited chips	1	1	1	1
Cost	1	2	3	2+

*Rank order: best = 1.
†Rank order depends on thermal design of substrate; 1+ = design with thermal vias and heat-spreading planes, 2+ = design without thermal vias.
‡ Rank order depends on leadframe material used: 1– = Alloy-42; 3 = copper alloy.
§Rank order depends on package type; 1– = PBGs (overmolded), 2 = PPGAs.

doubtedly continue, with ever increasing silicon capability. Laminated Al_2O_3 ceramic packages dominated the high-performance packaging technology up to the mid 1990s, but factors such as its high dielectric constant, modest thermal conductivity, and high cost have forced the industry to consider other packaging materials, including laminated plastic packages.

Requirements for higher packaging density on the PWB level are driving package designs towards smaller perimeter lead pitches and array type packages. To illustrate the differences in mounting areas of various package technologies, a 208-I/O package was used for comparison (Fig. 2.30).

The demand for BGAs has increased over the past few years because of their robust construction, ease of assembly to the PWB, and area advantage over peripheral lead type packages such as QFPs. Figure 2.31 shows that, above 100 I/Os, the BGA, with solder balls on 0.050-in (1.27-mm) pitch and fully populated, uses less area on the PWB than the QFP with peripheral leads on 0.020-in (0.50-mm) pitch. At a peripheral lead pitch of 0.012 in (0.30 mm), the crossover is at approximately 300 I/Os. BGAs are challenging the fine-pitch-perimeter packages, which are difficult to handle and have a greater board assembly yield loss than the array type packages.

To further increase the board densities, the drive is to use direct chip attach, chip scale packaging, stacked die BGA packaging, and SiPs (described below).

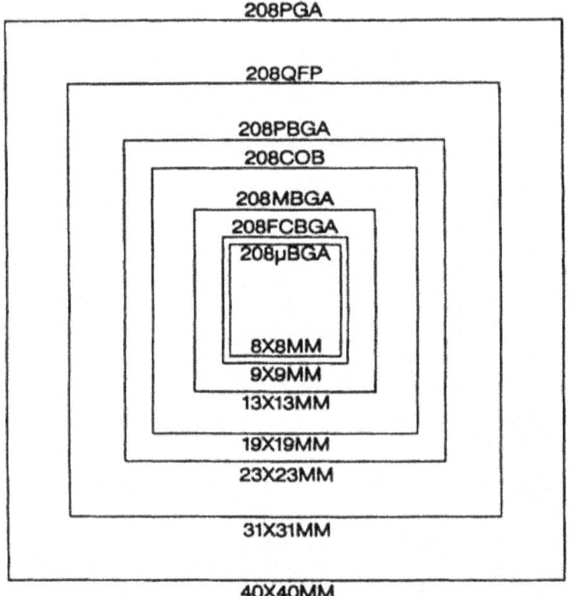

Comparisons based on the following:

208 I/O die size	= 7.6 × 7.6 mm max.
208 PGA	= 40 × 40 mm, 100 mil pin pitch
208 QFP	= 31 × 31 mm, 20 mil lead pitch
208 PBGA	= 23 × 23 mm, 1.27 mm solder ball pitch
208 COB	= 19 × 19 mm, wire bonded, 0.23 mm pad pitch
208 MBGA	= 13 × 13 mm, 0.8 mm solder ball pitch
208 FCBGA	= 9 × 9 mm, 0.5 mm solder ball pitch
208 μBGA	= 8 × 8 mm, 0.5 mm solder ball pitch

Figure 2.30 Mounting area comparison of IC packages.

As feature sizes of ICs decrease, demands are placed on interconnect technology capabilities for connecting to very closely spaced bond pads. Wirebonding will need improvements in placement accuracy to meet the demands of tighter bond pad pitches. It will also be necessary to provide the capability of longer wire spans with controlled wire dressing to bridge the gap between bond pads and the second bond targets while avoiding short circuits between adjacent wires in the dense wire fan-in pattern. Wirebonding of high-I/O die will require a Class 10,000 or better clean room environment to optimize automatic wirebonder performance and achieve acceptable bonding yields.

An alternative to wirebonding for high-I/O and high-performance devices is flip chip (FC) to package interconnection. FC has the advan-

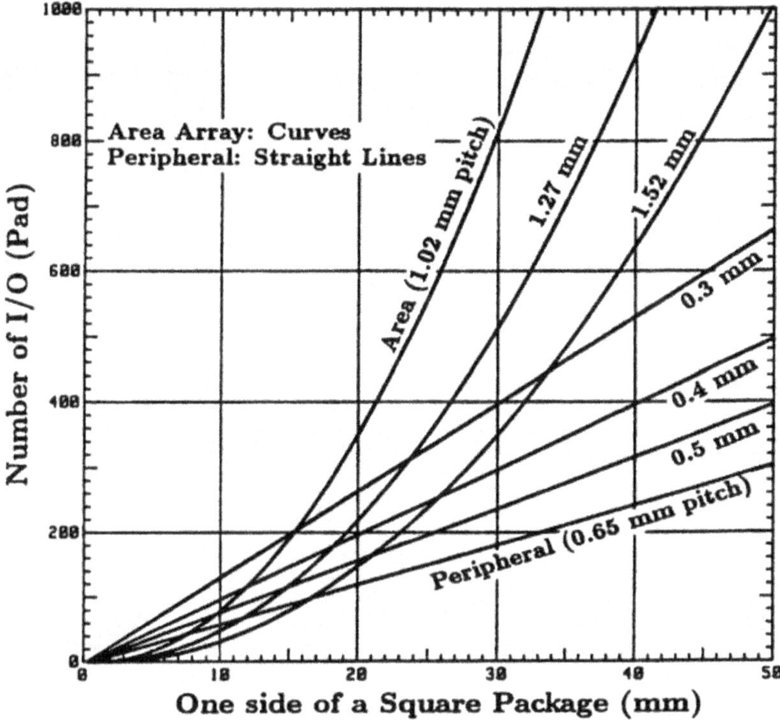

Figure 2.31 BGA and PQFP comparison: pin counts vs. package size, assumes full array for BGAs. *(After Lau.[3])*

tage of very low power and ground inductance connections to the package and the best capability in terms of the density of interconnection (using an area array) to the device compared to wire or TAB.

The electronics industry's demand for functional integration in smaller size packages, to gain performance and lower cost, has given rise to the system in a package (SiP) concept. The SiP is composed of a BGA substrate containing several ICs, discretes, and other elements to form a given functional block. Mixed chip technologies can be handled, i.e., flip chip devices alongside wirebonded devices. The SiP benefits are (1) smaller size packages than when ICs are packaged individually and (2) higher electrical performance, because the critical ICs can be placed closer to each other. The SiP also has the potential of reducing the number of layers and complexity of the PWB.

Newer packaging materials, with improved material properties, are being developed (see Chapter 8) to achieve better performance, better reliability, and lower costs. Designers will make more use of package enhancement features; for example, heat spreaders in molded plastic packages, ground/power planes, and heat-spreading copper slugs in laminated plastic packages.

The worldwide market share of IC packages, for the years 2000 through 2002, and the expectations for 2007, are shown in Table 2.7.[1] Although CSP, QFN, and BGA packages are growing at a faster rate than other packages, their overall usage in the industry is insignificant. The peripheral, leaded small outline package (SOP) still dominates the market share (50 percent) of all IC packages used. Wirebonding will continue to be the interconnection of choice for IC packages (91 percent of all package types) in 2007.

2.6 References

1. Prismark Partners LLG, "The Electronics Industry Report—2004," November 2003.
2. Harper, C., Ed., *Electronic Assembly Fabrication,* New York: McGraw-Hill, 2002, Chap. 3.
3. Lau, J. H., *Ball Grid Array Technology,* New York: McGraw-Hill, 1995.
4. Cohn, C., R. M. Richman, L. S. Saxena, and M. T. Shih, "High I/O Plastic Ball Grid Array Packages—AT&T Microelectronics Experience," in *Proc. 45th Electronic Components and Technology Conf.,* May 1995, pp. 10–20.
5. "Design Guidelines Multilayer Ceramic," Catalog CATD6, Kyocera America Inc., San Diego, Calif., 1995
6. Cohn, C., N. V. Gayle, L. S. Saxena, "Design and Development of Plastic Pin Grid Array Packages," in *Proc. IEEE 36th Electronic Components Conf.,* May 1986, p. 100.
7. Tessera Inc., data sheet, "The Tessera® µBGA™ Package—A Low-Cost Alternative to High Pin Count Packaging," 1995.
8. ASAT, 1999 Product Guide.
9. Amkor Technology Online Data Sheet for Stacked CSP, July 2003, available at www.amkor.com.

Table 2.7 Worldwide IC Package Market Share, 2000–2002 and Expectations for 2007 *(After Prismark[1])*

Package type	2000	2001	2002	2007	2002–07 CAAGR*, %	Percent of total IC 2000	Percent of total IC 2007
DIP/SOT	12.70	9.96	8.77	5.54	–8.8	14.7	4.9
LCC	2.10	1.49	2.12	1.61	–5.4	2.4	1.4
SOP	46.65	35.94	42.60	54.73	5.1	53.9	48.1
QFP	14.38	11.15	12.72	16.61	5.5	16.6	14.6
QFN	0.30	0.35	0.85	7.60	55.0	0.3	6.7
Wirebond CSP	1.37	1.47	2.59	8.23	26.1	1.6	7.2
Wirebond BGA/PGA/LGA	1.20	1.02	0.97	1.79	13.0	1.4	1.6
Flip chip CSP	0.00	0.01	0.02	0.77	119.8	0.0	0.7
Flip chip BGA/PGA/LGA	0.28	0.26	0.37	0.99	22.0	0.3	0.9
COB (wirebond)	4.75	4.02	4.27	6.90	10.1	5.5	6.1
DCA (flip chip)	1.52	1.63	1.92	6.47	27.5	1.8	5.7
Wafer CSP (FC)	0.00	0.20	0.28	0.84	25.0	0.0	0.7
TAB	1.25	1.05	1.11	1.67	8.6	1.4	1.5
IC total	86.50	68.65	78.56	113.73	7.7	100.0	100.0

*Compound average annual growth rate.

Chapter

3

Device Reliability

John W. Osenbach
Weidong Xie
Jason P. Goodelle
Brian A. Sensenig

3.1 What is Reliability?

The word *reliability* conjures up images of solid, dependable, robust, performance; long life; low maintenance; and high quality. Although we all understand these images, they leave the engineer and/or scientist with a lack of specificity and therefore a lack of understanding of the concept. For the scientist and engineer, reliability must be more concretely defined. We define reliability as the ability of a given component (mechanical, electrical, magnetic, optical, and so forth) to meet the performance specifications of the user, for a user-defined length of time, at user-specified operating conditions. Thus, reliability by definition requires one to have a well defined set of specifications including, but not limited to, the electrical, thermal, mechanical, optical, and other performance requirements. These specifications must also include the expected operating environmental conditions, the expected time period of operation, and a well defined failure and/or repair rate. Using this definition, it is easy to appreciate the fact that different end-user applications that involve the same device function might require much different device designs, processes, materials, and/or testing procedures to ensure that different reliability specifications are met for a particular application. An example of this would be two different system designs requiring a device to perform the same read/write function. In one application, this function is needed for a hand-held electronic game; in another operation, the function is needed in a space or ocean-bottom application. Because the costs associated with repair and/or downtime would be very high for a system used in space or at the ocean bottom, reliability specifications for devices used in such applications are always much more stringent than those for a hand-held electronic application. Because of the very different reliability requirements, there can be (and almost always is) a significant cost adder associated with devices used in space and undersea applications as compared to the same device being used in a consumer electronic game application.

Once the detailed application specific requirements are defined, the objective for the engineer is to quantitatively specify the reliability of the device (e.g., the percentage of devices that are expected to fail over a given time interval). Unfortunately, reliability is statistical in nature. Simply put, the statistical nature of reliability is tied to statistical variations in processing, materials, and environmental stresses, as well as the statistical nature of defects in the individual materials o which the devices consist. Thus, statistical sampling techniques are required for identifying the correct number and type of devices that should be evaluated in any reliability study. If, for example, all of the devices that are used for reliability studies are taken from one processing sublot, using only the nominal processing and assembly conditions, processed by or with engineering supervision, and tested to nominal specifications, then it is unlikely that the results of the study will be representative of the device population as a whole. Furthermore, it is likely that the results will provide for an optimistic (e.g., the best-case scenario) reliability estimate.

A more rigorous and proper technique for identifying the correct number and the type of devices to be used in the reliability would be, for example, the use of statistical sampling techniques to identify the correct number of samples from each processing lot. These devices would then be subjected to long-term environmental testing. At a minimum, the environmental testing would include a condition that represents the electrical, mechanical, and thermal stresses the device would be expected to experience in the field. These stresses would be applied for a time period equal to that specified by the end user. If the statistical sampling is done correctly and is based on devices that are truly representative of the overall population of manufactured devices, then the results of such testing will accurately define the reliability of the device. There are a few device applications where this type of procedure might be a viable option. In particular, those applications that are related to life and death (e.g., medical applications such as pacemakers, artificial hearts and lungs, and so forth). Additionally, those applications involved in space or undersea travel might be candidates for such procedures. However, for most applications, this technique has two very problematic constraints that make it impractical.

1. The cost is extremely high and is generally prohibitive.
2. The device cannot be used until the reliability tests are completed.

For many high-reliability applications, the time could be in excess of 20 years. That is to say, there would be no revenue for the device manufacturer or designer for a long period of time. Clearly, this is prohibitive for all but a very few high-end applications.

Thus, one is left with determining a practical way of accurately predicting what will happen to a population of devices after years of operation, in many cases 10 to 25 years in the future, with information gathered in the present, typically in less than 6 months, on devices sampled from processing lots made in the beginning of the product's life. This approach appears to be counter to established scientific methods. These methods clearly state that data should not be extrapolate outside of the range in which data is taken. And if such extrapolation is required, then one should never extrapolate the data many orders of magnitude beyond the range in which data is taken. Fortunately, it has been demonstrated that procedures that include sound statistical sampling techniques and accelerated stress testing can and do work if they are employed with careful hypotheses and experimentation.[1–10]

There are, however, a number of cases in which such sound procedures were not followed, leading to catastrophic consequences. The most notable one may be the space shuttle, Challenger, which exploded during takeoff in 1986.[11,12] To paraphrase the late Prof. Feynman, in spite of known variation in the reliability data taken from different sets of O-rings, which clearly indicated a range of potential outcomes, the officials acted as if they had a well defined physical understanding of the potential risks prior to takeoff. As the final outcome demonstrated, the bold assumption of a well founded physical understanding of the possible outcomes was in fact incorrect. Subsequent investigation demonstrated that the error occurred because the data taken at temperatures >60°F did not describe the physics of the situation at temperatures approaching 31°F. That is to say, if the starting hypothesis is incorrect, then the predicted outcome will be incorrect. If it involves life or death, then it is always better to err on the side of caution.

3.2 Accelerated Aging

The basis of the accelerated stress testing/extrapolation technique includes a carefully and thoroughly defined process. In general, this process includes, at minimum, the following steps:

1. The development of hypotheses for any and all potential degradation and/or failure mechanisms
2. The development of accelerated aging tests and procedures that are directed at identifying the hypothesized potential failure mechanisms
3. The development of statistically appropriate device sample sizes and device types
4. The collection of data and careful analysis of the data
5. The extrapolation of the data and reliability prediction
6. The verification of steps 1 through 5

However, as demonstrated above, if the methodologies used to define the hypotheses, tests, and data analysis are flawed, then the outcome of the extrapolation will also be flawed. Clearly, a flawless process would include a detailed understanding of all potential degradation mechanisms and how the kinetics of degradation associated with each mechanism depended on the accelerant and process control. For example, the accelerants of an integrate circuit (IC) package would include temperature, mechanical shock, thermal shock, humidity, voltage, current, and radiation, and others. And its process control would include variations in die attach, wirebonding, soldering, molding, and encapsulation, and so on. If a detailed understanding exists prior to design, then, in principle, one could design a device using materials and processes that prevent any of these failure mechanisms from occurring during the device lifetime. Furthermore, one could develop accelerated test methodologies that could be used to quickly verify the hypotheses. By "quickly," we mean in seconds to hours of testing. Unfortunately, there are a number flaws with this type of approach, including but not limited to the following:

1. Defects associated with the materials and processes used to produce the presumed failure-immune device can be presented; thus, the assumption of a well understood set of failure mechanisms is no longer valid.
2. For many device technologies, there is a relatively poor physical/chemical understanding of the details of material degradation, at least at a molecular level, and their relationship to device failure. This is especially true of emerging technologies, where new materials, processes, and designs are implemented for which little or no historical reliability or degradation data exist.
3. New applications of existing technologies and designs are emerging continuously in which environmental conditions and operating stresses fall outside of the well understood "normal" ranges.

4. Statistical variations that occur as a result of unknown process and materials variation and/or unknown and unwanted electrical, mechanical, and thermal stresses, lead to statistical uncertainty.
5. Some degradation mechanisms are caused by stochastic events (i.e., lightning strikes, ESD, EOS, flooding, earthquakes, and so on), and neither the occurrence nor the magnitude of the stress can be predicted *a priori*. Because of the stochastic nature of these events, it very difficult (if not impossible) to define the kinetics of this type of degradation based on first principles.

3.3 Failure Mode and Effects Analysis

To mitigate risks, failure mode and effects analysis (FMEA) should be applied to each and every new process, technology, material, design, and application space of existing proven technology and designs.[13–15] A similar technique can also be applied to identify potential root causes of failure for devices that have failed a qualification/reliability test or for field failures. The technique is a disciplined, systematic, analytical procedure that brings together designers, technologists, users, and subject matter specialists with expertise in all of the particular areas of interest, with the aim at mitigating risks *a priori*. Over the past few decades, being involved with FMEA, the authors have observed that the primary technical FMEA team should be inclusive but not too large. When teams get substantially larger than 10 people, the efficiency tends to drop of significantly. However, it has also been observed that, with careful technical assessment and written input, a larger group of individuals can be very helpful in moving the process to a successful end in a timely manner. As such, we suggest that a consulting team also be formed for additional technical assessment. The information the assessment team reviews should include, at minimum, the following:

1. Design layouts and modeling
2. Process flow charts
3. Process control charts
4. Testing procedures and measurement errors
5. A list of the piece parts used
6. A list of the critical performance parameters
7. A list of processes and/or materials perceived to be the most critical to device function and reliability
8. Any knowledge about the technology and materials used in the design that have been employed by the company in existing devices

9. Expected operational conditions
10. Beginning and end-of-life requirements
11. All potential or known failure modes
12. If the design, process, and/or materials have been used before, how often and by whom (internal or external to the company)

The team members then assess the information with respect to potential risks to design quality and reliability. The team distills these assessments and combines them into a comprehensive assessment. The document should then be used to drive the project work. Once the work is completed, it should be documented and revised again by the team. The final documentation should include, at minimum, the following:

1. The risk items
2. The individual assigned to address each of the items
3. The proposed action plan, including the hypothesis used to develop the plan and expected outcomes prior to execution of the plan
4. Any experimental procedure
5. The results of execution of the plan
6. Analysis of the results
7. Conclusions and potential additional work that should be continued to further address the issues

The authors have found that a FMEA process that is carefully planned and executed as suggested above not only leads to a more robust device and process technology (which ultimately brings in more revenue for the company and puts the customer at least risk), it also provides a means of documentation of the data. In addition, this documentation can be used in future designs and/or FMEA assessments to both improve the efficiency and mitigate potential issues prior to use.

3.4 The Kinetics of Degradation

Accelerated aging is a procedure that effectively speeds up the lifetime clock of a device by accelerating degradation mechanisms in a controllable, reproducible, and predictable manner that can be empirically related to actual operating conditions. Temperature, humidity, and temperature cycling are known to accelerate many physical and chemical processes such as plastic deformation, creep, leakage, diffusion, delamination, corrosion, stress corrosion cracking, and so on, all of which can lead to device failure. For example, degradation due to local temperature changes that occur during power on/power off cycles can occur as a result of plastic strain-induced work hardening of solder joints. The strain hardening eventually leads to fracture of solder joints. This process occurs rather slowly under normal operating conditions. As such, it takes a long time to observe this type of degradation as it relates to a measurable change in the device performance characteristics. Therefore, for all practical purposes, an accurate assessment of the degradation cannot be made in short (less than one year) time periods. That is to say, noise dominates the measurements of degradation for a long period of time.

Fortunately, it is possible to artificially increase the magnitude of the temperature change experienced by the device during power on/power off cycling. In this manner, one can increase plastic strain energy per cycle that is dissipated as a result of thermal fluctuations such that measurable changes in device performance can be found in weeks of thermal cycling, i.e., no longer noise dominated. If one can find a reproducible, physically plausible, empirical relationship between the degradation measurements made as a result of a larger change in temperature than that which would occur in during normal operation of the device, then the measurements under accelerated conditions can be used to estimate reliability under normal operating conditions. This type of procedure falls into the general category of accelerated stress testing (i.e., accelerated aging). The relationship that defines in degradation rates between the artificially created accelerated stress test to that which would occur at nominal operating conditions is referred to as the acceleration factor, *AF*, which is defined as

$$AF = \frac{\text{Time to fail at condition 1}}{\text{Time to fail at condition 2}} \tag{3.1}$$

where condition 2 is the operating condition (e.g., the customer use condition), and condition 1 is the accelerated condition. Using the *AF* and the statistical distribution of the data taken at the accelerated stress condition, one can estimate the reliability of the device. Thus, the objective of accelerated aging test development is to determine the *AF* functions for all of the relevant failure mechanisms that might occur during operation of the device. Note that an implicit assumption in the definition of *acceleration function* is that the same dominant failure mechanisms occur under both conditions. It should be noted that the acceleration factor is sensitive to the definition of how failure is defined.[5] Thus, we suggest that *failure* be defined in direct relationship to the end users' requirements whenever possible.

To a first order, the kinetics of failure (i.e., the degradation rate) is based on the concept of transforming the system from one state to another state and is conveniently expressed in statistical mechanics.[16,17] Figure 3.1 shows a schematic of a simple statistical mechanics model of the state of transformation. In this theory, the system is in an equilibrium state, and atom M_1 is in its lowest energy state when no additional energy stimuli are applied to the system. The application of a directional energy source* leads to a decrease of the energy state in lattice position 2 and an increase of the energy state in lattice position 1. Because thermodynamic law states that systems want to go to the lowest energy state at all times, the application of the external energy source provides the driving force that might lead to atom M_1 moving from position 1 to position 2. This can and does happen if the applied energy is sufficient to overcome the energy barrier, E_a, often referred to as the *activation energy.* It turns out that the rate for this type of reaction process follows a simple exponential dependence of temperature,

$$rate = A \times e^{\left(-\frac{E_a}{kT}\right)}$$

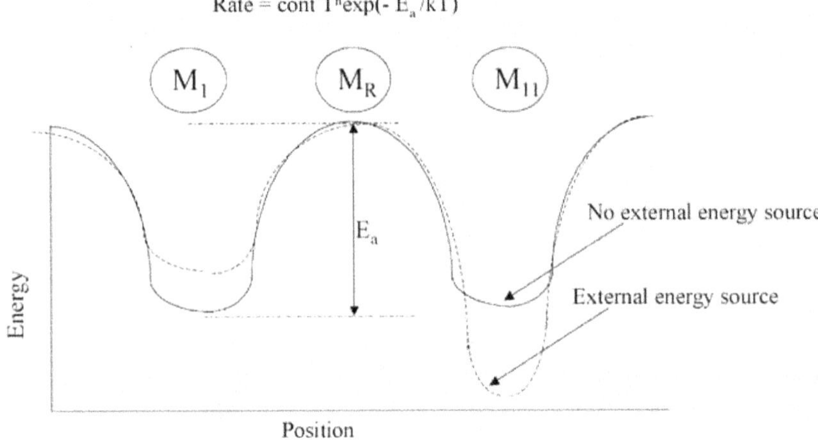

Figure 3.1 Schematic of the energy diagram used in the development of the Arrhenius expression.

---

*Energy sources include voltage (V), current (I), chemical gradient ($\Delta C/\Delta x$), thermal gradient ($\Delta T/\Delta x$), and strain ($\Delta x/x$). In standard reliability literature, the energy source is usually referred to as a *stress.*

where A = a constant
k = Boltzmann's constant
T = the temperature, in Kelvins, at which the degradation is occurring

This type of exponential relationship is referred to as an *Arrhenius dependence.* Although this theory is well established, there are still some troubling issues with the theory, most notably that if one assumes a reasonable jump distance between M_1 and M_2 (lattice spacing), then one calculates an E_a that is more than 10 electron volts (eV), which is substantially higher than those typically measured. (Note that typical measured activation energies are less than 2 eV.) Similarly, if one assumes a reasonable $E_a < 2$ eV, then the jump distance can exceed 100 nm, which is substantially larger than a typical lattice spacing of a less than 1 nm. Clearly, both of these conditions are not representative of the actual physical situation. In spite of this discrepancy, for more than a century, the empirically collected data for many chemical and physical processes have been found to be well represented by the basic Arrhenius expression. In fact, it is this relationship between temperature and time that is the core of accelerated life testing methodologies.

The "activation energy" in the Arrhenius expression can be thought of as the energy required for a change (reaction/degradation) in the system to occur. In principle, each reaction (degradation) mechanism has a uniquely defined characteristic activation energy. In reliability studies, an energy source or stress (temperature, temperature cycling, temperature humidity bias, and so on) is applied to a device for some period of time, and resultant changes in device characteristics are measured. If the characteristics fall outside of a predetermined range of acceptable performance, the device is considered to have failed. The resultant failures are due to the changes in the device performance characteristics, but not necessarily changes in M. These changes in performance characteristics need not be caused by a single reaction mechanism. Therefore, an "activation energy" that is related to the changes in device characteristics need not be representative of a single reaction mechanism. Thus, we prefer to call the energy required for degradation of a particular device the "effective" activation energy.

Implicit in the activation argument is that the mechanisms responsible for device failure at one set of accelerated stress conditions are the same as those found under normal operating conditions. If this is not true, then the acceleration factor is not valid. This becomes problematic on two fronts. There are a number of published, well established acceleration functions for common failure mechanisms in devices made with well characterized and exercised process technology, materials, and device designs, when operated in well established environmental conditions. These acceleration functions are the basis of the standard qualification tests (Table 3.1). In general, the acceleration functions take the following form:

$$AF = F(\text{temperature, temperature cycle, voltage, current, humidity, etc.}) \quad (3.2)$$

When applied correctly, these standard *AF* functions can be used to estimate the device lifetime, and therefore reliability, for given stress conditions. However, the functions and fitting parameters (E_a, current exponent, voltage exponent, humidity exponent, stain energy exponent, and so on) are valid only for the specific degradation mechanisms responsible for failures of the devices as produced with existing technology and materials sets and operated under well established environmental conditions. If, however, these *AF* functions are not applied with great caution, the resultant estimated reliability could in fact be seriously flawed, as attested to by the Challenger explosion. The unfortunate outcome of this is that passing a standard reliability test may not mean that the device is reliable. This is especially true for new, unvalidated design methodologies, processing technologies, manufacturing technologies, and operational conditions that have been used to create or operate the device.

Table 3.1 Common Reliability Tests and Conditions Used to Determine Device Long-Term Reliability

Stress test	Condition
High-temperature storage (HTS)	150°C, 1000 hr
High-temperature operating bias (HTOB)	100 to 150°C operating bias, 1000 hr
Temperature humidity bias (THB)	85°C/85% RH/bias, 1000 hr
Temperature cycling (TC)	–55 to 125°C, 1000 cycles –65 to 150°C, 1000 cycles 0 to 100°C, 3500 cycles

If one thinks of degradation or acceleration factors and tests used to uncover these degradation mechanisms in terms of atomistic processes that lead to device failure, then the chance of applying the incorrect stress tests or acceleration functions is minimized. When viewing in this manner, the mechanisms responsible for failure at all stress levels become explicit. At atomic/molecular level, degradation can be defined in terms of a stress-induced rate of change of a material M with time.

$$\frac{\Delta M}{\Delta t} = F(\text{temperature, himidity, bias, current, strain, etc.}) \tag{3.3}$$

Because of random variations within the device structure, M has some defined spatial density distribution, SDD_M. That is to say, M has a distribution in x, y, and z within the device. Device failure will occur when the change in material property, M, causes one or more of the device performance characteristics to fall outside of the required/specified range. Such a change would occur when the change in M intersects with a sensitive area of the device. Therefore, the device lifetime, t_{fail}, can be defined in terms of a time rate of change in M, the spatial density distribution of the device, SDD_M, and the spatial distribution of the sensitive areas of the device, SDD_D.

$$t_{fail} \propto \left(\frac{\Delta M}{\Delta t}\right)^{-1} \times SDD_M \times SDD_D \tag{3.4}$$

Assuming that, on average, the spatial distribution of the material M that changes is independent of the applied stress, the acceleration factor (AF) can be written as the ratio to $t_{fail\,1}$ to $t_{fail\,2}$, or

$$AF = \frac{t_{fail\,1}}{t_{fail\,2}} = \frac{\left(\frac{\Delta M}{\Delta T}\right)_1^{-1} \times SDD_{D1}}{\left(\frac{\Delta M}{\Delta T}\right)_2^{-1} \times SDD_{D2}} \tag{3.5}$$

For typical devices, there are n such material properties that could change with time under an externally applied stress and intersect with m sensitive areas of the device, any or all of which could lead to device failure. Thus, there are a minimum of $n \times m$ individual acceleration functions, and therefore acceleration factors, for any given device. Each factor has its own unique stress and time dependency. Fortunately, only those material property changes that lead to device failure in the shortest time at nominal operating conditions must be considered when developing acceleration models and tests. Because each material property has its own unique $\Delta M/\Delta t$ function and spatial distribution, one has to be extremely careful to ensure that only those $\Delta M/\Delta t$ functions that can lead to device failure in the field are involved in device failure at the accelerated stress test conditions. One could extend these arguments to early life failures that are often considered to be related to manufacturing defects. In this case, it is possible that additional $\Delta M/\Delta t$ mechanisms will be introduced by defective processing. Furthermore, these processing defects could also lead to increases in both SDD_M and SDD_D. Thus, one has to add additional acceleration functions to account for defective devices.

In the next few paragraphs, we explore what these concepts mean in terms of reliability stress testing and extrapolation. Let us assume that failure occurs as a result of a diffusion process. This could be the case, for example, for the development of thick Cu/Sn intermetallics at a solder joint as a result of exposure to high temperatures for extended periods of time, strain (work) hardening of a solder joint during temperature cycling, porosity and voiding in a metal line or solder joint as a result of electromigration, popcorning and/or warpage as a result of moisture ingress and outgress, growth of Au/Sn intermetallics at the solder Ni interface, and so on. Although the activation energy for diffusion is sensitive to material type, composition, microstructure, and the technique used to produce the material, there are some general rules of thumb that can be stated about the magnitude of E_a for different diffusion mechanisms that can occur in polycrystalline materials. If the activation energy for diffusion through crystalline lattice structures is E_a (typically between 1 and 1.5 eV), then the activation energy for grain boundary diffusion is of order 0.5 E_a and for surface diffusion is of order <0.35 E_a. The activation energy for interfacial diffusion is typically characterized by 0.35 to 0.7 E_a, and that for moisture diffusion though polymeric systems is of order 0.4 to 0.5 eV. For metal and ceramic systems, the transition from one dominant mechanism to another can be roughly defined in terms of the ratio of operating temperature, $T_{operating}$, to the melting point of the material, T_{mp}, when temperature is defined on an absolute temperature scale (i.e., Kelvins). This ratio is often referred to as the homologous temperature. When $T_{operating}/T_{mp} \leq 0.35$, the diffusion is surface diffusion dominated; when $0.35 \leq T_{operating}/T_{mp} \leq 0.7$, it is grain boundary diffusion dominated; and when $T_{operating}/T_{mp} \leq 0.7$, it is lattice diffusion dominated.

Let us consider a diffusion dominated failure mode such as strain hardening during temperature cycling. If the solder is eutectic lead-tin in composition, then the approximate temperatures at which surface, grain boundary, and lattice diffusion control the rate are $T_{operating} <$ –113°C, –113°C < $T_{operating}$ < 46°C, and $T_{operating}$ > 46°C, respectively. Note that the melting point of the lead-free ternary alloys that contain Sn, Ag, and Cu, the binaries containing Sn and Ag, and pure Sn are within +35°C of the eutectic alloy. The actual transition temperatures will most likely be a little higher as a result of the incorporation of additional elements such as Ni/Cu and/or Au into the solder. Because we are usually interested only in temperatures greater than –55°C for typical device applications, for all intents and purposes, we can ignore degradation resulting from surface diffusion processes and consider only grain boundary diffusion and lattice diffusion processes when degradation occurs.

Figure 3.2 is a schematic Arrhenius plot of device lifetime (inverse degradation rate) versus temperature assuming two different dominated degradation mechanisms: mechanism 1, δM_1, with 1-eV activation energy, and mechanism 2, δM_2, with 0.5-eV activation energy. We further assume that the lifetimes for the two mechanisms are equivalent at 60°C. In the case of strain hardening, both lattice and grain boundary diffusions play a significant role, because the temperatures of interest during accelerated temperature cycling lie above and below 60C.*

Clearly, if all of the data needed to arrive at Figure 3.2 can be collected, then one can extrapolate the lifetime measured at any accelerated stress condition to any stress condition of interest without introducing any error. This would clearly be the idea case. It is more likely that the best-case scenario is as shown in Figure 3.3. In this case, data are taken at three temperatures. The solid line represents the best-fit line using least-square analysis. The dotted line represents the two different curves for two different mechanisms of degradation shown in Figure 3.2. As shown, the best-fit line introduces only small errors if the estimated lifetimes are taken at temperatures within the range at which measurements were taken. However, as the lifetime estimates are made at temperatures far outside of the temperature range where degradation rate measurements were taken, the error is

---

*Deformation of materials can occur by a wide variety of different mechanisms: (1) dislocation glide (typically referred to as plasticity), (2) dislocation climb (power law creep), and (3) diffusional flow due to point defect in the lattice. In general, at low temperatures and high stresses, deformation is dominated by dislocation glide [rate $\propto$ $(\text{stress})^2 \cdot \exp(E_{aglide}/kT)$]. In this case, it is the dislocation motion past obstructions such as localized precipitates, other dislocations, and solutes. At higher temperatures and moderate or higher stresses, deformation is dominated by dislocation climb (rate $\propto$ $(\text{stress})^n \cdot \exp(E_{aclimb}/kT)$. Note that the dislocation glide mechanism is also operating but, for the most part, is not the dominant mechanism. In this case, it is dislocation climb (movement out of the bulk of the film toward a surface or interface). At the highest temperatures and low stresses, deformation is dominated by motion of point defects either by grain boundary or lattice diffusion (rate $\propto$ (stress) $\cdot \exp(E_a/kT)$. Thus, this simplified model, which deals only with temperature and its influence of degradation mechanisms, does not very well capture the true mechanism. That is to say, additional errors are easily introduced above those related to temperature, and that makes it even more difficult to predict *a priori* what might happen relative to device degradation and ultimately device reliability.[18] The actual functionality becomes much more complicated when voltage, current, and the magnitude of the temperature cycle are included in the degradation model.

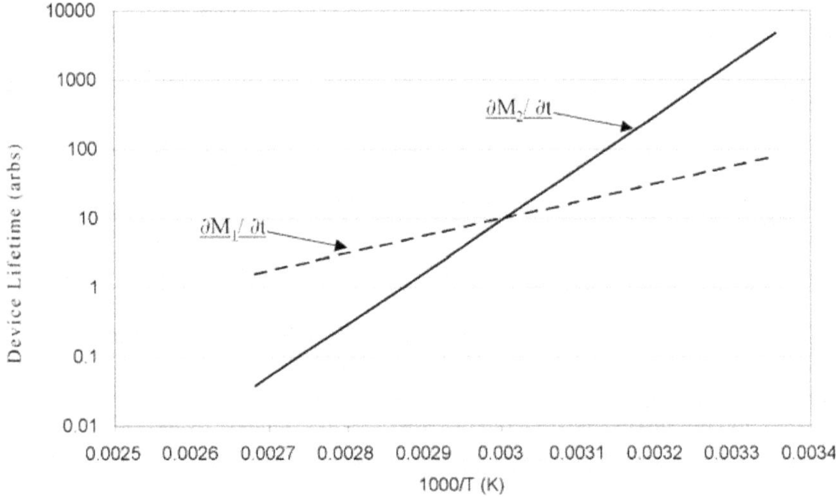

Figure 3.2 Arrhenius plot of the device lifetime vs. temperature. The two curves represent two different mechanisms of degradation with reaction rate activation energies of 0.5 eV and 1 eV, respectively.

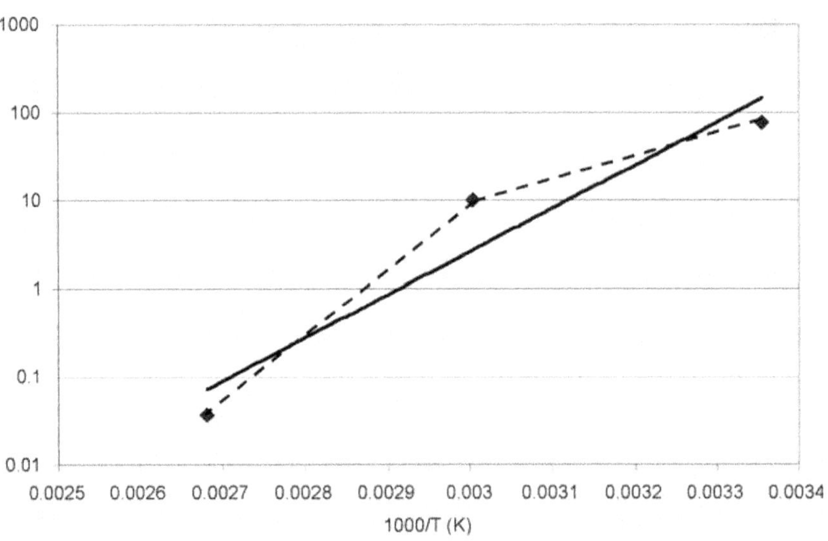

Figure 3.3 Arrhenius plot of the device lifetime vs. temperature. The solid line represents the best-fit curve to the three data points measured. The dotted line represents the two different curves for two different mechanisms of degradation shown in Fig. 3.2.

magnified. Reliability estimates at higher temperatures than were taken would be extremely conservative in that the estimated lifetime would be significantly lower than the actual lifetime. An estimate like this might preclude the use of the device in the field whereas, for reliability estimates at lower temperatures than were taken, the estimate would be optimistic in that the estimated lifetime would be significantly longer than the actual lifetime of the device. Such an estimate might lead to disastrous consequences if the device were field deployed in an application were life and death were at stake. Furthermore, because of the exponential dependence, the estimated lifetime becomes grossly in error as the gap between the temperature of interest and the maximum and minimum temperatures at which data were actually collected increases. Because it is common to use only one temperature or temperature range (as defined by the standard qualification tests) for device reliability qualification testing, the actual situation could be substantially worse than portrayed in this example.

The potential errors involved in extrapolation of one data point that might be determined via a standard acceleration function are further demonstrated in Fig. 3.4. In this case, the measured device lifetime was taken to be 100 hr at 120°C. The data were then extrapolated from the measurement temperature to other temperatures using three different effective activation energies, 0.5, 1.0, and 1.5 eV, respectively.

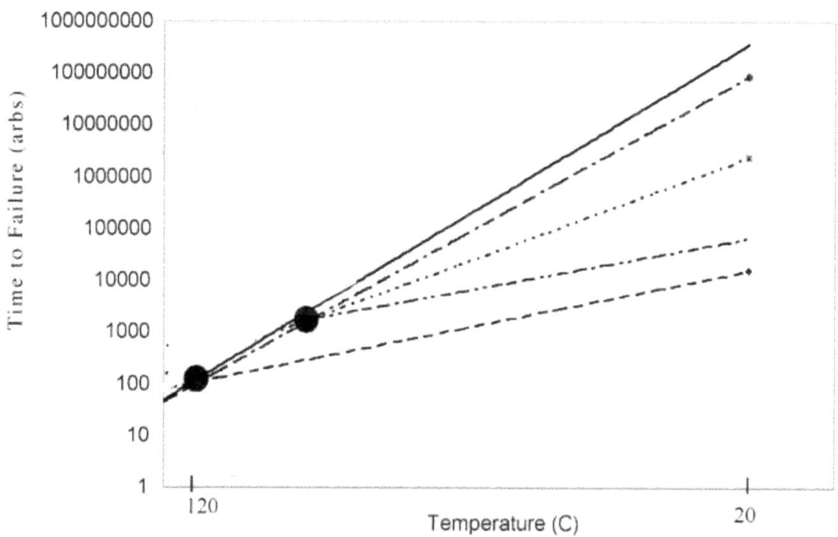

Figure 3.4 Plot of time to failure vs. temperature for data collected at 120°C and extrapolated to other temperatures using 0.5-eV (dash), 1.0-eV (dot-dot-dash), and 1.5-eV (solid) activation energies, respectively. In addition, the 1-eV extrapolation at 85°C was further extrapolated sign 0.5 eV (dot-dash-dash) and 1.5 eV (dot dash).

As shown, the lifetime estimates for the device at 20°C differ by more than two orders of magnitude, depending on the assumed effective activation energy used for the extrapolation. To further illustrate this issue, we assumed that the actual a median lifetime at 85°C is equal to that extrapolated from 120°C using 1 eV as the effective activation energy. We then apply the three different effective activation energies, 0.5, 1.0, and 1.5 eV, to extrapolate this data point to other temperatures. As shown, the differences at 20°C among the predicted lifetimes for the three different activation energies are smaller than that found for 120°C extrapolation. However, the difference still approaches two orders of magnitude. Given that one does not know the correct activation energy for extrapolation based on one data point, it is reasonable to argue that any of the three values used in this example is correct. Thus, without any additional information and/or critical analysis of potential failure mechanisms, there is no reason to believe one extrapolation over another.

As shown, the magnitude of the variation in extrapolated lifetimes depends on both the assumed activation energy and the magnitude of the temperature difference between the point at which the data are taken and at which the device will actually operate in the field. Note that the closer the field temperature is to the accelerated life testing temperature, the less variation is found between the extrapolation based on different activation energies. Therefore, the extrapolation is better. The lesson here is that, if one is introducing a new device process technology, material set, or other concept, then reliability data should be collected as close to the field operation temperature as possible. Furthermore, more than one temperature should be used for the reliability qualification accelerated tests.

3.5 Hazard Rates

The field failure rate is generally referred to as the *hazard rate (HR)* and is simply defined as the percentage failure per time interval of interest.[9]

$$HR = \frac{\text{Total failure percentage in time, } t}{\text{time, } t} \tag{3.6}$$

Because the failure rates for commercially available electronic devices are very low, the units of hazard rates are typically defined as *failures in time (FITs)*, which is 1 device failure in 1 billion device hours. That is to say, if one had 1 million devices in the field, each of which had a 1-FIT hazard rate, then 1 device would fail, on average, every 1000 hr (6 wk) of operation. Equivalently, an average hazard rate of 1 percent per year is approximately 1160 FITS. The rule of thumb the authors use is that an average hazard rate of 1000 FITs is roughly equivalent to 1 percent failure per year. Given the error in the extrapolation as the statistical nature of failures, we have found that this rule of thumb captures the essence of the true hazard rate.

Figure 3.5 illustrates how the data shown in the examples in Fig. 3.4 translate to actual estimated average field failure rates (hazard rate) at 120°C. In this case, the hazard rates were determined by extrapolation of the 120°C data to 45°C. Finally, estimates were made using the three different activation energies of the device. Furthermore, the failure distribution was assumed to be lognormal with a lognormal standard deviation (dispersion) of 0.8. As shown, the estimated (i.e., predicted) field failure rate is very sensitive to the assumed E_a. In fact, one would most probably not want to use this device, even for handheld applications, if the 0.5-eV extrapolation was truly representative of the field reliability of this device. Without any other information, one can only hope the estimate using 1.5-eV as the activation energy was not used, especially if the device is used in a high-reliability application such as a mechanical heart.

This example was meant to illustrate the difficulties in data collection, analysis, and extrapolation. It also demonstrates to the user of the devices that is important to critically examine device manufacturers' reliability data and analysis techniques, including all of the assumptions used in the extrapolation, to ensure that the hazard rate predictions are reasonably representative of those observed in the field.

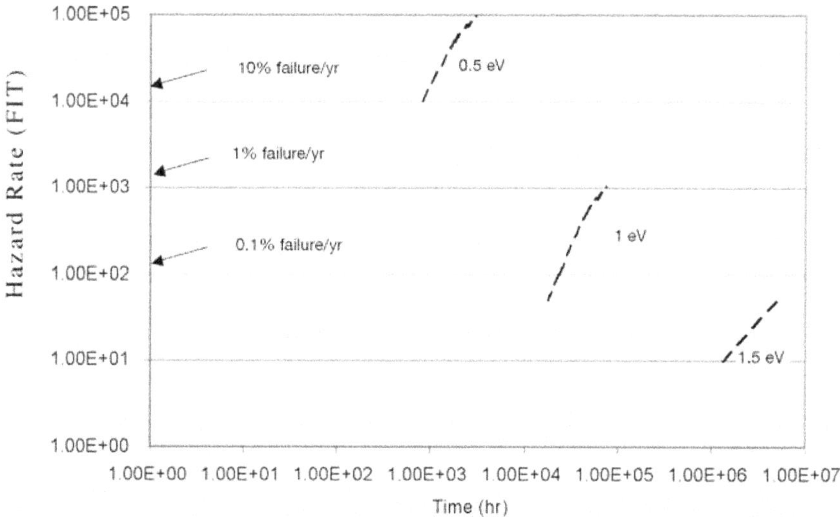

Figure 3.5 Plot of the average hazard rate for the 120°C data shown in Fig. 3.4.

3.6 Common Mathematical Functions Used for Estimating Reliability

The mathematical functions and the nuances of the applications of these functions are dealt with in great detail in many books[3,5,6,8,10] and will not be developed in this chapter. The reader is referred to these books and the references therein for further study and explanation. However, it is important from a physics and chemistry of failure point of view to have a general understanding of these functions. Device reliability in time is often refereed to via the *bathtub curve* (Fig. 3.6). Its origin is taken from analysis of human mortality data.[10] The data indicate that the risk of death decreases over the first few years of life (infant mortality), then becomes roughly constant over the next few decades of life (constant failure rate), followed by an increasing death rate in the last few decades of life (wear-out). This general bathtub curve functionality of mortality is well known and understood by the large majority of the population. Remarkably, a similar function has been found to represent the "mortality" of electronic devices as well as other commonly used objects.

Figure 3.6 is a schematic of a bathtub curve for a typical device type. The plot shows the hazard rate (e.g., the normalized failure rate per time) versus time the device is in the field. Also shown in Figure 3.6 are the commonly used mathematical statistical distribution functions that describe the different regions of the curve and their respective

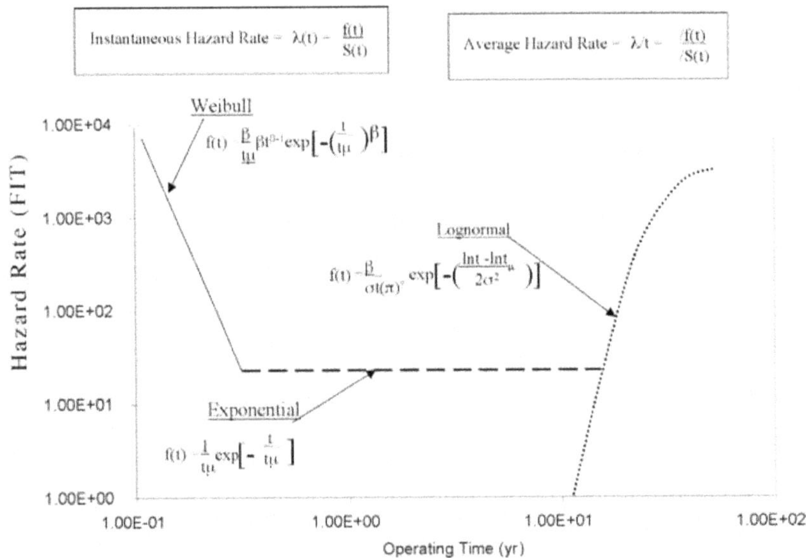

Figure 3.6 Hazard rate bathtub curve for a typical electronic component.

failure rate functions. Note that it is the object of the reliability, design, process, and quality engineer to develop a detailed understanding of the possible causes of device failure in each section of the bathtub curve. In this way, the infant mortality can be driven to zero, the constant hazard rate region can be driven toward zero and extended in time, and the where-out can be pushed out in time such that it effectively never plays a role in the device reliability.

The infant mortality region of the curve is thought to be the result of some type of pre-existing "defect" or flaw that, over time, effectively wears out (i.e., degrades). Usually, it is thought to be related to a processing or material-handling defect. Examples of such defects are cold solder joints and thin metallization and/or dielectric regions in the package substrate. In general, the hazard rate for defective devices decreases with time. The physical reason for this time dependence is that devices with the largest defects fail first, followed by devices with smaller defects, until all of the defective devices are removed from the population. These defects can be removed from the population using burn-in. Burn-in is a test and/or stress technique that is used to weed out all weak (defective) devices in very short periods of time. Each device is subjected to burn-in so that all defective devices are eliminated from the population prior to shipment of the devices. In this way, the defect population region of the hazard rate curve is eliminated. Unfortunately, the burn-in technique adds significant cost to the device. Because of this, and the fact that this cost will always be present if nothing else is done to eliminate defective devices, there is an economic driver for identification of the root cause of failure for the defective devices and the development of process and/or design fixes that lead to an elimination of the defects. This process of identification and elimination of defects is sometimes costly, but the authors have found that it often pays for itself in a very short period of timed, especially when the percentage of devices that exhibit these weaknesses is in excess of a few percent.

The constant hazard rate region of the curve is thought to be the result of some stochastic unpredictable event, such as a lightning strike or earthquake. The wear-out section of the curve is thought to be the result of material fatigue in the device, such as work hardening of the solder joint and solder fracture. The object of the device manufacturer is to push this section of the curve to a region of time in which the device will no longer be in service. If we go back to the previous section and remember that failures occur when a change in a material property of the device intersects with a sensitive area of the device, then we come to the conclusion that it is important to develop a fundamental understanding of the degradation kinetics of all possible failure mechanisms so as to improve the overall reliability of the device.

3.7 Cost of Reliability

What is the "cost of reliability"? The cost of reliability is simply the cost incurred by a company to complete any type of reliability evaluations, including testing and burn-in, plus costs incurred as a result of product failures in the field. The cost related to field failures depends on the stage of the life cycle in which the product failure is detected and resolved. For example, if a product failure occurs early in the R&D phase of the life cycle, the cost is minimal. However, if the failure occurs near the end of the product life cycle, the cost could be very high, especially if the failure involves any type of recall. The cost of reliability is not always tangible. Even with a complex algorithm, it is not always possible to estimate it. It is not the same for all companies or for all products. There is no standard cost adder to apply for a specific industry or product.

Figure 3.7 illustrates the normalized costs incurred by the device manufacturer at various stages of the product's life cycle. All products go through their own life cycle at their own speed. The speed of progression from one phase to another phase is not fixed and varies depending on the complexity and "newness" of the product. Clearly, as shown in Fig. 3.7, the cost will be significantly lower if reliability is-

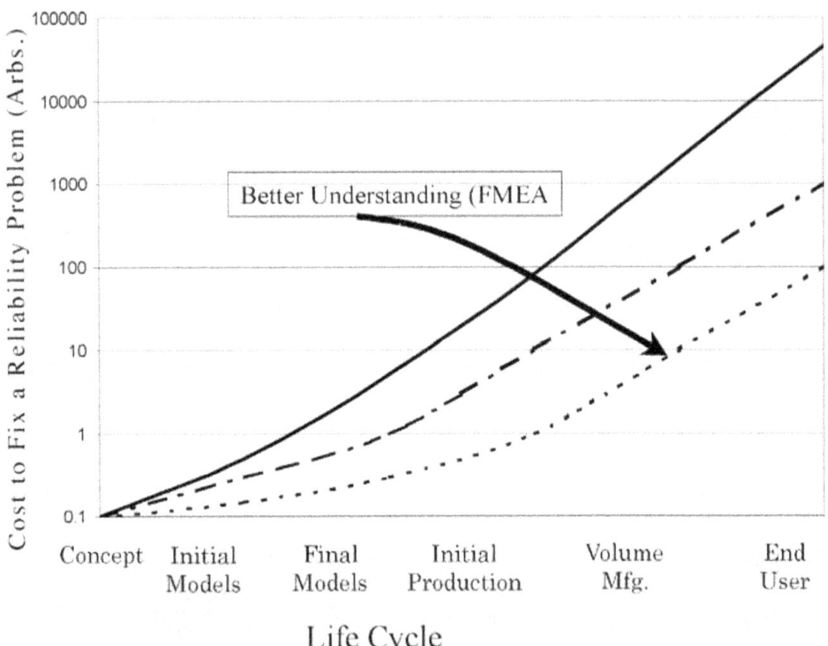

Figure 3.7 Schematic of the cost to fix a failure throughout the life cycle of the product.

sues are detected early in the product life cycle as compared to those reliability issues found near the end of the product life cycle. Furthermore, early detection will avoid costly recalls or field returns and potential legal issues that could arise depending on the product and its end application.

Figure 3.7 shows a basic (common) life cycle of a product. This example shows six phases but, in actuality, you may have more or fewer, depending on your business model or product and customer requirements. The life cycle begins with a concept phase. In this stage, there is no tangible product, but an idea of the product and a team of research and development (R&D) engineers and scientists brainstorming and designing a new product or process technology. This phase is the ideal stage in which to identify potential reliability concerns and issues, with little to no cost impact on the company. The major costs incurred at this stage are the resources' time. This time has a cost associated with it, but the overall cost impact is low as compared to the other stages. As the product moves into the initial and final models, the incurred cost remains relatively low.

The model stage is followed by early production. At this stage, the cost impact begins to grow, because failures involve a larger amount of product. At this stage, failure in the customers' hands could also lead to a poor reputation. This could have a large impact on the overall business picture.

The final stage in this example is when the product reaches the customer. This is the most costly stage if any reliability issues are encountered. This stage involves not only costs to the company but missed or delayed shipments to the end customer as well. Reliability issues detected in this stage may require massive recalls. Such recalls can have a large financial impact and resource strain on both the user's company and the device manufacturer's company.

Clearly, the concepts discussed in the previous sections of this chapter speak directly to minimization of potential reliability problems. If they are applied carefully and diligently during the R&D and early models phase of the product life cycle, then cost is minimized. A further benefit of doing an effective job in the R&D phase would be to establish the company's reputation for avoiding recalls and reliability problems with the devices in the field. This might help sales. In contrast, a poor reputation for reliability and quality almost always leads to a loss of sales.

3.8 References

1. Lawless, J. F., *Statistical Models and Methods for Lifetime Data,* John Wiley & Sons, New York, 1982.

2. Desu, M. M., and S.C.Narula, "Reliability Estimation Under Competing Causes of Failure," in *The Theory and Applications of Reliability*, vol. 2, Ed. C. P. Tsokos and I. N. Shim, Academic press, New York, 1977.
3. Lewis, E. E., *Introduction to Reliability Engineering,* John Wiley & Sons, New York, 1987.
4. Nash, F. R., "Making Reliability Estimates When Zero Failures are Seen in Laboratory Aging," *Mater. Res. Soc. Sym. Proc.* 184, 31, 1990.
5. Joyce, W. B., K. Y. Liu, F. R. Nash, P. R. Bassoud, and R. L. Hartman, "Methodology of Accelerated Aging," *AT&T Tech J.* 64(3), 717, 1986.
6. Nelson, W., *Statistical Models, Tests Plans, and Data Analysis,* John Wiley & Sons, New York, 1990.
7. Amster, S. J., and J. H. Hooper, "Statistical Methods for Reliability Improvements," *AT&T Technical J.* 65(2), 69, 1986.
8. Klinger, D. J., Y. Nasada, and M. A. Menendez, *AT&T Reliability Manual,* Van Nostrand Reinhold, New York, 1990.
9. Goldthwaite, L. R., Failure Rate Study for the Lognormal Lifetime Model, *Proc. 5th Nat. Symp. Reliab. Qual. Conf.* 208, 1961.
10. Nash, F. R., *Estimating Device Reliability: Assessment of Credibility,* Kluwer Academic Press, Boston, 1993.
11. Feynman, R. P., *What Do You Care What Other People Think,* Bantam Books, New York, 1989.
12. Dalal, S. R., E. B. Foulker, and B. Houdley, "Risk Analysis of the Space Shuttle: Pre-Challenger Prediction of Failure," *J. Am. Stat. Assoc.* 84(408), 945, 1989.
13. "Introduction to Failure Mode Effects Analysis (FMEA)," America Supplier Institute, Dearborn MI, 1993.
14. Menon, H. G., *TQM in Product Manufacturing,* McGraw Hill, New York, 1992.
15. "FMEA Software," Ford Motor Company FMEA Headquarters, Plymouth, MI.
16. Sheng-Keng Ma, *Statistical Mechanics,* World Scientific, Singapore, 1998.
17. Tolman, R. C., *Principles of Statistical Mechanics,* Oxford University Press, London, 1962.
18. Verhoeven, J. V., *Fundamentals of Physical Metallurgy,* John Wiley & Sons, New York, 1975.

Chapter

4

Physics and Chemistry of Failures in Packaged Devices

John W. Osenbach
Weidong Xie
Jason P. Goodelle

4.1 Introduction

The underlying physics of failure is discussed in Chapter 3. In this chapter, we review the most prevalent microscopic failure mechanisms responsible for most of the device failures. As discussed in Chapter 3, failure results from changes in the material via physical movement of some or all of the material under applied energy, i.e., stress. Failure occurs when this movement causes a shift in device performance. This material movement is driven by the thermodynamic law that all systems tend toward the lowest energy state to attain equilibrium. Thus, given infinite time, the system will change its state as a result of the application of an external energy source. Fortunately, the rate at which the system is transformed from one state to another is controlled not by thermodynamics but by kinetics. Under normal operating conditions, most reaction kinetics for typical packaging material systems are so slow that the final thermodynamic equilibrium state will never be attained in times that are important to the end user. Thus, most reactions that can lead to device degradation do not impose a substantial risk of failure.

In this chapter, we deal with a number of physical phenomena that could lead to device failure in times that are important to the end customer. First, we discuss solid-state mass transport effects, followed by liquid phase mass transport effects. We then discuss current-driven mass transport effects and follow up with a discussion on thermal mismatch effects. Finally, we review moisture-driven degradation effects.

4.2 Mass Transport Effects

4.2.1 Solid-State Reactions

Several reliability- and quality-related issues are resultant from reactions among two or more dissimilar packaging materials. In particular, all electronic packaging technologies require the use of a number of dissimilar materials to form physical and electrical connections from the IC to the outside world. Electrical joints (connections) that fall into this category include Au wire to Al bond pads and solder (Pb-containing and Pb-free) to Cu, Ni, Au, or Pd. The formation of a "good" metallurgical joint requires, at a minimum:

1. Chemical reactions between the two materials that are used to form the joint must occur in a controllable and reproducible fashion when the joint is formed.
2. Any additional reactions that might occur after the joint is formed and deployed in the field must be sufficiently slow that no significant degradation occurs.

In item 2, "significant" is defined as a change that results in a shift in the performance of the device. The reactions typically involve the formation of binary, ternary, and other compounds or solid solutions of the individual metals that ultimately form the joint. If such reactions do not occur, then the joint is generally of poor quality, i.e., it has poor mechanical strength as well as poor electrical and thermal conductivity. Furthermore, if the reactions occur in an uncontrollable (optimal) fashion at a given set of manufacturing or post-assembly conditions, then these joints could be mechanically and electrically weak (e.g., in a nonoptimal, degraded state). It is likely that both early life and wear-out failures will occur more frequently on devices with poorly formed joints than that on devices with well formed joints. This, for example, could be the situation if, during wirebonding, the time, temperature, ultrasonic energy, and/or the applied force and the spatial uniformity of the energy sources are set incorrectly.*

For illustrative purposes, let us consider what happens when two dissimilar materials are brought into contact with each other. As discussed in Chapter 3, if a reaction between the two materials leads to lower energy, then the reaction will occur. Furthermore, if the temperature is below the melting point of either of the two materials, and if any liquid

* The "energy source" includes the ultrasonic energy and the applied force and temperature. By "spatial uniformity," we mean that the energy sources are applied to the materials such that it is has a uniform distribution along the entire interface between the two materials being joined.

formation region that exists in the binary system, the rate of reaction will be controlled by solid-state diffusion. Given infinite time, the reaction will result in the lowest energy state—the thermodynamic equilibrium state of the system. Fortunately, for the package technologist, the equilibrium states of most commonly used package metallurgies (a function of composition and temperature) have been empirically determined. The equilibrium states can be conveniently shown in phase diagrams (see, for example, Refs. 1 and 2). Figure 4.1 schematically shows the phase equilibrium diagram of a binary system, composed of A and B, that has a complete solid solubility, i.e., complete miscibility. Complete miscibility, by definition, implies that, if equal concentrations of A and B were placed in contact with each other at room temperature and given an infinite time to stabilize, the system would end up being a random mixture of 50 percent A and 50 percent B.

One common package system with complete miscibility is the binary system of Cu and Ni. The rate at which the solution would migrate to a random 50/50 mix is given by the diffusion constants of A in B and B in A. A schematic of its functionality is shown in Figure 4.1*b*. Note

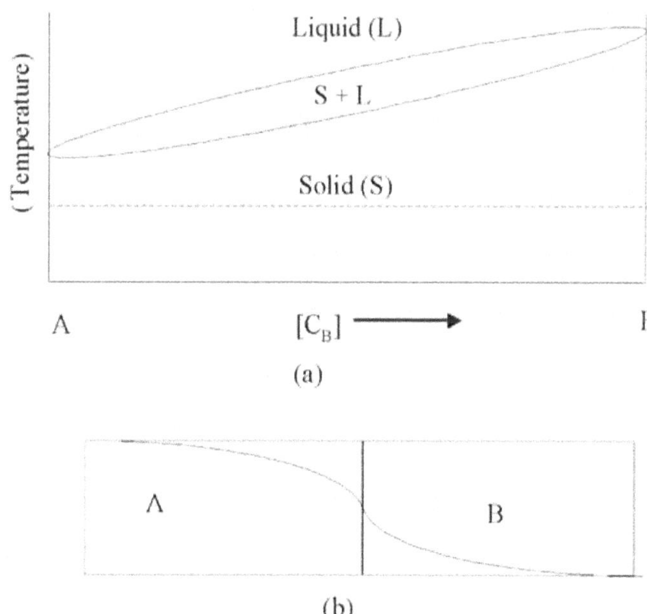

Figure 4.1 (*a*) A schematic of a phase equilibria diagram of a completely miscible two-component material system and (*b*) a schematic of a diffusion profile at time, *t*, taken along the dotted line in (*a*). Note that, in the limit of infinite time, the system would go to a 50/50 mixture of A and B, and the diffusion profile would be horizontal at this concentration.

that there are no physical or chemical requirements for the diffusion coefficients of A in B and B in A to be equal. That is to say, the concentration gradients of A into B and B into A need not have the same spatial or time functionality.

Considering a more complicated system that has a stable binary phase, δ (Figure 4.2*a*). The diffusion-driven concentration profile is schematically shown in Figure 4.2*b*. The binary system of Sn and Mg is such a system. In this case, diffusional transport occurs between the different materials forming the stable phase as well as between the end point materials and the stable binary phase. On either side of the binary compound, δ, there is a diffusion tail where element B is diffusing into element A from the δ phase, and element A is diffusing into element B from the δ phase. The exact shape of the profiles of elements A and B usually will be given by the respective diffusion constants in the stable binary compound. As can be seen, the situation with a simple binary compound is much more complicated than that for a completely miscible system. In fact, many of the commonly used packaging metallurgies have more than one binary phase, and often-

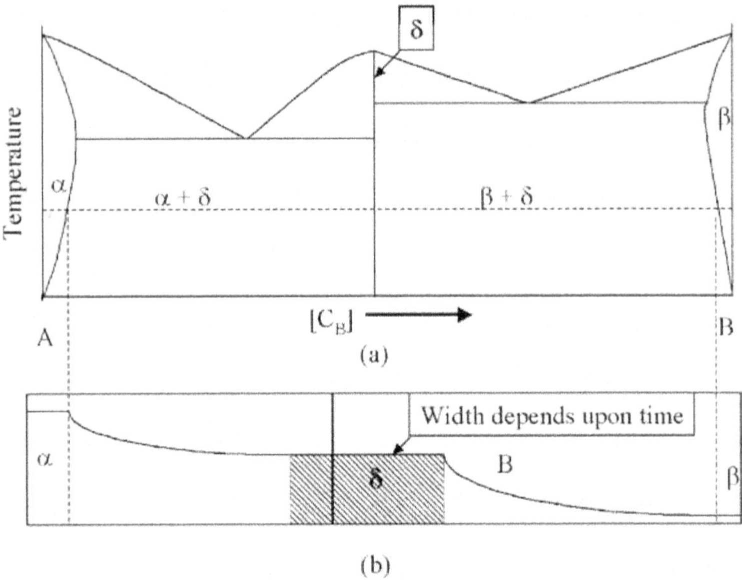

Figure 4.2 (*a*) Schematic of a binary phase diagram that has one binary intermetallic phase, δ. In this case, there is some solubility of B in A, phase denoted as α, as well as some solubility of A in B, phase denoted as β. (*b*) Schematic of the diffusion profile at time t along the dotted line in (*a*). Note that the concentration of B in A, α, and A in B, β, are both pinned at that composition shown by the dashed line in the figure. The concentration of the intermetallic phase, δ, is also held constant. The width of the intermetallic phase increases with time at the expense of the end component phases.

times there are ternary and even quaternary phases that form. Thus, the actual situation for commonly used packages is usually much more complicated than these simple examples. For example, the commonly used Au wire to Al pad (Al/Au system) has five intermediate stable binary phases in the technologically important temperature range. Similarly, the Cu/Sn system has three such phases. However, the general physics of the situation is not altered with the increased phase complexity—just the overall concentration profile and time evolution of it.

The stable binary phases are generally referred to as *intermetallic phases of compounds.* (In the literature, these phases are commonly referred to by the acronym IMC.) Many are characterized by a well defined singularity in composition, usually referred to as a *line compound* because they are only stable at one particular composition. This is because they have a well defined, closed packed crystal structure with lattice parameters that essentially prevent even low levels of solid solubility of either end point element from occurring in the structure. These line compounds are often mechanically strong and brittle* and have higher thermal and electrical resistance than the individual end components. One notable exception is the Sn/Ag IMC, Ag_3Sn, which is somewhat ductile. Finally, these compounds typically have a different, often lower, coefficient of thermal expansion than the end point elements.

As discussed previously, diffusion is thermally activated, and the diffusion mechanism that controls the diffusion rate depends on the ratio of the temperature of interest to the melting points of the individual materials. In most cases, the diffusion coefficients (e.g., rate constants) of the different end point elements or materials will be different. Therefore, the rate of reaction will be controlled by the element that diffuses at the lowest speed. For technological and illustrative purposes, let us take the widely used Al/Au system as an example. In this system, Au diffuses into Al faster than Al into Au[3],† which means that, for an infinite thickness of Au and Al, at temperatures of technological interest, more Au will diffuse into Al than Al into Au in any given time period (Figure 4.3*a*). For diffusion behavior that occurs via specific lattice site to specific lattice site, the movement of an atom results in an open atomic site. This site is referred to as a *vacancy*, and this type of diffusion is referred to as *vacancy diffusion.* For such a

*In this context, we define a brittle material as having a strain to failure of less than 5 percent.

†It is interesting to note that, in this case, the rate-controlling species is the material with the lower melting point (T_{mp}), Al (660°C), rather than the material with the higher melting point, Au (approximately 1063°C).

system, assuming the volume of the system remains constant, the neutral plane of mass (original interface) will effectively move toward the slow diffusion species, the Al side of the joint. This effect is known as the *Kirkendall effect.*[4] As the concentration of vacancies builds up in one location, they can coalesce and form a void. This build-up of vacancies is referred to as *Kirkendall voiding.* Darken was the first to predict such an effect mathematically.[5] These two effects are illustrated in Figure 4.3*b* and *c*. Finally, if the Kirkendall void density becomes large enough and/or coalescence of the voids occurs, then Kirkendall voids will degrade the strength of the joint. This, coupled with the brittle nature of the joint and the difference in thermal expansion between the IMC and the material on either side of the reaction zone, could lead to device failure.

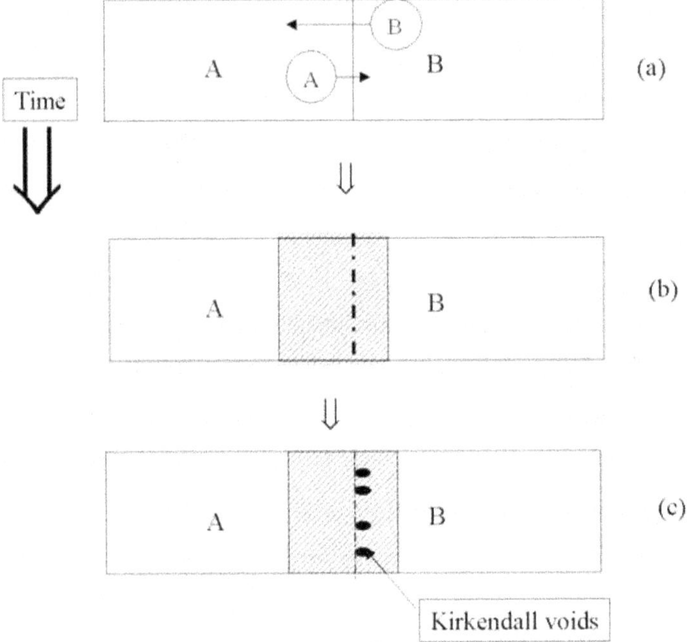

Figure 4.3 The time evolution of the location of the boundary between A and B as well as the development of Kirkendall voids is shown schematically. The length of the arrows in (*a*) represents the relative magnitude of the diffusion coefficients of A and B, respectively. As shown, B has a larger diffusion coefficient than A. The gray-hatched region in (*b*) represents the reaction zone as time progresses. The actual interface is denoted by the dark, solid line in the figure. Its original position is shown by the dashed-dotted line in (*b*) and (*c*). The formation of Kirkendall voids is also shown in (*c*). In this case, they form at the original interface near B, because more B diffuses than A at any given time, thereby leaving vacancies that coalesce into voids.

It should be noted that, although the intermetallic compounds between Al and Au are brittle, they are not usually responsible for device failure. Similarly, Kirkendall voids are normally not the root cause of failure of Au wire to Al pad joints in modern wirebonding processes and assembly processes.[6] Failures, however, do in fact occur if there is poor welding (poor contact), which can occur if the contact interface between the ball at the tip of the wire and the aluminum pad is not spatially uniform. Figure 4.4*a* and *b* show examples of a good joint and

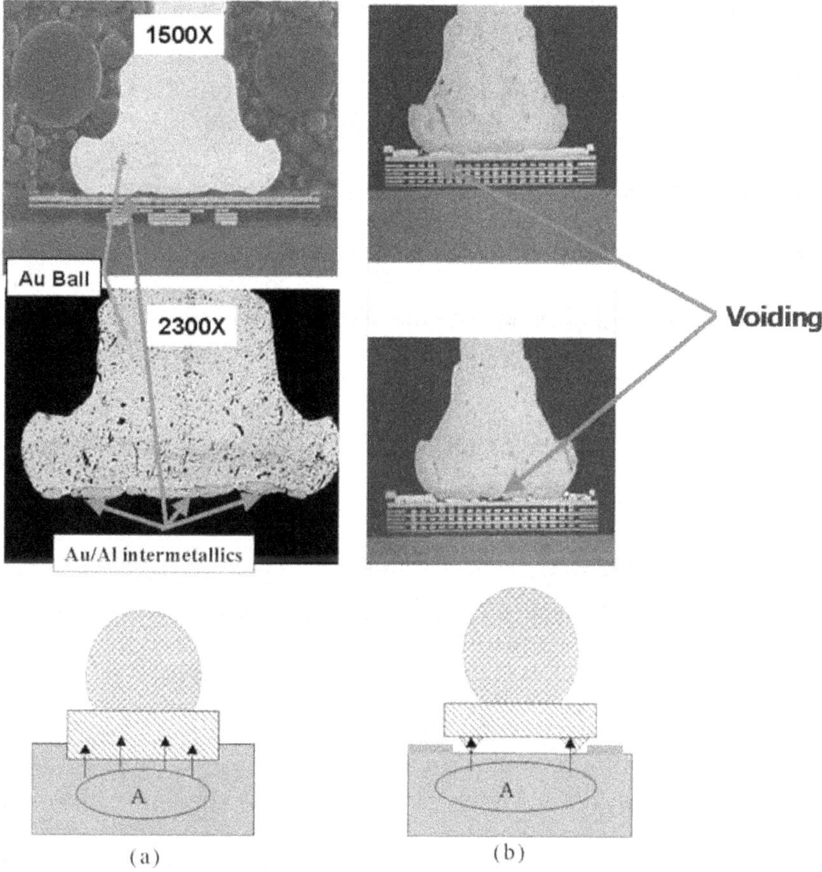

Figure 4.4 (*a*) Two cross sections of a well formed gold wire to Al bond pad joint. In this case, the bond is made on a Cu/low-k IC. As shown, the Au/Al intermetallic phases are relatively uniform, and no large voiding is observed. In contrast, (*b*) shows two micrographs of a poorly formed bond on a similar device. As shown, there are large voids in the joint. These voids are a result of poorly formed interface caused by microwelding. The schematics of the diffusion front for each case are shown under the respective micrographs. As shown, the diffusion of A, denoted by the arrow, is uniform through the entire interface for case (*a*). For case (*b*), there are only point contacts through which matter can move.

a bad joint between a Au wire ball bond, 25-μm diameter, and a 1.2-μm thick Al pad, on Cu/low-K 130-nm silicon technology, respectively. Also shown are schematics of the respective interface reactions. In the case of a good interface (Fig. 4.4*a*), the formation of the Al/Au IMC leads to the consumption of most of the Al layer under the ball bond, but there is no sign of large voiding. Furthermore, the IMC is somewhat uniform and continuous across the entire interface. In the case of the poor interface (Fig. 4.4*b*), the Al/Au IMC consumes most of the Al layer under the ball bond, but there are regions where large voids are present. In addition, there are regions along the interface where little or no IMC is observed. The large voids are not Kirkendall voids in the classic sense. We call this effect *contact limited diffusion induced voiding.* In this case, the initial contact welds between the Au ball and the Al pad were made only in localized regions where the Au reacted with the Al, consuming most of the Al pad. In this case, a large void is left behind where the Al once was. Other failures of the Al/Au interface have been traced to impurities in the joint as well as specific mold compound flame retardants.[7,8] However, in general, except in cases where the temperatures used to assemble electronic devices exceed 300°C or the operation temperatures approached 300°C, Kirkendall voids resulting from the faster diffusing rate of Au as compared to Al have not been identified as the root cause of wirebond failures in packaged devices.[3]

Kirkendall voids are common in most multicomponent systems, as each component in such systems usually has a different diffusion rate. Therefore, Kirkendall voiding can (and often does) occur in other types of package joints. Thus, laboring with a lack of information, one could always revert to indicting Kirkendall voids as the root cause for many package joint failures. In one such case, it was reported that Kirkendall voiding was the root cause of failure of Pb/Sn solder joints to an electroless Ni-immersion Au pad.[9] Figure 4.5 shows the different layers found in such a joint in the cross section. For more detailed explanation of the reaction between the solder and the electroless Ni-immersion Au system, the reader is referred to Refs. 9 and 17 and the references therein. Note that, in addition to Sn/Ni IMC, electroless Ni, and Ni_3P, there are Kirkendall voids in the joint. In this case, the Kirkendall voids form at the Ni/Ni_3P interface and have been indicted as one of the root causes of failure for electroless Ni/Au solder pads.[9] Our preliminary results, although not conclusive, indicate that although Kirkendall voids can and do lead to a weakening of the joint, Kirkendall voids in themselves are in fact not the root of this failure in this system.[10] This is not to say that Kirkendall voids do not or cannot lead to failure of this system or any other systems, but just to say that, at present, there are very

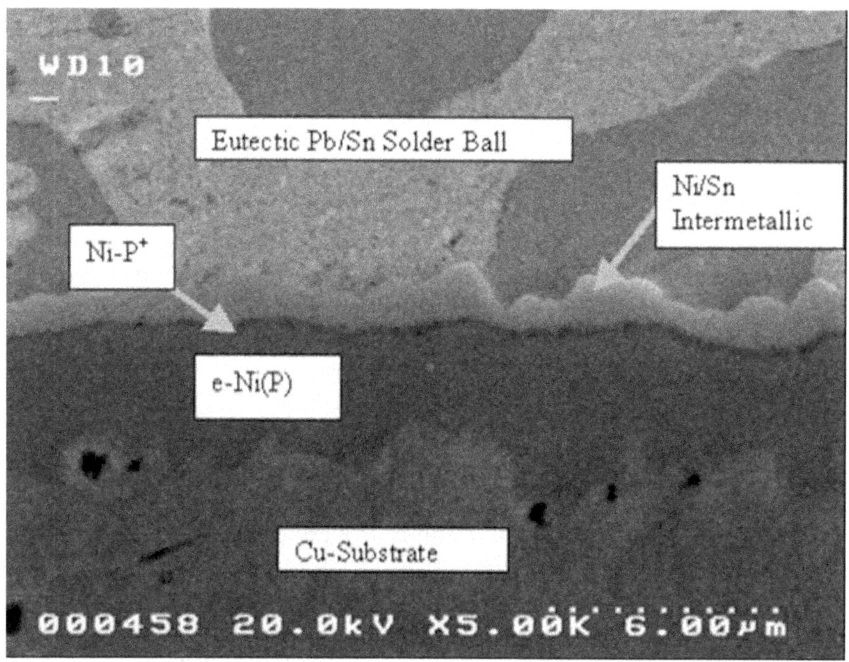

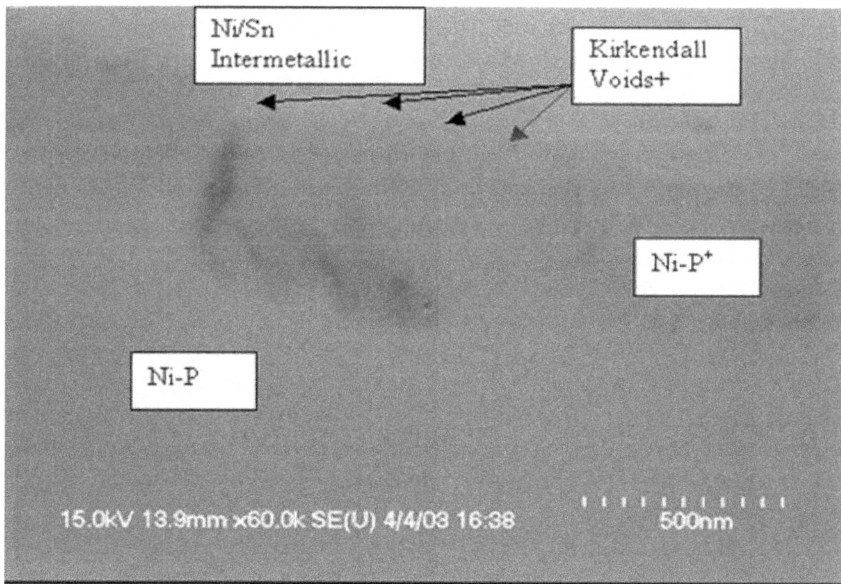

Figure 4.5 Micrographs of a solder joint between electroless Ni/Au and eutectic Pb/Sn solder. Note that Kirkendall voids are observed between the Ni/Sn intermetallic layer and the Ni-P^+ region of the joint.

few cases indeed in which failures have been directly traced to this effect, at least over the last decade.

4.2.2 Liquid-State Reactions

When a device is soldered to a lead frame, substrate, board, or other apparatus, the solder melts and begins to dissolve the metal on the two sides of the joint. In most cases of interest to the package engineer, the solder is an Sn/Pb based solder, the most common being eutectic Sn/Pb or near-eutectic compositions. In some cases, additional elements such as Ag are added to the solder to improve the solder characteristics in some way. Solder characteristics that are affected by adding the additional elements include solderability, processability, and mechanic strength. The solder plus the two piece parts are brought to a temperature at which the solder melts. At least one of the elements in the solder, in this case Sn, reacts with the metallization of the two piece parts, usually leading to the creation of Sn-based IMCs. Typical substrate metallizations include, but are not limited to, Cu, Ni, and Au. The solder process involves melting of the solder followed by dissolution of part or all of piece part metallization and the formation of IMCs, when liquid solder becomes supersaturated with the pad metallization. This is followed by solidification of the solder when the temperature drops below the melting point. The system contains at minimum three different metals and often as many as four or five. By analogy to the discussion above, one concludes that the soldering process leads to substantially more complicated reactions than that found in solid-state reaction processes between two dissimilar metals. As such, although it has been used for many centuries, the soldering process is extremely complicated and difficult to understand and model.

Because solder has been and continues to be widely used to join metals, there is a large body of literature on the subject. The reader is referred to Refs. 11 and 12 and references therein for further reading. Fortunately, the process can and has been empirically optimized to yield highly reliable joints, as evidenced by the fact that solder is used to join the large majority of packages to system-level printed circuit boards. There are, however, some interesting mass transport effects that occur during and after soldering. They can and sometimes do change the characteristic of the joint and hence its long-term reliability.

The first two that are possibly most innocuous are the redistribution of Sn and Pb from the supersaturated grains of Pb and Sn, respectively. Because the melted solder is a homogeneous mixture of Pb and Sn, on average, the local composition will be that given by the ratio of Sn and Pb in the solder when it is in the liquid state. When it cools below the melting point (the eutectic point for 63/37 wt.% Sn/Pb solder), the system equilibrium dictates the occurrence of a phase separation of Pb and Sn such that, at room temperature, there is much less than 1 percent Sn in the Pb phase and 1 percent Pb in the Sn phase. However, the separation of Pb and Sn into their respective phases is diffusion controlled, and the time constant associated with it is much larger than the available time during cooldown. Thus, there will be an excess concentration of both Sn in the Pb and Pb in the Sn in the joint. Over time, even at room temperature, Pb will diffuse out of the Sn phase into the Pb phase, and Sn from the Pb phase into the Sn phase. In addition, the overall energy of the system is reduced by the elimination of small grains of Sn and Pb and by the growth of large grains of Sn and Pb. This process is called *grain growth and recrytallization*. The combination of these two solid-state, diffusion-driven reactions leads to a net decrease in the shear strength of the joint. Fortunately, it also leads to an increase in the strain energy needed to cause solder fatigue. The reader is referred to Ref. 21 for more detailed explanation of the grain growth and recrystallizing processes.

The influence of these effects on the mechanical properties of the joint is shown in Figure 4.6. This figure is a plot of the shear strength of 635-mm eutectic Sn/Pb solder balls on a Cu metallized substrate

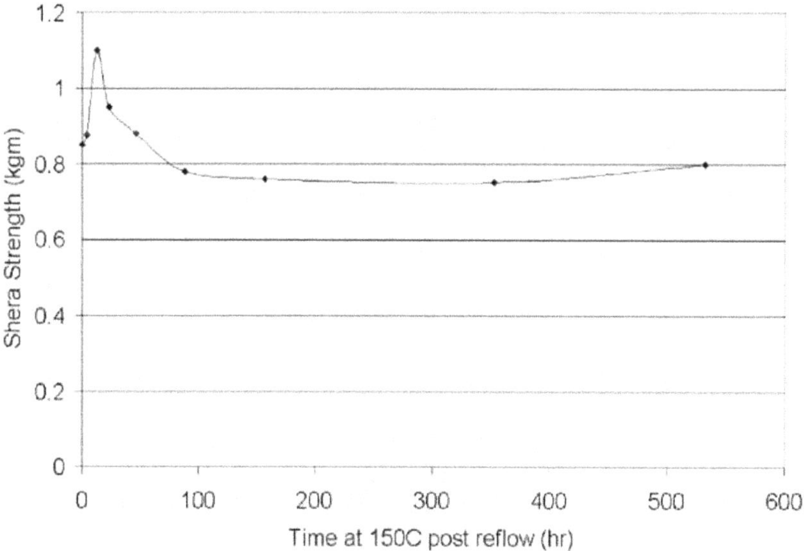

Figure 4.6 Shear strength of 635-mm diameter eutectic solder balls on Cu solder pads. The time-zero data point was taken within 2 hr after a simulated solder board attach with a peak temperature of 220°C. The subsequent data were taken after various instances of exposure of the same set of packages to 150°C.

versus time at 150°C. In this case, the joint was made with a standard board solder attach process in nitrogen ambient with a peak temperature of 220°C. Note that there is an initial increase in shear strength over the first tens of hours of storage at 150°C, which is followed by a decrease in strength over the next hundred hours and, finally, a stabilization of the shear strength. This change in the shear strength of the solder joint is a direct result of redistribution of Sn and Pb from the supersaturated grains, grain growth, and recrystallization. Because the mechanical properties of most materials strongly depend on the grain size and the local chemical composition, the effects of redistribution of Sn and Pb recrytallization lead to changes in the mechanical stability of the joint over time that can either improve or degrade reliability. Fortunately for the Pb/Sn solder system, the changes described here usually lead to slight improvements in reliability so that, for eutectic solders, these mechanisms in themselves do not play a significant role in the overall reliability of the joint. The same, however, cannot be said for the lead-free replacement materials, as there is too little data to arrive at such a conclusion.

After solder joints are made, other solid-state reactions can also lead to degradation in reliability. If Au is part of all of the substrate metallization, then care has to be taken to ensure that no Au/Sn intermetallic embrittlement occurs. Sn dissolves and reacts with Au to form Au/Sn IMC, the most important of which is the Sn-rich $AuSn_4$. Foster has determined that in bulk samples more than 3 to 5wt.% Au in Sn leads to embrittlement of the material.[13] Thus, for the past three decades, 3 to 5 wt.% or less has been quoted as the limit for Au in a Sn-based joint to prevent embrittlement and premature failure. A few studies have indicated that much lower concentrations of Au can lead to embrittlement over time as a result of precipitation and growth of the $AuSn_4$ IMC at the Ni interface.[14–16]

Figure 4.7*a* shows an example of a joint that was embrittled with Au and one that was not. The effects of Au embrittlement are problematic, because the exact concentration needed to cause embrittlement of a particular joint not only depends on the time at temperature but can also depend on the joint geometry—especially the cross-sectional area of the solder-to-substrate interface. In general, we have found that the Au content should never exceed 0.8 wt.%, and for some geometries less than 0.8 wt.%. The effect reinforces the notion that solid-state reactions that occur at used conditions can and often do influence the reliability of the device. Thus, care has to be taken so as not to discount possible degradation mechanisms that at first seem unreasonable but, after further study, are identified as important degradation mechanisms. That is to say, care has to be taken when changing the metallurgy of a joint to ensure that reactions that might occur over a long

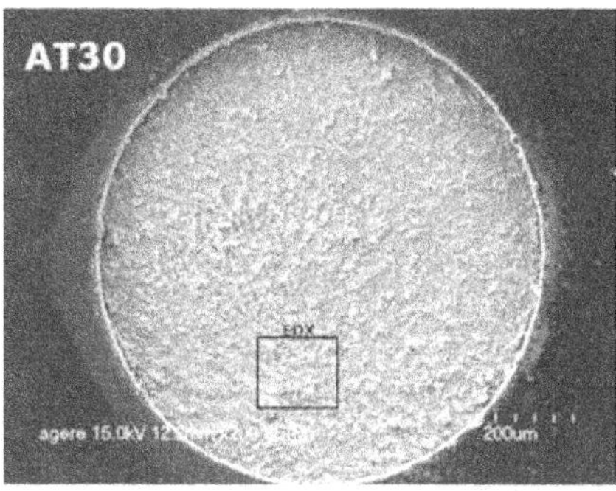

Brittle fracture mode is shown. Fracture surface is near Cu pad/solder ball interface!

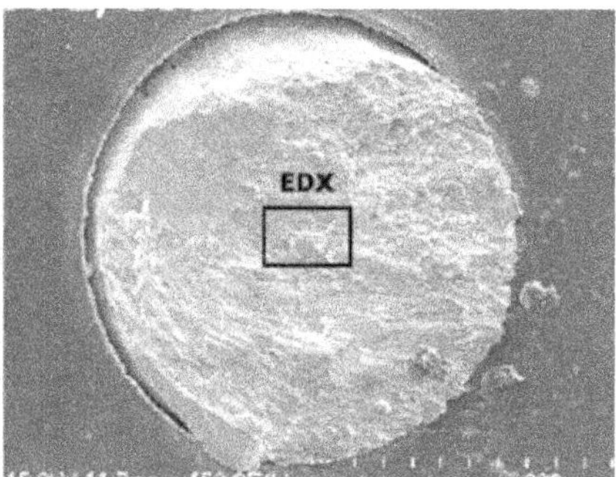

Ductile fracture mode is shown. Fracture surface is in solder ball.

Figure 4.7 Micrographs of the electrolytic Ni/Au pad after a eutectic solder ball was sheared from them. As shown, there are two types of fracture: a brittle fracture mode resulting from Au embrittlement, and a ductile fracture mode where no Au embrittlement is observed. Brittle failure was found on a sample with 1 wt.% Au, ductile was found when the Au content was below 0.8wt.%.

period of time, but that are not yet identified or understood, will be found in standardized reliability testing.

Because Cu is a fast diffuser in many materials, the structure and thickness of the Cu/Sn intermetallic layers at such a joint depend not only on the reflow process, or multiple reflow processes, but also on the post-processing thermal excursions—even those that occur at room temperature. The thickness and structure can also be influenced by strain-induced diffusion. Given that the reliability, e.g., device wearout lifetime, decreases with increasing Cu/Sn thickness, this effect is undesirable. As such, care should be taken to minimize the thermal excursions that a joint experiences so as not to degrade the device lifetime unwittingly. Finally, although Sn/Ni and Sn/Cu joints are known to have adequate reliability when properly formed, it has been reported that ternary mixtures of Sn, Ni, and Cu produce joints with less than optimal reliability. This is particularly true for Sn-based solder joints connecting a Cu pad to an electroless Ni-immersion Au pad.[17,18] As the packaging community moves toward Pb-free solder attachments, the thermal excursions that the solder joints will be exposed to will increase by at least 20°C and as much as 40°C. In addition, the most common lead-free replacement solders, including those in the Sn, Cu, Ag family, contain significantly higher levels of Sn than eutectic solder. Since the reaction rate is, at a minimum, proportional to the Sn content and the exponential in temperature, the combination of these two effects might prove to have a significant effect on the reliability of lead-free solder joints. Of particular interest are flip chip packages.

For flip chip ICs, it is common to use an Al/Ni/Cu under-bump metallurgy (UBM) onto which the solder is electroplated or stencil printed. Note that the UMB layers are thin and, in particular the Cu layer is typically less than 1 μm as compared to a minimum of 15 μm on a typical PWB or flip chip substrate pad. For the case of stencil printed solder, the solder is reflowed twice after printing, and the surface tension causes it to form a truncated sphere on top of the UBM. The solder on the IC is referred to as a *bump*. During the solder reflow, the Cu layer of the UBM reacts with the solder, forming a Cu/Sn IMC. The thickness of the IMC is determined by the time-temperature profile of the reflow process and the solder composition. The bump is exposed to a third reflow process when the die is attached to the package substrate. This process leads to additional consumption of the thin Cu layer. The bump is exposed to at least two additional reflow process thermal excursions by the time the package is attached to the PWB, one when the balls are attached to the substrate and one when the package is attached to the PWB. It is possible, for PWB rework reasons, for the bump to be exposed to at least two additional reflow processes. Each of these additional thermal excursions will cause additional consumption of the Cu layer. This could lead to complete consumption of the Cu layer. Figure 4.8 shows examples of the solder bump/UMB interface of a flip chip silicon die after it was exposed to ten simulated solder board attach reflow processes with a peak temperature of 260°C.[19] These devices had lead-free (Sn/Ag/Cu, 97.5 wt.%/1.5 wt.%/0.5 wt.%) solder bumps. As shown, there is evidence that the Cu UMB layer is breached, and the solder is beginning to react with Ni as well as the Cu. Thus, the interface contains a ternary mixture of Sn, Cu, and Ni. In the end, the additional thermal excursions that are required for Pb-free processes might be the biggest problem to overcome for flip chip die for lead-free applications.

4.2.3 Electromigration

The final topic we cover in solid-state transport effects is electromigration. This should not be confused with electrochemical migration, which is a corrosion-driven process. Electromigration is a mass-transport phenomenon that occurs as a result of momentum exchange between the flowing electrons that carry current (in the Franklin convention, the electron current is negative current) in an electrical interconnect and the host metal lattice.[4] Effectively, the host atoms are coupled to the electron wind of the current flow and are dragged along with the electron wind. The energy that drives this process is derived from the current-driven chemical gradient that this momentum transfer produces. The gradient leads to diffusion of the atoms in the direction opposite to the flowing current, i.e., in the direction of the

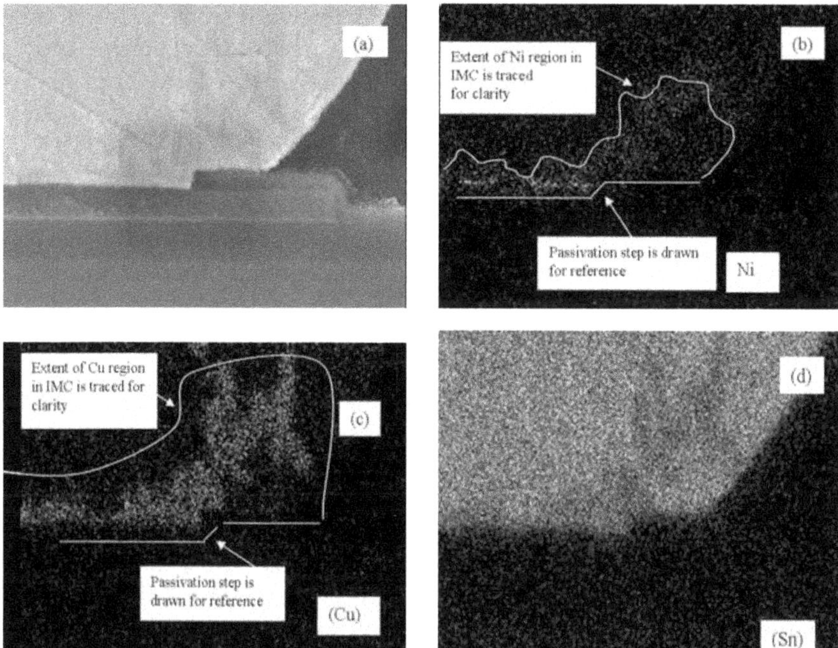

Figure 4.8 Elemental maps and cross section of lead-free solder bump exposed to ten simulated solder board attach reflow processes.

electron flow. The displaced atoms are effectively dragged along with the electron current and deposited at some other location along the interconnect metallization. This leads to a restoring force that tends to resist additional electromigration. However, over time, when electromigration occurs as a primary failure mechanism, there is a build-up of additional metal in the metal trace somewhere downstream in the direction of electrons. This causes a build-up of vacancies where those atoms once sat. The vacancies can coalesce into a void. The development of voids decreases the cross-sectional area of the metal. This leads to an increased I^2R thermal rise locally. This then leads to an increased rate of void formation. Thus, void formation leads to a feedback circuit that increases the rate of void formation with time. If and when the void grows large enough to breach the metal line, an open circuit occurs. In addition, in the area where the atoms are deposited, there is metal build-up. This leads to a localized stress. If the stress is large enough, metal extrusion can occur to relieve this localized stress. This could lead to a short-circuit failure. In the end, failure can occur as a result of an open in the line (effectively coalescence of the vacancies to form a complete open), or a short (coalescence of the metal in one position and stress-driven diffusion leading to a short).

Furthermore, the mass transfer of atoms from one location to another leads to an increase in resistance of the interconnect path. As the speed of the device increases, small changes in resistance that previously did not cause performance degradation might do so. This may well become a very important issue for high-speed flip chip devices in the near future. In general, this problem can be treated as an Arrhenius rate problem with a pre-exponential that is current density dependent.[20] As such, all of the diffusion constraints and concerns discussed in the previous chapter must be included in the analysis.

4.3 Thermal Mismatch Effects

This is perhaps the most technologically important and scientifically difficult quality and reliability challenge for electronic (and optical) packaging. In addition, it could very well be the most difficult to quantify and predict *a priori*. The importance of degradation resulting from thermal mismatches can be judged by the thousands of published papers on the subject as well as the fact that it is addressed in some detail in most electronic packaging text books. Differential thermal expansion effects influence the assembly process as well as the lifetime of the device.

As discussed in Chapter 2, an electronic package consists of the integration of a rather large variety of different materials, including the silicon device itself. Table 4.1 lists the common packaging material types and the range of coefficient of thermal expansion (CTE) and the modulus of elasticity that characterizes these type materials. Note that we have included the properties of silicon and its metallizations and dielectrics, including Cu and low-k.

Before discussing brittle fracture or fatigue, it would be proper first to review the elastic properties of typical materials used in packaging. A simple time stress-strain plot, Fig. 4.9, can represent the elastic properties of different materials. Note that the increase in stress near the fracture point occurs because of a reduction in cross-sectional area. For engineering stress-strain of materials, the stress would actually decrease near this transition, because the cross-sectional area reduction (necking) would not be taken into account. That is to say, if held at the stress near the transition, the joint would fail. A characteristic stress/strain curve of brittle/rigid materials (e.g., silicon, inorganic glasses, and ceramics) indicates that there is essentially a linear dependence of stress on strain. That is to say, these types of materials follow Hook's law. The slope (modulus) is high, and the strain to failure is typically less than 0.1 percent. These types of materials have, for all intents and purposes, no memory in that the stress-stain curve is invariant with number times the materials is subjected to stresses

Table 4.1 Coefficient of Thermal Expansion and Modulus of Common Package an IC Materials

Material	~CTE @ RT (ppm/°C)	~Modulus (GPa)
Cu	17	400
Al	23	65
Si	3	150
Au	19	170
Ag	10	110
SiO_2	0.5	150
FSG	3	70
Low K	17	20
Al_2O_3	6	250
Mold compound	8–20	10–30
Organic substrate	10–25	5–30
Die attach epoxy	8–25	10–40
FC underfill	15–50	2–10
Eutectic solder Pb/Sn	24	35
Pb-free solder	20–25	50
High-Pb solder (95Pb/5Sn)	26	24

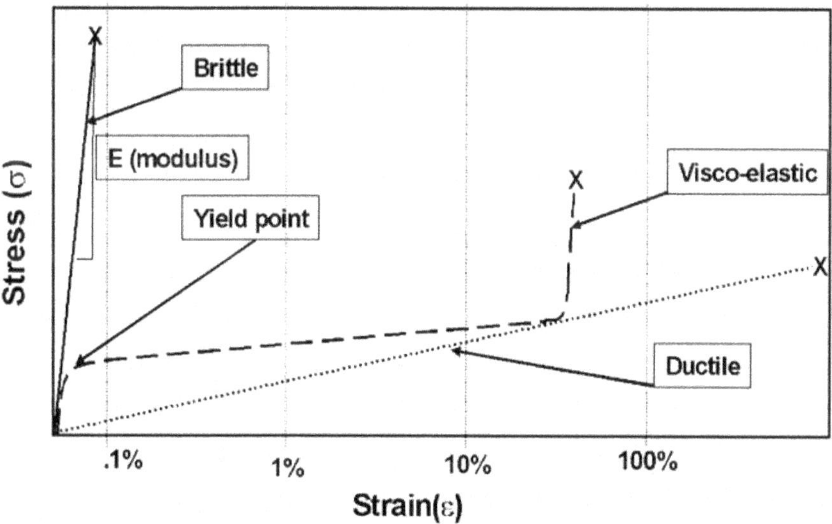

Figure 4.9 Schematic showing the stress-strain behavior of brittle, viscoelastic, and ductile materials.

below the fracture point, denoted by an X in the figure. The implicit assumption here is that any flaws will not grow under the cycling stress. The characteristic stress/strain curve of ductile materials (e.g., solders, Cu, Au, and Al) is nonlinear. There are three regions of interest for ductile materials.

1. There is a linear dependence of stress on strain with a high slope at low strains. In this region, the stress-strain dependence is for all intents and purposes completely reversible, and the material behaves in a manner similar to a brittle material in this region.
2. The linear reversible region is followed by a region where the slope rapidly decreases by three to four orders of magnitude. The transition from the high slope to the low slope is roughly defined as the yield strain (stress). In this region, the stress-strain dependence is not reversible. That is to say, the stress strain dependence has some hysteresis in this region.
3. The low slope region is followed by a transition to a second, higher slope region. In this region, the slope increases with increasing strain, and this continues until the material ruptures.

A characteristic stress-strain curve for a fully compliant material (e.g., silicones) indicates that there is essentially a linear dependence of stress on strain over all strain space. The slope (modulus) is many orders of magnitude lower, and the strain to failure is typically many orders of magnitude higher than that found in typical brittle materials. As is the case for a brittle material, the stress-strain dependence of the ductile material is completely reversible. Note that the ductile material exhibits classic linear-elastic behavior at low strain, typically less than 0.1 percent for a polycrystalline material, and with the exception of reversibility (i.e., no hysteresis), ductile behavior for strains up to tens of percent. The yield stress is dependent on the ambient temperature. In general, the yield stress increases as the melting point of the material increases. Similar to many other thermomechanical properties of materials, the yield stress of a ductile material is a function of the material type, the microstructure, the film thickness, the grain size, the thermal and mechanical history, the contaminants or dopants, the surface treatments, the density, and the grain orientation.[21] Furthermore, the yield strength also increases with increasing strain rate.

In general, the yield stress decreases as the strain rate decreases. The standard terminology for this effect is *creep*. Qualitatively, one can think of creep as having a characteristic time constant which is diffusion limited. The time constant depends on the dominant diffusion mechanism of creep. If the strain rate exceeds the diffusion rate for the dominant creep mechanism, then creep cannot occur. In general, one can separate creep for polycrystalline, nonglass, phase-containing materials into three main mechanistic types,

1. Dislocation glide
2. Dislocation climb
3. Point defect diffusion

Note that two of the three mechanisms require dislocation motion. This, plus the fact that, in general, grain boundaries contain a higher density of dislocations and point defects than a single-crystal material, indicates that polycrystalline materials tend to be more susceptible to creep than single-crystal materials. As discussed in Chapter 3, the creep mechanisms can be further differentiated by the specific atomic diffusion mechanism that is operating in the temperature/strain rate space of interest. Furthermore, because the creep behavior is diffusion limited, the stress-strain characteristics are temperature dependent. In general, as the temperature increase, the modulus and yield point of ductile materials decreases. This is qualitatively easy to appreciate in that the characteristic time constant for creep relaxation as well as the bond strength of the material will decrease with increasing temperature. For creep, the temperature dependence is given by the Arrhenius expression. For noncrystalline brittle materials, a transition between brittle and ductile or compliant behavior occurs in and around the glass transition temperature, T_g. Polycrystalline brittle materials become more plastic as temperature increases.

Note that the stress of a ductile material is essentially pinned at the yield strength for strains between the yield point and the strain where the second high slope region occurs (i.e., near the elongation to rupture strain). Furthermore, the yield point depends on the temperature and the strain rate. That is to say, the material cannot be treated as a linear elastic material with one modulus.

Finally, the stress-strain characteristics of typical epoxy-based polymeric packaging materials strongly depend on temperature. This is shown schematically in Fig. 4.10. As shown, the modulus, slope of the stress-strain curve, is high at a low temperatures and decreases by a few orders of magnitude as the temperature is raised above the glass transition temperature, T_g. Thus, one can define polymeric material as being brittle below T_g and essentially compliant above T_g. Note that the transition does not occur at one unique temperature as one would find for a simple first-order phase transition but has some finite range of temperature over which the modulus decreases by a few orders of magnitude. The width and slope of the transition depend on the particular material of interest. In general, the width increases with increasing con-

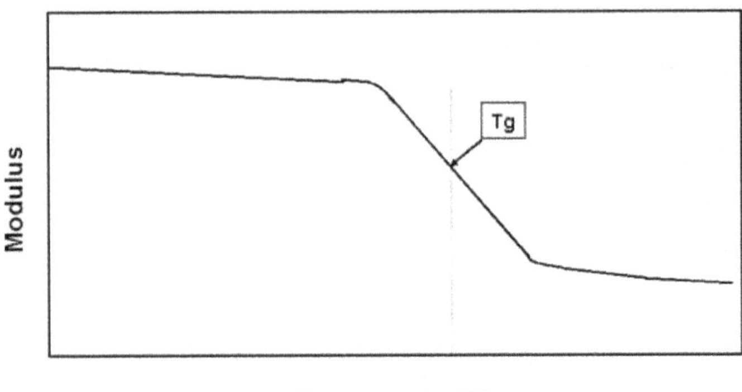

Figure 4.10 Schematic of the modulus vs. temperature for a typical underfill material.

centration (loading) of second phase filler materials such as silica spheres or glass fibers and as the degree of crystallization decreases. The behaviors of ductile metals and polymers are often not appreciated or included in modeling the stress and strains that the different element of the package experience during assembly and field use. Often, the temperature and strain rate dependencies are not well characterized.

Figure 4.11 is a schematic of a flip chip package with no underfill. The strain induced by thermal expansion mismatch occurs below either the melting point of the solder or the glass transition temperature of the polymeric materials used in the package. Because the stress is driven by differences in thermal expansion of the materials in the system (e.g., strain), and the stress is not always linearly dependent on the strain (see above), we prefer to look at the problem as a strain-driven process as opposed to stress-driven process.* In general, the more rigid (i.e., high elastic modulus) the interface material is, the higher the strain on the substrate and the die is, the less rigid [i.e., compliant (low elastic modulus)] the interface material is, the higher the strain on the interface material is, and the lower the strain on the die and substrate will be. The strain/stress configuration resulting from differential thermal mechanical properties of the different parts of the package is very complex, but the general principles related to the differential thermal mechanical properties can be developed for a

*How to simplify the analysis of a package depends on the primary evaluation target. For instance, to estimate stresses of a silicon die in an underfilled FC package, it is proper to treat the system as a stress-driven problem, because the cured underfill material that locks the silicon die to the substrate is rather brittle and can be reasonably considered as a linearly elastic material.

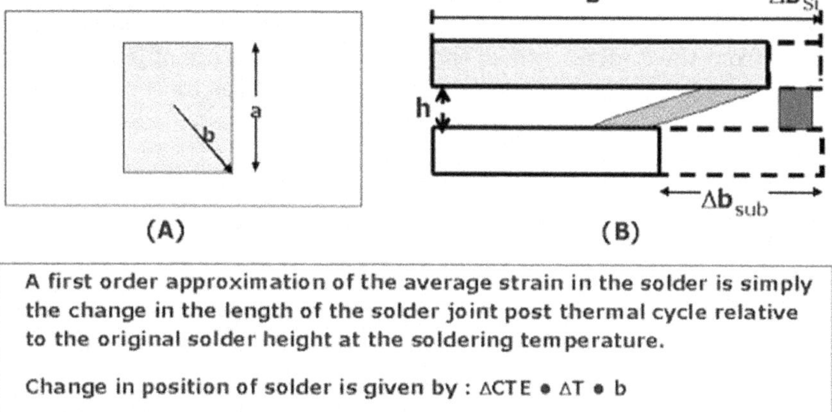

Figure 4.11 (*A*) Schematic of a planar view of the die (gray) on the substrate. Also identified are the length (*a*) and distance from the neutral axis (*b*). (*B*) Cross sectional schematic of the die on substrate at temperature and after it is cooled to room temperature.

much simpler case. We consider, for illustrative purposes, the stains/stresses that develop between an isotropic substrate and an isotropic die connected via one bump (see Fig. 4.11*a* and *b*). For a more detailed analysis, the reader is referred to Refs. 4, 11, and 12 and references therein.

To further simplify the situation, we assume that the temperature of the system is uniformly changed in such a way as to ensure thermal equilibrium at all times. That is to say, there are no thermal gradients present in the system at any time. Finally, we assume that there is perfect alignment of the die pad and substrate pad at the melting point of the solder, and all differential changes occur during cooldown after solder attachment is complete. Something approaching this would occur, for example, when the system is cooled from the soldering temperature to room temperature after bump attach. If the die and substrate are infinitely rigid, then the solder bump will effectively absorb the differential thermal expansion via straining. This is shown schematically in Figure 4.11*B*. In this case, the CTE of the substrate is substantially larger than that of the die. The global difference in the location of the central axis of the solder bump on the substrate relative to that on the die is given by $\Delta CTE \cdot \Delta T \cdot b$, where ΔCTE is the difference in thermal expansion coefficient between the die and substrate,

$\Delta T = T_{mp\text{-}solder} - T_{ambient}$ is the difference in temperature between the melting point of the solder and the ambient, and b is the longest distance from the neutral axis of the die (for a square die of length a, $b = \sqrt{2}\,a$). In this case, the solder bump length increases to

$$\sqrt{\Delta CTE \cdot \Delta T \cdot b^2 + h^2}$$

where h = height of the solder bump at $T_{mp-solder}$

Therefore, the average strain the solder experiences is given by

$$\gamma = \frac{\sqrt{(\Delta CTE \cdot \Delta T \cdot b)^2 + h^2} - h}{h}$$

The vertical component (tensile) and horizontal component (shear) of the strain are given by the geometry of the system. Clearly, the strain is a sensitive function of h. That is to say, as h approaches infinity (e.g., long bumps or equivalently very thick bond lines for standard die attach), the strain on the bump arising from CTE mismatches goes to zero and is essentially tensile. On the other hand, as h approaches 0 (e.g., short bumps or equivalently thin bond lines for standard die attach), the strain is maximized for the particular design in question and is essentially shear. The magnitude of the strain depends on the particular geometry and the thermal mechanical properties of the materials used in the design and the difference in temperature between the solder melting point and the ambient temperature. A schematic representation of the strain the bump would experience if no underfill were present versus temperature for two different solder bump heights (see Figure 4.12*a*).

The addition of an underfill component to the system leads to a somewhat more complex strain configuration. The exact nature of the strain state depends on the glass transition temperature, CTE, and elastic modulus of the underfill. Figure 4.12*b* shows schematically how the strain on the bumps might change with temperature in such a system. At temperatures below the melting point of the solder, but above the T_g of the underfill, the strain is mainly taken up by the solder joint as discussed above, and shown in Fig. 4.12*a*. Below T_g, where the underfill becomes rigid, the strain is mainly taken up by the underfill (assuming that the underfill has a high enough modulus to be considered rigid) but is significantly lower than that of either the substrate or the die. Note that, in general, a slope change occurs in both cases shown in Figure 4.12*b*. The magnitude of the slope change depends on the bump height and thermomechanical properties of the underfill. The temperature at which the slope change oc-

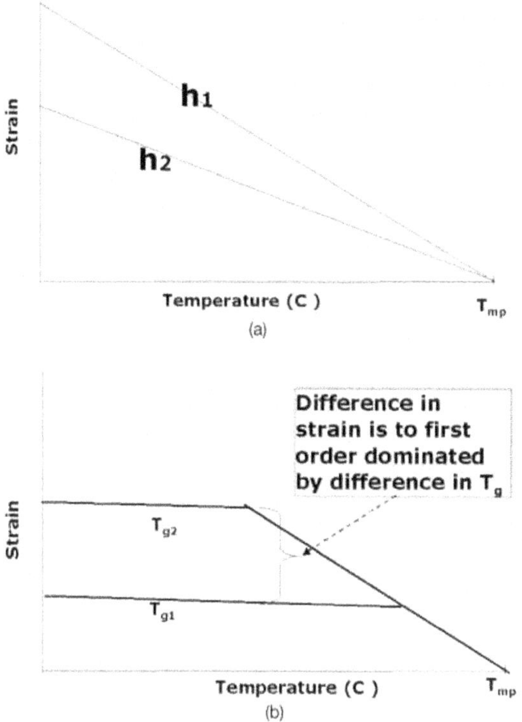

Figure 4.12 Schematic of the strain that would be imposed on a bump as a function of temperature (*a*) with no underfill and (*b*) with underfill. Note that we have included two different solder bump heights and two different T_g underfills.

curs depends on the T_g of the underfill, and the width of the transition depends on the width of the transition of the underfill from a rigid material to a compliant material.

Clearly, the system becomes more and more complex as one adds additional elements to it, including different T_g polymers, multilayer substrates with alternating metal and polymer layers, metal vias in substrates, nonsymmetric metal/polymer density, substrate warpage, and so on. Such complexity is difficult if not impossible to tackle with a series of simple 2-D models. Fortunately, these elements can be included into a stain/stress analysis in 3-D with finite element analysis.[22] This technique is discussed in detail in many publications and books, and a number of available commercial software packages provide a means to perform the analysis. It is beyond the scope of this chapter to go into FEA, but the reader is encouraged to read the literature on this subject, as FEA is a powerful tool for designing and evaluating the reliability and quality of packages. The reader should be aware that a successful FEA usually requires intensive engineering work on both modeling and results interpretation. For example, the input of correct boundary conditions and material properties are critical. The boundary conditions include, but are not limited to, the geometry of all of the elements of the package, the mesh size used in the computation, and the temperature and strain rate dependent mechanical properties of the different materials. The authors have encountered a number of situations in which the boundary conditions were incorrectly specified so that the resultant calculation led to improper designs. Therefore, it is often useful to perform model validation, before deploying the complex and costly full 3-D analysis. This can be done using empirically measured data, closed-form solutions of the problem if available, or even simple 2-D model results. The analytical model discussed above is a good example of a simple means for checking the correct FEA modeling. Adaptation of this technique can avoid, or at least reduce, some computational errors.

The importance of understanding the strain configuration under different conditions applies in two areas: failure as a result of assembly and field failure. In general, failure will occur when the strain energy of the system exceeds that required for material rupture. The strain energy required for failure is just the area under the stress-strain curve for the materials of interest. For rigid/brittle materials, we have found that, if the strain energy imposed on the particular materials of interest is less than a factor of 5 of that required for fractures, then this material generally will not fail in the field. One exception to this is stress corrosion cracking induced failures.[23] For ductile materials, if the strain imposed exceeds the yield strain but is below that required for rupture, then a fraction of the strain energy above the yield strain is stored in the ductile material as potential energy.* Each time the ductile material experiences a strain higher than the yield strain, an additional sum of potential energy is stored in the solder. This would be the case, for example, during power cycling of the device. If and when the cumulated potential energy in the solder joint exceeds that required for rupture of the material, then the energy is released as kinetic energy, and failure occurs.

We note, however, that subjecting a device with a strain/temperature characteristic such as shown in Figure 4.12*b* to a standard accelerated temperature cycling stress test, and then applying standard acceleration factors to the measured test data, would in general pro-

---

* In this formulation we assume that no relaxation such as grain growth recrystallization can occur in the material to reduce this stored potential energy.

duce reliability estimates that would not be valid. If, however, one had a reasonable model of the temperature-strain behavior and the stress-strain characteristics of the materials used in the system, then one could modify the acceleration factors so as to account for the slope change and thus have a much more accurate estimate of device lifetime.

4.4 Humidity Effects

Two of the most technologically important humidity effects are

- Moisture ingress and saturation of the polymeric materials used in the package (e.g., mold compounds, die attach, substrates, and so on)
- Moisture-induced electrochemical corrosion[4,12,24,25]

Moisture adsorption and the induced hygroscopic stresses that result have been shown to affect both the structural integrity of the package and the warpage behavior of the packages. Moisture absorption in packaging materials can lead to a degradation of the adhesion between polymeric encapsulants and the structural elements to which they are bonded via hydrolysis of the polymer. Furthermore, moisture can lead to softening and lowering of the T_g of the polymeric materials. Reduction of the T_g of a given polymer will generally lead to decreased adhesion strength. This effect must be carefully evaluated when new materials are being considered.

Moisture absorption can also lead to package cracking and delamination via *popcorning*. When the package is subjected to a reflow process used for board assembly, the temperature is rapidly raised above the T_g of the polymer, as well as above the boiling point of water. Moisture absorbed in the polymer can vaporize at these temperatures, resulting in a volume increase of the moisture and thus increased pressure. If the pressure is greater than the adhesion strength of any of the interfaces or the intrinsic strength of any of the materials used in the package, then delamination and/or cracking between the various materials that compose the package can occur. In extreme cases, the package may crack in locations that are often difficult to see during visual inspection. C-mode scanning acoustic microscopy is a useful technique for identifying such issues. In addition to creating a crack, which might propagate over time and cause the packaged device to fail in the field, the loss of adhesion at interfaces makes the device susceptible to electrochemical corrosion effects, because external contaminants, including moisture, can easily enter the package and migrate to metal surfaces. The industry has developed a standardized set of categories and tests for these effects, as listed in Table 4.2. Effectively, the different moisture-sensitivity levels indicate how long a package can sit in room ambient without additional bake outs prior to board attachment solder to reflow without encountering moisture-induced problems. Clearly, the higher the moisture level resistance, the fewer process controls have to be imposed on the device prior to board attach, with the result that the assembly cost is reduced. To minimize these moisture effects, the industry has gone to moisture bagging of sensitive parts. In this case, the devices are baked to a dry state and then packaged in a moisture-resistant, ESD-protective bag that includes a desiccant and a moisture monitor. When the bag is opened, the user can inspect the moisture monitor to determine if additional baking prior to assembly is required.

Table 4.2 JEDEC/IPC-Defined Moisture Levels (J-STD-020A)

Level	Storage life out of bag
1	Unlimited @ ≤ 85%RH
2	1 year @ ≤ 30 C/60%RH
2a	4 weeks @ ≤ 30 C/60%RH
3	1 week @ ≤30 C/60%RH
4	72 hours @ ≤ C/60%RH
5	48 hours @ ≤ 30C/60%RH
5a	24 hours @ ≤ 30C/60%RH
6	Bake before assemble

Recently, moisture has been shown to play a role in warpage of the package. This effect can be significant as the industry moves toward lead-free processing. In this case, the temperatures used for reflow are 20 to 40°C higher than that used for Pb/Sn eutectic solder attach. It is the additional temperature that induces the moisture-dependent warpage. It has been found that warpage in many cases is a bigger issue to deal with than popcorning in modern-day polymeric packaging materials.

Corrosion, although largely eliminated in state-of-the-art plastic IC packages, is still an important potential failure mode that must be carefully considered. Because there is a galvanic cell between the Al pad and the Au wire, wirebonds are susceptible to corrosion. In particular, Al readily reacts with halogens such as Cl, Br, and F. In the presence of moisture, Al reacts with these elements to form water-soluble compounds such as $AlCl_3$. These soluble corrosion products are then hydrolyzed, liberating the halogen. The halogen can then participate in further corrosion. As an extreme case, it would take only three Cl atoms to corrode the entire Al pad, given enough time. Bromide flame-retardants that are added to the molding compounds can also lead to corrosion-type degradation of the aluminum pad. They do so, however, by promoting additional Au/Al intermetallic compounds that lead to joint embrittlement and early life failure.[8] Although corrosion-induced failures of the leadframe or solder bumps or pads are very rare, they can occur. Cracks or opens in the Cu- or Sn-based leadframe, resulting from poor plating quality or high bending stresses imposed during trim and form, can expose the less-noble base Cu metal to a corrosive ambient. Furthermore, if delamination or crack formation occurs, then a path for corrosive materials to enter the package and possibly corrode the base metals is formed. All of these types of corrosion-induced problems can be avoided by wisely selecting mold compound type and additives, properly controlling contamination prior to plating or molding, and carefully determining the allowable moisture level below which delamination does not occur.

4.5 References

1. Brandes, E. A., and G. B. Brook, Eds., *Smithells Metal Reference Book,* 7th ed., Butterworth Heinemann, Boston, MA.
2. Hansen, M., and K. Aderko, *Constitution of Binary Alloys,* 2nd ed., McGraw-Hill, New York, 1958.
3. Harman, G., *Wire Bonding in Microelectronics Materials, Processes, Reliability and Yield,* 2nd ed., McGraw-Hill, New York, 1997.
4. Ohring, M., *Reliability and Failure of Electronic Materials and Devices,* Academic Press, New York, 1998.
5. Gupta, D., and P. S. Ho, Eds., *Diffusion Phenomena in Thin Films and Microelectronic Materials,* Noyes Publishing, New York, 1988.
6. White, M. L., J. W. Serpiello, K. M. String, and W. Rosenzweig, "The Use of Silicon RTV Rubber for Alpha Protection on Silicon Integrated Circuits," *Proc. 19th IRPS,* 43, 1981.
7. Horsting, C. W. "Purple Plague and Gold Impurities," *Proc. 10th IRPS,* 155, 1972.
8. Thomas, R. E., V. Winchell, K. James, and t. Scharr, "Plastic Outgassing Induced Wire Bond Failure," *Proc. 27th IEEE Electronics Components Conference,* 182, 1977.
9. Gayal, D., T. Lane, P. Kinzie, C. Panichas, K. M. Chong, and O. Villalobs, "Failure Mechanism of Brittle Solder Joint Fracture in the Presence of Electroless Nickel/Immersion Gold (ENIG) Interface," *Proc. 52th IEEE Electronic Components and Technology Conference,* 254, 2002.
10. Osenbach, J. W., and A. Amin, unpublished research, 2003.
11. Hwang, J. S., *Modern Solder Technology for Competitive Electronic Manufacturing,* McGraw-Hill, New York, 1996.
12. Tummala, R. R., E. J. Rymaszewski, and A. G. Klopfenstein, Eds., *Microelectronic Packaging Handbook,* Kluwer Academic, Boston, MA, 1997.

13. Foster, G., "Embrittlement of Solder by Gold From Plated Surfaces," in *Papers on Soldering,* ASTM STP 319,13.
14. Benerjio, K., and R. F. Darveaux, "Effects of Aging on the Strength and Ductility of Controlled Collapse Solder Joints," in *Microstructures and Mechanical Properties of Aging Materials,* P. Kiliaw, R. Viswanathan, K. L. Murty, E. P. Simenen, and D. Fear, Eds., The Minerals, Metals, and Materials Society, 431 (1993).
15. Zhang, C. H., S. Yi, Y. C. Mui, C. P. Howe, D. Olson, and W. T. Chen, "Missing Solder Ball Failure Mechanism in Plastic Ball Grid Array Packages," *Proc. 50th IEEE Electronic Components and Technology Conference,* 242, 2000.
16. Xiong, Z. P., J. W. Osenbach, L. Tok, and K. H. Chua, "Degradation of Pb/Sn Solder Balls on Electrolytic Ni/Au Substrates as a Result of Post Assembly Heat Treatments," *Proc. 5th IEEE Electronic Packaging Technology Conference,* 2003.
17. Yokomine, K., N. Shimizu, Y. Miyamoto, Y. Iwata, D. Love, and R. Newman, "Development of Electroless Ni/Au Plated Build-Up Flip Chip Package With Highly Reliable Solder Joints," *Proc. 51st IEEE Electronic Components and Technology Conference,* 441, 2001.
18. Personal communication with K. Yokomine (2004).
19. Moyer, R. S., unpublished research (2003).
20. Poate, J. M., K. N. Tu, and J. W. Mayer, Eds., *Thin Films—Interdiffusion and Reactions,* John Wiley & Sons, New York, 1978.
21. Verhoeven, J. D., *Fundamentals of Physical Metallurgy,* John Wiley & Sons, New York, 1975.
22. Lau, J. H., *Low Cost Flip Chip Technologies*, Chap. 12, 13, 14, and 15, McGraw - Hill, 2000 and references therein.
23. Ulig, H. H. and R. W. Revie, *Corrosion and Corrosion Control,* John Wiley & Sons, New York, 1985.
24. Shook, R. L., J. J. Gilbert, E. Thomas, B. T. Vaccaro, A. Dairo, C. Horvath, G. L. Libricz, D. L. Crouthamel, and D. L. Gerlac, "Impact of Ingressed Moisture and High Temperature Warpage Behavior on the Robust Assembly Capability for Large Body PBGAs," *Proc. 53rd IEEE Electronic Components and Technology Conference,* 1823, 2003.
25. Osenbach, J.W., "Corrosion-Induced Degradation of Microelectronic Devices," *Semicond. Sci. Technol.,* 11, 155, 1996.

Chapter

5

Strategies for Locating Failures

Jason P. Goodelle
Weidong Xie
Albert Seier

5.1 Interpreting Electrical Test Results

In most cases, package defects will be revealed as electrical contact failures by automatic test equipment (ATE). These failures are usually contact opens, meaning that the current forced into the failing package pin has a broken path to the chip. Once the device is screened on ATE and opens are reported, a number of laboratory analysis tools are used to locate the failure and find root cause.

Several nondestructive and destructive failure analysis techniques are used in an attempt to locate defects. Among them are acoustic microscopy, X-ray analysis, curve trace analysis, time domain reflectometry (TDR) analysis, decapsulation, and optical inspection. Details of these techniques can be found in Chapter 6. The flow of a device through the failure analysis process is affected by several variables such as device pin-count and package style—the most significant being flip chip packages, in which exposing the die surface while maintaining electrical integrity requires a method other than standard decapsulation.

In the case of wirebonded packages, X-ray analysis can quickly reveal gross opens, such as wires that have disintegrated as a result of electrical overstress. The opens can then be quickly correlated to the ATE data log. When dealing with devices reported as contact opens, or shorts that are not found in X-ray analysis, curve trace and TDR techniques quickly provide data that aid in the isolation of defects. A curve tracer applies a voltage to package signal pins to bias the PN junctions of the buffer protection diodes, resulting in a current-versus-voltage (IV) curve (see "DC Electrical Characterization," Chap. 6). The absence of current draw is evidence of an open circuit. TDR (see the TDR section of Chap. 6) can then be used to further isolate the open to determine if it is package related, die related, or related to the die-to-package interface, i.e., wedge bond, wirebond, or wire.

Once this is accomplished, the package can be decapsulated to expose the die while retaining electrical integrity. If TDR results indicated a die-related open, evidence of electrical overstress may be evident on optical inspection of the die surface. In the case of a package open that was previously determined through TDR analysis, TDR results can be confirmed by electrically probing between the wedge connection (for non-flip chip packages) and the package pin. A lack of current flow could then serve as confirmation of the package open. Destructive package parallel lapping or cross-sectioning can then be performed to locate the package defect.

5.2 Using Package Design Files as Failure Roadmaps

The basic blueprint for any package construction is the design file, which is used for the attachment of the silicon, substrate fabrication (or leadframe selection), and package assembly. For the most part, the package design files consist of a whole package of schematic wiring diagrams that provide the details of how the external power, ground, and signals from the device level are connected to the silicon and other active or passive components. These files are typically used as the primary template for the manufacture of the packaged components and can subsequently be used to aid in the failure analysis of reliability issues, combined with proper information from the testing platform. Tools available to the failure analysis professional, such as those mentioned in the previous section, can help isolate the device failure to general regions of the assembly. Also, information from the electrical test can indicate general locations of the anomaly, unless a logic failure is indicated, which can come only from the silicon. In most cases, it is easily associated with specific regions of the die, based on the type of information from the tester. However, in many cases, the only means to isolate the true nature of the failure in a packaged device will be to perform a destructive analysis. Since destructive analysis involves a physical removal of portions (in many cases, large portions) of the package, the progress of the process will ultimately depend on visual references and knowledge of the structure to aid in the slow process of removing small layers of material to physically locate the source of the failure (which, in many cases, can be identified only by knowing what the structure should "look" like).

The most common process for destructive analysis is to perform a planar (from top or bottom of the devices) or cross section using fine-grit polishing techniques that remove fractions of a micron (1×10^{-6} m) of material at a time. (This is usually performed after unaffected portions of the device are removed using sawing or decapsulation techniques.) It is very important during this process to recognize both the relative location in which the plane of the section lies and the locations of nearest neighboring features to determine the amount of material to be removed to expose the next feature. Design files are uniquely capable of telling the FA professional not only what type of failure to expect (open, short, and so on) from the ATE output, but also which conductive pathway the failure lies on and how to physically reach the failure, through sectioning, once the general location is identified. The following sections will outline the various types of design files commonly used in the failure analysis process.

5.2.1 Introduction to Package Design File Structures

Figure 5.1 shows the generalized high-level flow typically used in the design stages for both the silicon and package sides of an IC product, detailing the nature and types of files provided at each step. This process is initiated based on a customer request (providing specifications and requirements). The design files are provided in a format specific to the tool used for the design (e.g., Cadence, Synopsis, Encore), which will subsequently be used for the manufacture of the silicon and for the package substrate. (Lead frames usually do not require complicated design files.) Since the focus of this chapter is to outline methodologies used to locate a failure in a defective device, and the failure is assumed to be package related, the discussion will focus mainly on package-level design files. However, because it often becomes necessary to use the silicon design file in conjunction with the package files, some mention of top-level interconnect structures on the silicon will be made.

5.2.2 Wirebond Diagrams

In simplest terms, the wirebond diagram is the means by which the silicon is connected to the rest of the package. This connection will, of course, depend on the functionality of the silicon and will be influenced by how the signals must be routed through the device and the placement of the device on the circuit board. The wirebond (or bump map in the case of a flip chip package) constitutes the primary interaction of the package and the silicon and will be used in the assembly of the device. In the event of a failure in the packaged device, the wirebond diagram will provide an important guide to how the chip circuitry is connected to

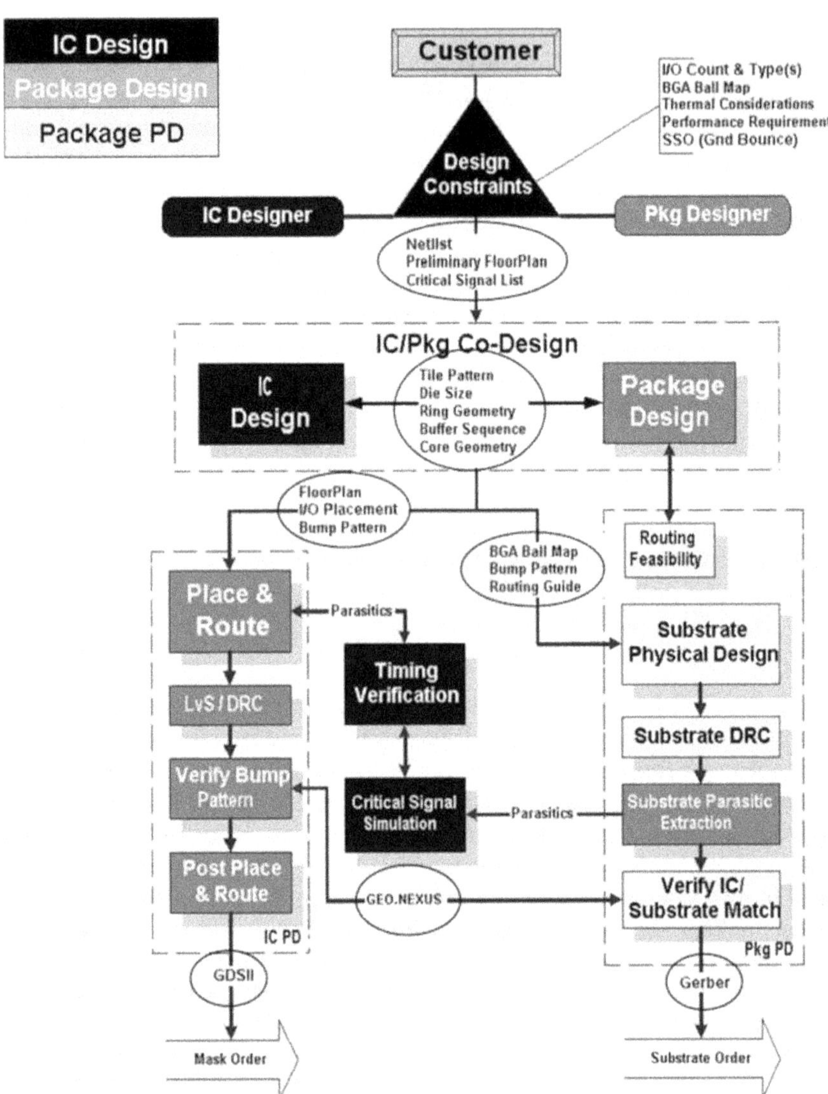

Figure 5.1 Generalized silicon and package codesign process. Output files are circled.

the carrier. The failure analysis professional will use the wirebond diagram as a first means to troubleshoot a package failure. If, during a package test, a short or open is discovered in a circuit, the wirebond (and wires leading to the bonds within the circuit) will often be the first elements checked for integrity. Many times during package stressing, a wire will sever, dislodge, or otherwise become compromised, leading to an "open" reading on the test equipment. In other cases, crossed wires, severed wires touching adjacent wires, or contamination bridging adjacent wires will result in a "short" reading from the test equipment. Armed with the data provided by the test equipment (as described in previous sections), the FA specialist can use the wirebond diagram to aid in the employment of nondestructive techniques (such as X-ray or TDR) to pinpoint the failure. In destructive evaluations, the bond diagram can assist in the removal of package materials by providing visual clues or acting as a reference to determine the proper location of the wires while progressing with the sectioning process through the device. Figure 5.2*a* (full view) and 5.2*b* (close-up view) shows typical wirebond diagrams for a single-chip plastic overmolded BGA package. Although the actual wire structure may differ slightly from the diagram, modern wirebond design tools mimic the actual process very closely such that the diagram resulting from the design process (Fig. 5.2) will be very useful during FA to understand what the wire structure should look like and the geometric relationship from wire to wire.

5.2.3 Flip Chip Bump Maps

Much as with the wirebond diagrams, the flip chip bump diagram provides a visual tool for reference when evaluating the linkage between the silicon and the package in a flip chip device. Again, this interconnection region provides the primary interface between the silicon and the package. The bump diagram typically will be a highly customized arrangement of the solder connections used to perform chip attach. This diagram not only will aid in the assembly of the package, it can also serve as a very useful tool in aiding failure analysis of the package after a defect is identified (also similar to the wirebond diagram). In many cases, a failing circuit can be identified quickly (at least at the identification stage) by marking the affected bumps. This identification will help the failure analysis professional plan for destructive analysis. This process will also help to identify patterns to the failures, which can then potentially be traced back to a process issue. For example, if a flip chip device is subjected to an incoming test and identified as an open, and all of the failures are traced back to a particular grouping of solder bumps (all one side or all centered), there is a strong possibility that the solder bumps were not completely fused as a result of problems during the chip attach process. Figure 5.3 shows an example of a flip chip solder bump map, with bumps registered as incoming failures filled in as a solid bump. The process by which these bumps (or wires in the wirebonded case) are linked to the entire package's conductive closed loops from one BGA to another (often termed *nets*) is the topic of the following sections.

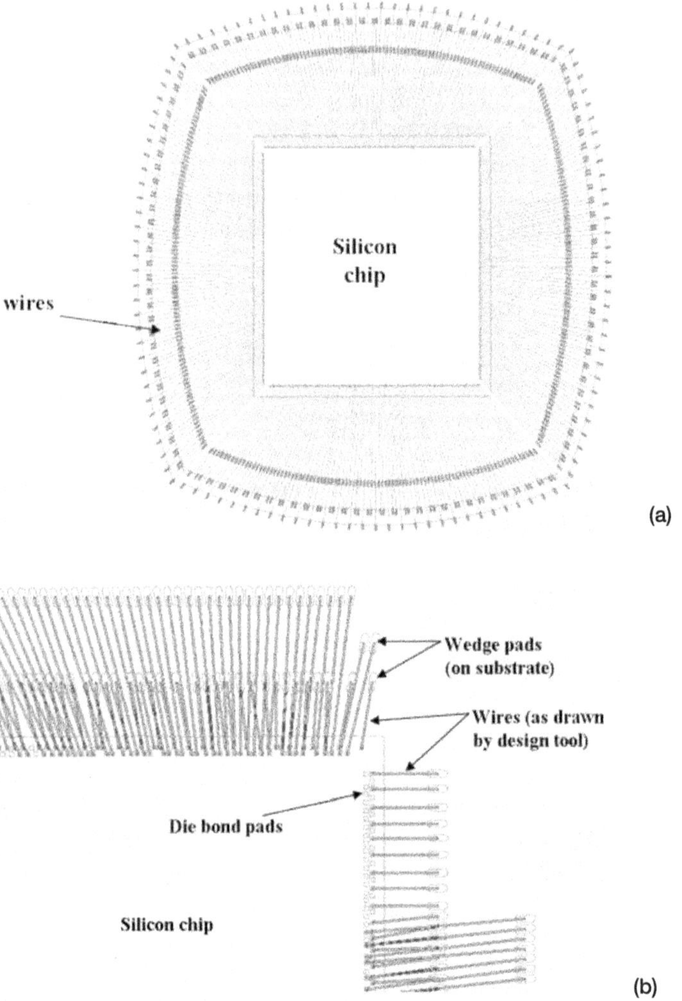

Figure 5.2 (*a*) Example of a single-chip wirebond diagram and (*b*) a different single-chip wirebond diagram, close-up of bonding area.

5.2.4 Package Design Files

The most basic blueprint of a package structure is the design file associated with the type of package being produced. This could range from a simple CAD drawing of a limited-I/O ceramic leadless chip carrier, showing the electrical connection routes, to highly complex, multilayer laminate substrate design files. In some of the latter cases, the design files can have as many as 20 or more layers, depending on the applica-

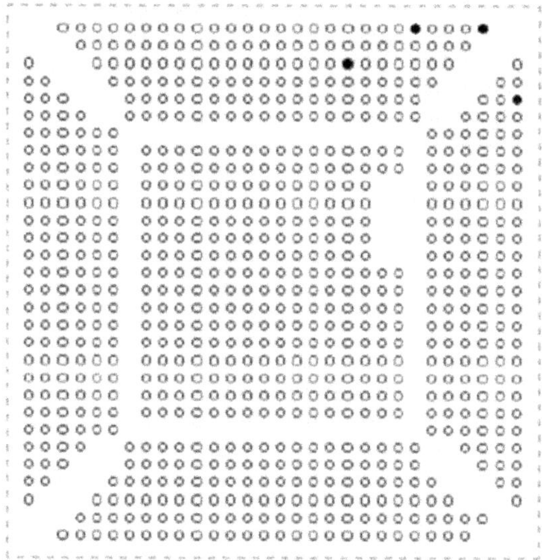

Figure 5.3 Schematic of a flip chip bump map with incoming failures identified by darkened bumps associated with the failed nets.

tion or the substrate technology. For plastic laminates, each layer can contain detailed conductor, dielectric, drilling, via, or bond pad information. For ceramic substrates, each design layer can yield conductor, via, or dielectric information. For the purposes of failure analysis, these files contain valuable geometric clues that will be extremely helpful during destructive analysis. Certainly, analytical tools such as TDR and curve tracers can help isolate the general location of the failure; however, in many cases, the tool does not have the resolution to isolate the exact location of the failure. If the testing process yields an issue (tagged as a failure) with the logic portion of the circuit, it is clear that the silicon must be scrutinized during later stages of the failure analysis (more on this in the section on interpreting automatic test error logs). However, if the tester simply indicates an open or short circuit, the entire package must be evaluated at a deeper level. Armed with the information concerning the failed nets and what is commonly referred to as a *net list* [which is a spreadsheet or text document that shows how a lead or BGA ball is connected to the chip (through wirebond or bump location) and then back out to a corresponding BGA or lead], it is possible to use the design file as a map for failure analysis. This is possible because, typically, the destructive analysis process (usually accomplished through grinding and polishing with constant visual inspection) is a layer-by-layer process that is highly dependent on visual references. Figure 5.4 is a simple schematic showing how several design file layers are combined to form a final package element. In this case, a flip chip BGA substrate is depicted, revealing top and bottom layers.

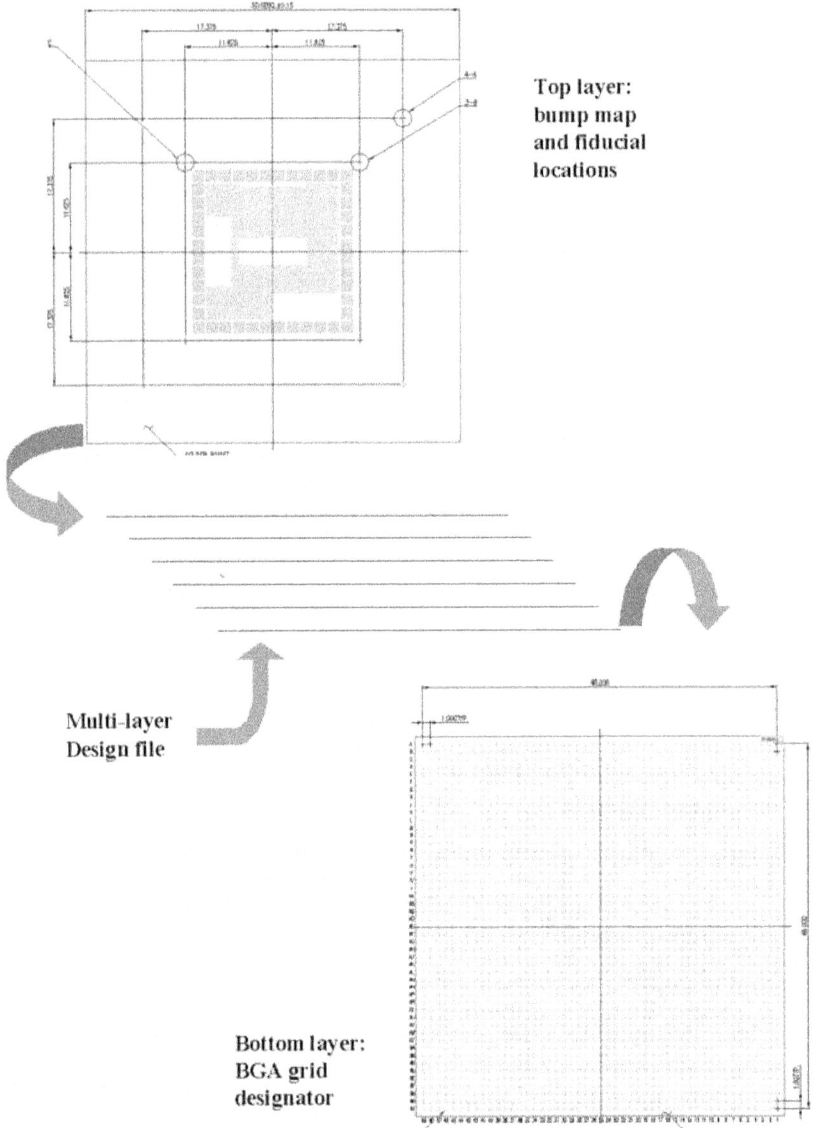

Figure 5.4 Schematic of a typical multilayered design file construction, in this case a flip chip BGA.

5.2.5 Package Net Lists

As mentioned in Section 5.2.3, the term *net* is commonly associated with the path from one BGA to another and all the functionality (including silicon) between them. The list of these critical signal connections will be one of the first design files that result from the co-design (package and silicon) process. From this file, the complete path from board-level connection to the silicon and back to the board again can be ascertained. Figure 5.5 shows a typical full net structure for both a wirebonded and a flip chip BGA package. For the wirebonded structure, one example of a complete net is a track of the path from BGA ball 1 to trace 1 to wedge pad 1 to wire 1 onto the chip through chip pad 1 and, then trace back though a similar path to BGA 2. A similar exercise can be completed for a flip chip device, in this case replacing wire connections with solder bumps. Signals on and off the die will have this general structure. However, in many cases, one of the connections to a complete circuit will consist of one connection to a common power or ground such that potentially only a few BGAs will service the entire common "plane" (referred to as a *plane*, since this usually is a separate plane of solid metal contained within the BGA laminate substrate) for power or ground. The complete net list will usually take the form of a spreadsheet that will link the BGAs (in the

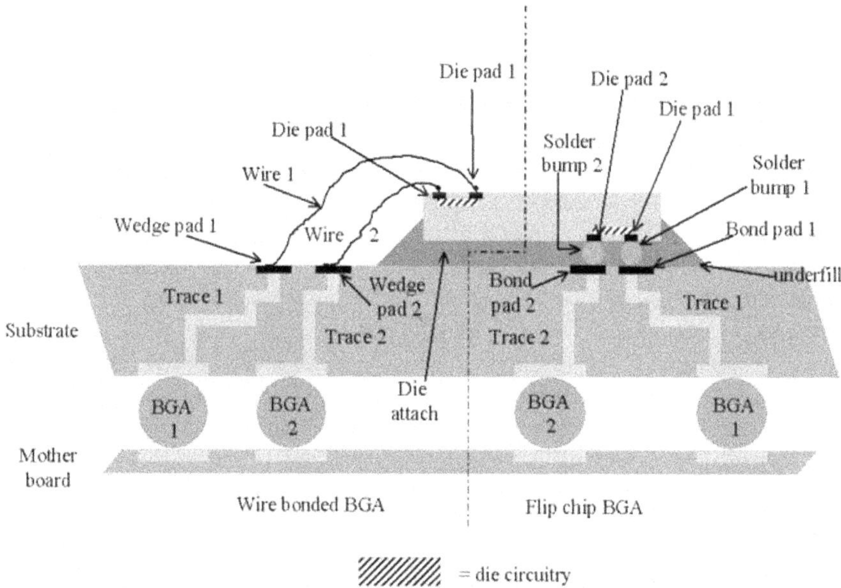

Figure 5.5 Schematic of the constituents of a complete package net for both wirebonded and flip chip BGA packages.

manner completed for the wirebonded structure shown in Fig. 5.2) to a connection to the die (through a particular metal trace) and then show how this input path exits the silicon and is recaptured by a single trace or plane serviced by one or many BGAs. This file is a very critical tool in the hands of a failure analysis professional since, in many cases, the failure from the automatic testing platform will provide information only about the BGA (or lead, keeping in mind that the whole argument commutes to simpler leadless packages with little modification) that was tested. If this BGA is "netted" to a trace that is captured by a common power or ground on the closed portion of the circuit, the only information about this hidden portion of the net will be provided by the net list. This will prove vital in defect isolation, since the failure can exist on the entire net, and only through proper manual probing (armed with the netlist) can a more precise location of the failure be revealed and exposed using destructive analysis.

5.3 Automated Test Equipment Failure Error Logs: Starting with the Failure Data

When a packaged device is tested during evaluation and qualification stages, visual techniques often will be used to determine whether the stress-testing process is affecting the structural integrity of the assembly. However, to truly understand the effect of accelerated testing on the critical structures of the device (such as new design rules in the processing of metal traces or through or blind vias in substrates or the effect of material changes on the silicon integrity), electrical testing is required, at least intermittently, during stress testing. The automatic tester will run a program that verifies the function of the logic portions of the silicon (for production devices or testers with fully functional logic regions) within allowed parameters. The tester will also verify circuit integrity within allowed parameters. For example, if a resistance level is obtained during a portion of the test that is higher than the allowed value, the circuit is flagged as an open. The same process can be performed to identify leaky circuits (i.e., circuits showing a voltage drop but not completely shorted) and shorted circuits.

The test equipment will commonly automatically reject the device once a portion of the test (usually testing hundreds of signal pins in addition to basic continuity, employing several types of testing in one program) fails. A list of the package nets that failed the specific tests will then be provided to the test engineer. The list of failed nets is commonly referred to as an *error log* or *failure log*.

The information provided by the test equipment will not be sufficient to isolate the failure in a defective device. The critical linkage comes when the entire net (from BGA to BGA, from BGA to common plane, and so on) is made. This process requires pairing the error log data with the net list information, which leads to successful destructive analysis once the design file reveals the physical location of the suspect net. This process of pairing the information is usually totally manual but is not usually so cumbersome as to halt the process entirely. This is because, in many cases, the failure will consist of only a few nets. Failures that consist of several nets often can be linked to specific issues that do not require destructive analysis, such as a flip chip device that has been attached incorrectly (all nets are likely to show up as failed) or a massive failure signature associated with a handling issue (e.g., damaged silicon). This process of failure "triage," once we are in possession of the failed net information, is a vital process that can be invaluable in a quick, successful failure-analysis process. This first step in failure analysis is sometimes referred to as *common-sense debugging,* and it is performed before any destructive analysis takes place. Unfortunately, this process is highly dependent on having experience with a particular device type in many use and stress conditions. However, the first step in building experience with failures with a particular device will start with linking failures from the error log to the nets within the device. Table 5.1 shows an example of the type of data provided by the automatic test equipment paired with nets from the net list information. The design file will contain the same net names listed in the net list, making identification through visual reference possible, especially during destructive analysis. The BGA location names found in Table 5.1 refer to positions on a grid, similar to that found in Fig. 5.4, where the bottom layer of a BGA device is shown. The BGAs are typically the starting point for verification of the failure and any subsequent destructive analysis.

5.4 Test Vehicle Design with Improved Fault Isolation Capabilities

Clearly, any advantage that can be provided to a failure analysis professional to quickly isolate the source or location of a package failure will be immediately useful. To this end, when presented with the opportunity to produce test vehicles specifically for the purpose of evaluating the performance of a package (representing an entirely new technology or containing new materials or process elements) in the presence of simulated or accelerated field conditions, several tactics can be employed in designing a device that provides more information or simplifies the FA process. The following sections will outline several possible methods that can be adopted to facilitate the package development process through more efficient, cheaper, and more effective test vehicles.

Table 5.1 Example Error Log Data Coupled With Corresponding Net List Data

Error log indicator (provided from test equipment)	Failure type	Failing net name (often different from error log name)	Failing BGA	Failing die pad	Failing wedge pad	Paired net to form complete circuit	Paired BGA	Paired wedge pad	Paired die pad
CHIRXDATA_B_2	Open	CHIRXDATA[2]	L19	U2–563	U2–57	CHIRXDATA_B_1	L20	U2–55	U2–561
CHITXDATA_D_6; DNDATAIN_D_5	Short	DS3NEGDATAIN[5]	U31, AP25	U3–140	U3–160	DDINCLK_D_4	AN25	U3–158	U3–138
DS3RXCLKOUT_C_2	Open	DS3RXCLKOUT[2]	AJ1	U4–158	U4–114	DPDATOT_C_3	AH–3	U4–116	U4–160
DPDATAOUT_A–5	Open	DS3POSDATAOUT[5]	C5	U1–176	U1–462	DS3DATAOUTCLK_A[4]	B4	U1–460	U1–174
DPDATAOUT_A–5	Short	DS3POSDATAOUT[5]	C5	U1–176	U1–462	DS3DATAOUTCLK_A[4]	B4	U1–460	U1–174
DPDATAOUT_A–5	Open	DS3POSDATAOUT[5]	C5	U1–176	U1–462	DS3DATAOUTCLK_A[4]	B4	U1–460	U1–174

5.4.1 Issues with Fault Isolation Using Production Packaged Devices

Typically, when performing package-level testing, the need for chip-level parametric logic evaluation is not needed to understand how the package will perform during thermomechanical stress testing. This is usually true because most failures that occur at the package level will result in a short or a break (open) in a conductive path. Even if the package causes damage to the die, as long as there is some level of connectivity between the package and the die, automatic testing using an applied bias voltage will detect it. It is a different situation when trying to determine the effect that a package will have on the silicon's functionality itself. In this case, evaluating parametric shifts in the silicon during exposure testing may be indicative of issues caused by package-induced stresses or the failure to prevent environmental attack. However, if the failure is indeed isolated to some issue with the logic functionality, failure analysis becomes very detailed and time consuming. Therefore, especially for large-scale package-level evaluations, simplicity in the testability of the vehicle is the key.

5.4.2 Cost of Doing Package Evaluations on Production Devices

The average cost of a full-production mask for an 8-in wafer is nearly $30,000, and the production run costs approach several thousands of dollars per wafer lot (25 wafers). Typical lead frame and substrate prices, while still lower than the silicon per-unit cost, can quickly consume the new product development test budget. In fact, for large body sizes, multilayer flip chip substrate costs are often more expensive than the product die. The average material evaluation can consume 500 to 1000 devices per experiment (depending on variables tested and sample sizes). It is clear that, to keep costs down, cheaper, more reusable approaches must be developed to maintain the level of experimentation that is often necessary to guarantee product quality. What is needed are inexpensively produced test vehicles that, while not themselves actual product, simulate all or a good portion of the current product portfolio. However, careful attention must be paid to ensure that, after the effort to drive costs down by using test vehicles to run evaluations, the tester still represents the product.

5.4.3 Benefits of Simplified Opens and Shorts Testing for Quick Evaluation of New Packaging

One means to an inexpensive test vehicle is the use of single-metal-layer testers that do not require complicated logic testing to evaluate the failure. While performing package-level evaluations, the presence of sensitive logic circuits often will unnecessarily complicate the test process, making defect isolation more difficult. Ideally, for package-level testing, using simple conductive circuits placed in sensitive, high-stress regions of the package (or package silicon interconnection region) is the best tactic for efficient evaluation of package performance. For example, to evaluate the performance of a new PBGA substrate with cutting-edge design rules, routing traces through the high-stress regions of the substrate from one BGA ball to another, using the die interconnection to complete the circuit, is an ideal method (employing manual or automatic testing equipment to test the continuity of the circuit) for understanding how the new substrate will perform in the presence of thermomechanical stress. The following sections describe some techniques for creating these evaluation vehicles. While the discussion revolves mainly around laminate PBGA package types, the techniques are applicable to any package technology with only a few slight modifications.

5.4.4 Generating Opens and Shorts Testers in Production Devices

The use of simple test vehicles certainly provides an advantage over the use of devices with full production silicon when evaluating new package technologies. This results from the lower cost and easier test strategies. However, in many cases, a production substrate is the only available resource, so some unique approaches must be employed to take advantage of the simplified testing, cost benefits, and easier failure analysis afforded by the use of what is often referred to as an *opens and shorts* tester. The name is derived from its simplified conductive loop structures wherein the only information generated during a test is whether the conductive path is open or shorted. An open conductive path usually results from cracking during stress testing. A shorted conductive path may find its root in contamination, corrosion, whisker formation, or some other artifact generated during stress testing. The following section details some techniques for creating simple single-metal-level testers for use in production laminates.

5.4.5 Single-Metal-Level Silicon Die as Drop in Opens and Shorts Testers

The term *single-metal-level* is used to describe testers of the sort presented here, which employ only a few mask layers from a normal silicon mask set (usually approximately 30) to keep the silicon production costs to a minimum. This process eliminates most steps required for transistor production such that what remains is only a conductive metal layer (Cu or Al, depending on the technology) and a substructure and passivation layer that will depend on the application. The conductive metal layer can be designed to perform many tasks, from complex structures such as resistive heaters and temperature gauges (some of which are described later) to simple shorts of complete resistive loops throughout the rest of the package being used for package-level evaluations. The latter is the focus of this section.

5.4.5.1 Wirebond structures. To take advantage of a simplified test in a production package structure, a limited number of options are available. The most effective option is to take the existing silicon and modify a mask layer to short out existing nets on the production substrate such that applying a voltage to these structures during test (and modifying the net list to facilitate test program production and assisting in failure analysis) allows us to verify the net's integrity. For a wire-bonded interconnection, the device is bonded in the same fashion as the production device (i.e., using the same wirebond diagram and bonder parameters) and relies on shorts created on the modified silicon structure to form the testable nets. The result of using a modified silicon structure is a cheaper, easier-to-test vehicle that is also easier to analyze if failures occur. The tester relies on only a single metal layer to complete the shorted nets, so a fully processed silicon structure is not required. This provides cost savings over the standard production silicon.

Test vehicles of this type are primarily used for opens and shorts testing without relying on logic testing in any fashion, which, as outlined in Section 5.4.3, is much preferred when trying to evaluate the robustness of new materials or processes in the presence of stress testing. The simplicity of this approach lies in the use of the production silicon bond pad layer and shorting adjacent signal nets that will be tested on the automatic tester in the same fashion as a production device (making use of sockets and pins within the socket that already exist, even further driving down the costs) but with a modified test program that applies only a small voltage (as mentioned above) to one side of the net and reads the output at the shorted side.

Figure 5.6 shows an example of the shorting structure, requiring only changes at the metal layer containing the bond pads that can be used to tie together adjacent signal wires, and the circuits within the substrate to which they are connected (including metal traces, wedge pads, and BGAs). After wirebonding, the structure is completely assembled, just as a production device would be, and requires no special handling other than modifications to the test program. Since the test device uses the same substrate design rules and materials, silicon dimensions, and other assembly materials as the production device, the performance of the test vehicle will be identical to that of the product evaluations with a high degree of certainty. If failures do occur, the

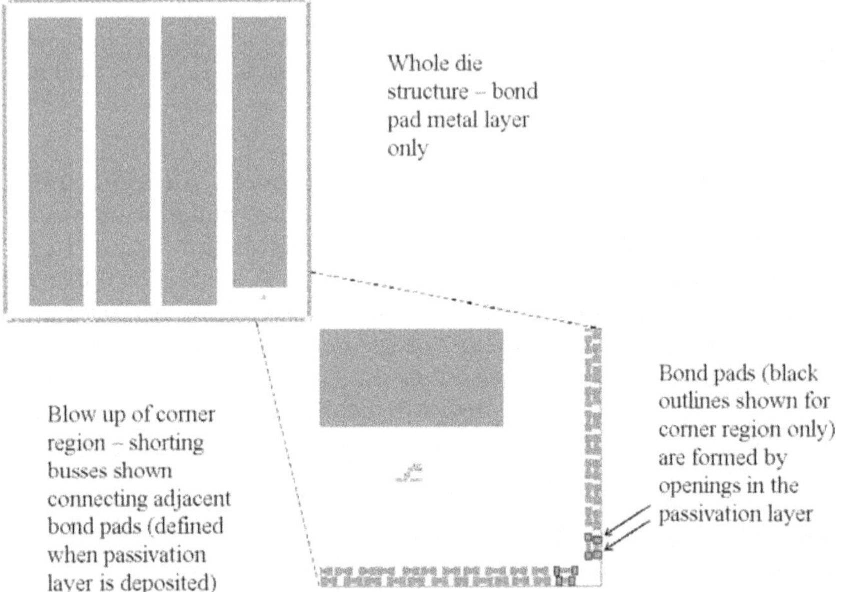

Figure 5.6 Schematic of a typical wirebonded drop in replacement silicon die used to convert a production device into a test vehicle. Shown is the bond pad metal layer on silicon with shorting busses to create independent opens and shorts that are testable in the device.

simpler net structure and more direct opens-and-shorts tester output help to debug anomalies more quickly and facilitate destructive analysis. The latter is true because, although the device has full coverage in terms of the number of nets coursing through the package, the number and density of the nets are much smaller than for a production device, thus providing more physical space in which to maneuver when sectioning the assembly for failure location.

5.4.5.2 Flip chip structures. In the same manner as outlined above, a flip chip version of the single-metal-level tester for drop-in use with production devices can be created. The only difference is that, rather than shorting two wires together, adjacent signal bumps will be shorted.

5.5 Case Study No. 1: Finite Element Analysis to Help Identify Root Causes

Up to this point, this chapter has focused on specialized techniques and package structures that help isolate and identify failures in IC packages, the intent being that root cause of the failure will be determined and corrective action will be implemented in a timely manner. However, with advent of sophisticated finite element modeling tools, detailed models of prototype devices can be constructed and evaluated for potential high-stress regions that may prove to be problematic during use or stress testing. FEA is often used to aid in the isolation of the root cause after a particular failure mode is identified.

As a complex material system, an electronic package normally consists of multiple materials with dissimilar thermomechanical properties, as is evident from much of the work described in this text. Thermomechanical stresses can build up inside the package during fabrication, assembly, qualification testing, and field use as a result of a CTE mismatch among dissimilar materials that are packed together. It is always important to understand how a material behaves in a package under given mechanical or thermal loading conditions. FEA, a robust and effective numerical method, is a useful tool for locating the critical sites, sorting out the significant factors, and identifying the root causes of a failure. In particular, FEA is versatile in doing a parametric study, which allows one to investigate different scenarios through simulation in a timely manner. FEA can be applied successfully in various stages of a product's life cycle, e.g., to choose important design parameters in the early design phase. This allows the developer to verify a design target from different designs without performing lengthy and costly verification tests in the prototype phase, to down-select materials in the assembly phase, and to identify potential solutions of a failure in the qualification phase. Benchmarking and verification of finite element models with available experimental data, or at least engineering intuitions, are vitally important for a successful analysis. The simulation results should always be judged and interpreted carefully, based on physical insights into the analyzed problems. The following case study is an example that shows how important it is to correlate FEA with the observations of failure analysis to find the root causes of a failure and identify potential solutions.

5.5.1 Package Information

The 1444 I/O FCBGA schematically shown in Fig. 5.7 is a test vehicle for high-density and high-speed flip chip substrate technologies. The package has a body size of 40 × 40 mm. A 12-mm square die is attached onto the substrate with 1700 bumps. Figure 5.8 shows the outline of the substrate, which consists of seven metal layers. There are 1444 (38 rows and 38 columns) solder balls attached to the substrate with 1-mm pitch.

An electrical open was detected after exposing the package for 300 cycles of condition B (–55 to 125°C) temperature cycling (TC). Time do-

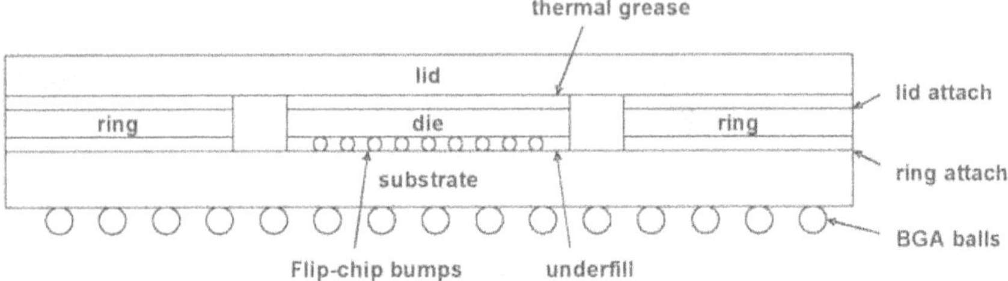

Figure 5.7 Schematic of the 1444 FCBGA package.

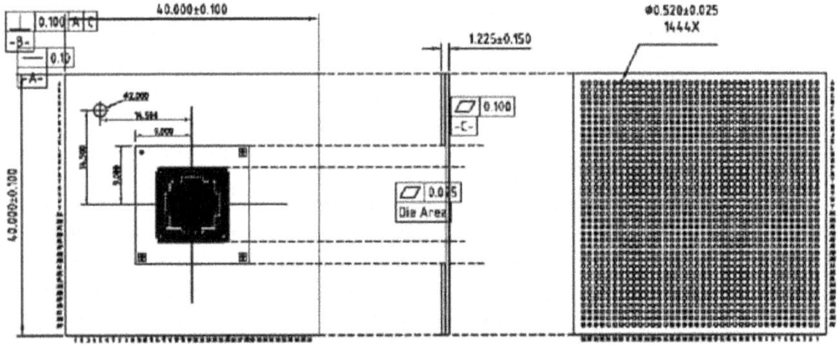

Figure 5.8 Outline drawing of the substrate with an attached stiffener.

main reflectometry (TDR) confirmed that the failure was within the substrate, close to a BGA pad. Cross-sectioning revealed that a signal trace on metal layer beneath BGA pad had been cut through by a crack in the substrate, resulting in the detected electrical open. The failure mechanism analysis (FMA) suggested that the crack was initiated from the corner of the interface between a BGA pad and the solder mask (SM) under the die corner. More details of this failure and the steps used to isolate and address it are contained in the next section. The initiated crack, denoted as portion A in Fig. 5.9, propagated into the substrate during TC and penetrated through layers as shown in Fig. 5.9 (portion B). In some cases, the propagating crack had been observed cutting through the copper traces on the adjacent metal layer, causing electrical opens.

Cross-sectioning of incoming bare substrates before assembly did not reveal any microcracks. The test sockets used for electrical test were first suspected as the root cause for existing cracks, as the mechanical clamp force of those sockets in electrical testing could have induced microcracks at the interface between the BGA pads and SM. But the same failure mechanism was reproduced in the follow-up TC

Figure 5.9 Crack initiated from the pad/SM interface (A) and then penetrated into the substrate (B).

with samples that had not been electrically tested. Therefore, the crack must have been generated in the packaging assembly or the follow-up temperature cycling. FEA was conducted to help understanding the failure mechanism and identify the root cause of failure.

5.5.2 FEA Modeling Considerations

Ideally, more accurate results would be expected if more detailed features of a package were modeled in the finite element model. In reality, a trade-off has to be made in consideration of the limitations of computing resources and computation time. Some modeling techniques can be employed to overcome the resource limitation. In this case, the technique used is *submodeling,* in which a relatively coarse-meshed global model is first used to capture the global deformation, and a follow-up, fine-meshed smaller local model is then used to compute the stress and strain distribution in the region where accurate results are desired.

Recall that Fig. 5.7 showed the schematic of package cross section. The nominal design parameters are listed in Table 5.2. All materials are modeled as linear elastic, and the properties are listed in Table 5.3. To reduce the model scale, one-eighth of the package has been modeled after employing geometry and loading symmetric conditions. Figure 5.10 shows the global model with lid attached, which was used for simulating a free-standing package under thermal loading as it was in the temperature cycling. The fillet shape and lay-ups in the substrate were modeled in the global model as shown in Fig. 5.11. As seen, the bumps and pads were not modeled in the global model. In-

Table 5.2 Nominal Dimensions

Die size	$12 \times 12 \times 0.724$ mm
Package size	$40 \times 40 \times 2.53$ mm
Lid and ring thickness	0.4943 and 0.645 mm
Substrate thickness	~0.5 mm
Pad diameter/SM opening	0.63/0.53 (1-mm pitch)

Table 5.3 Material Properties Used for FEA

Component	Young's modulus (GPa)	Poisson's ratio	CTE (ppm/°C)
Lid/ring attach	0.69	0.34	40
Die thermal	0.00035	0.45	232
Underfill	6.3	0.33	41
Silicon die	160	0.23	3.3
Cu foil	129	0.35	16.9
Solder mask	1.6	0.43	98
63/37 Sn/Pb bumps	17	0.43	23.2
Lid/ring	120	0.35	18.7
Dielectric	17	0.25	19

stead, a uniform layer with bulk material properties, estimated from the rule of mixture, was used. The glass transition temperature (150°C) was assumed to be the stress-free state, and the package was brought down to the lowest temperature point in temperature cycling (–55°C).

There was a concern that the lid attach process could induce microcracks at the corner of a pad and solder mask interface, because the

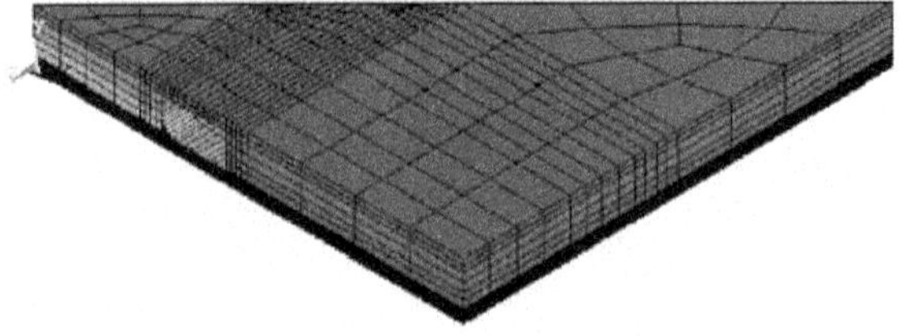

Figure 5.10 The global model with lid.

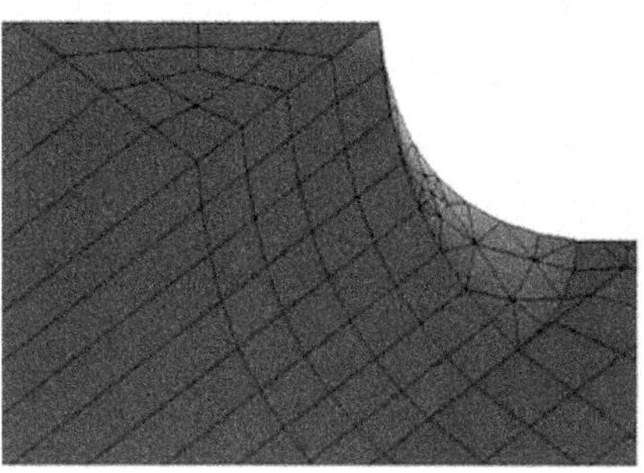
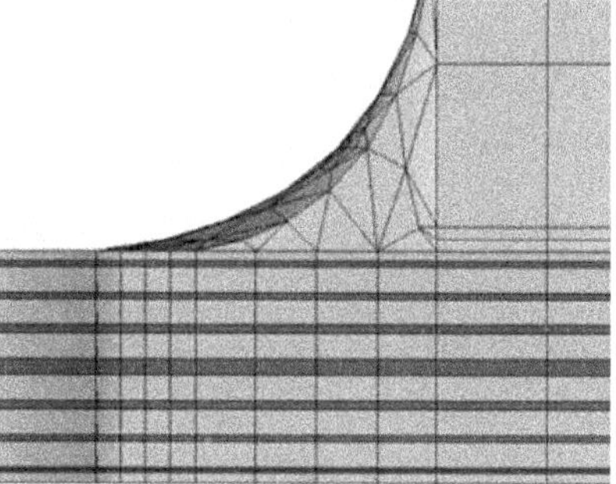

Figure 5.11 Close-up of the die corner.

rail used to hold the substrate supports only four corners of the holding substrate. A model was created, as shown in Fig. 5.12, to study the stresses inside the package with different supporting condition during the lid attach process. The clip force for lid attachment was applied as a uniformly distributed pressure on the top surface of the die. The amplitude of the clip force was 1 lb (0.37 kg). Two extreme bottom supporting conditions were simulated: fully supported and edge-only supported. These represent the best and the worst situations the package would encounter in the lid attach process.

The results from the global models showed high stresses along the die edges, as shown in Fig. 5.13, and those stresses reach peak values at the die corner during both the lid attachment process and temperature cycling, which is consistent with the fact that most failures were observed in the die corner area. Since a full 3-D model was proven to be infeasible due to limited computing resources, a submodel was built

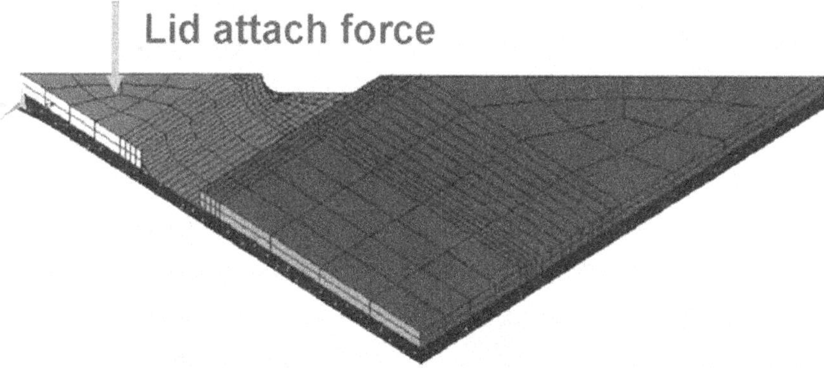

Figure 5.12 Lid attach simulation.

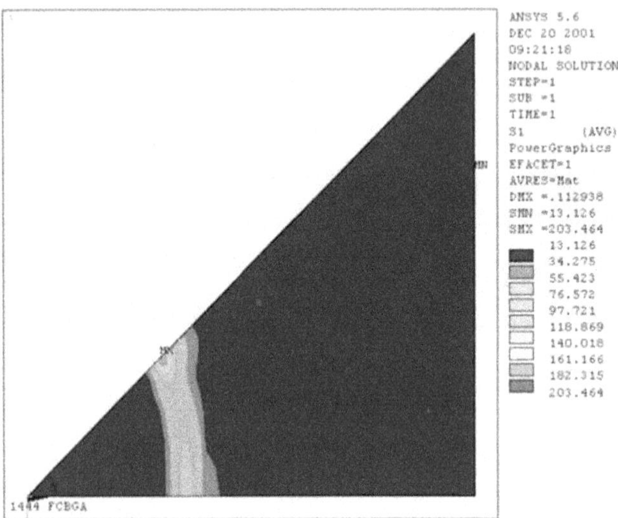

Figure 5.13 High stress shown along the die edge.

to mimic the detail features under the die corner where the failure was observed. The submodel covers the die corner area, including three rows and three columns of bumps as shown in Fig. 5.14.

5.5.3 Simulation Results

In the lid attach process, the stresses at the pad and solder mask interface are found to be largely dependent on the supporting condition provided by the holding rail. In a fully supported rail, the stresses at

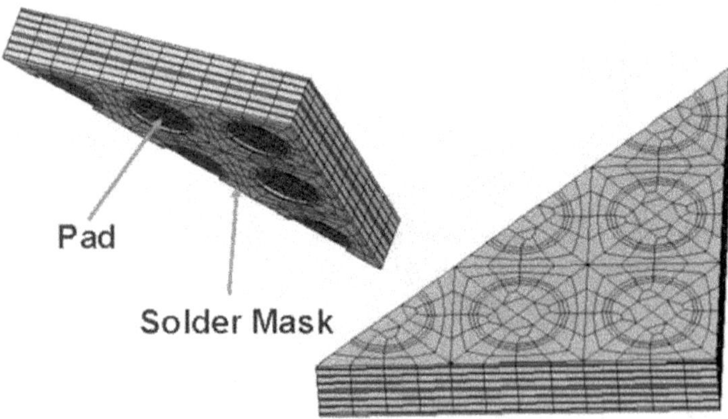

Figure 5.14 Top and bottom of the submodel.

those interfaces are negligible. Even a 10-lb lid attachment force was applied as shown in Fig. 5.15, which depicts the first principal stresses that evolve during this process. On the other hand, as seen in Fig. 5.16*a*, significantly large tensile principal stress was realized if only the edges of the substrate were supported with the normal 1-lb lid attachment force. These tensile stresses at the pad and solder mask interface (see Figs. 5.17 through 5.19) could initiate an interfacial crack, as the fracture resistance is generally weaker at the interfaces. Additionally, these tensile stresses would be the driving forces causing an initiated crack to propagate, penetrating the dielectric layer and eventually causing electrical open failures.

5.5.4 Discussion

FEA results revealed that the interface between BGA pads and solder mask (SM) of the substrate under the die corner area was the critical region where the highest stresses were found for lid attachment. The stresses in such an interface depend on the supporting condition while attaching a lid; a firmly supported substrate could significantly reduce the interfacial stresses and therefore reduce the possibility of microcrack initiation. Significantly fewer cracking failures were found in follow-up qualifications of parts built with newly designed rails that fully support the package during lid attachment.

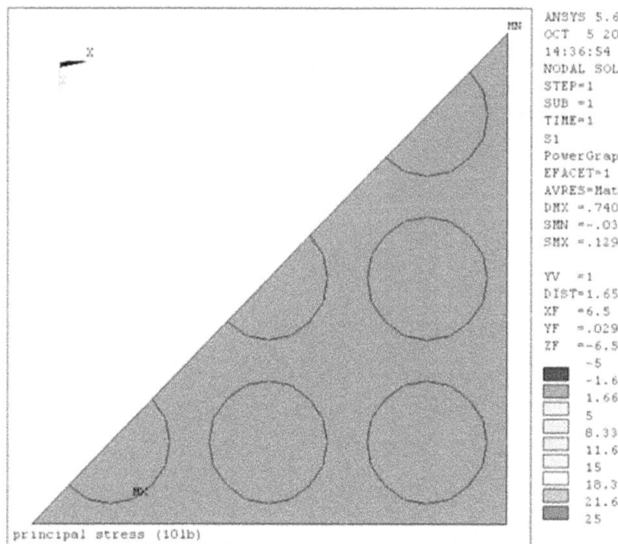

Figure 5.15 First principal stress on pads and in solder mask at the die corner when the bottom is fully supported (lid attach force = 10 lb).

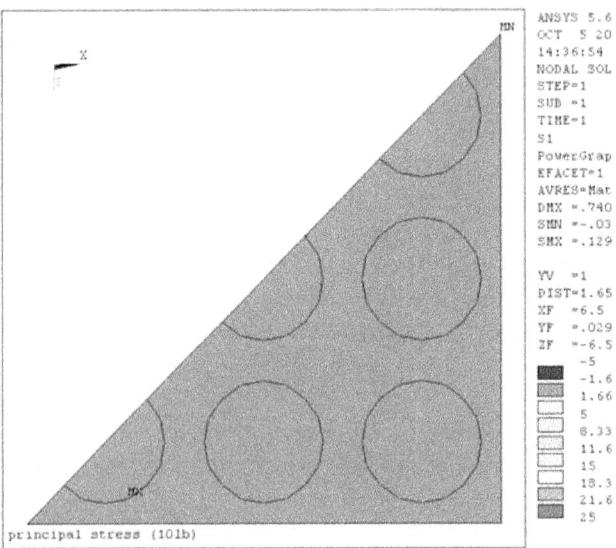

Figure 5.16 First principal stress on pads and in solder mask at the die corner when only the edges are supported (lid attach force = 1 lb).

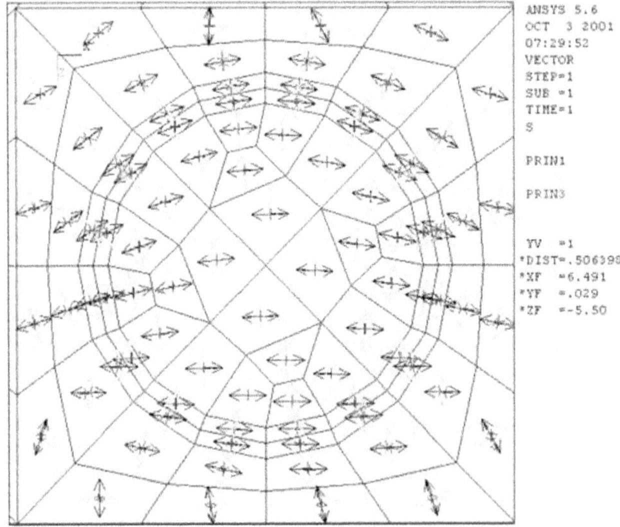

Figure 5.17 Direction of principal stresses along the pad and solder mask interface.

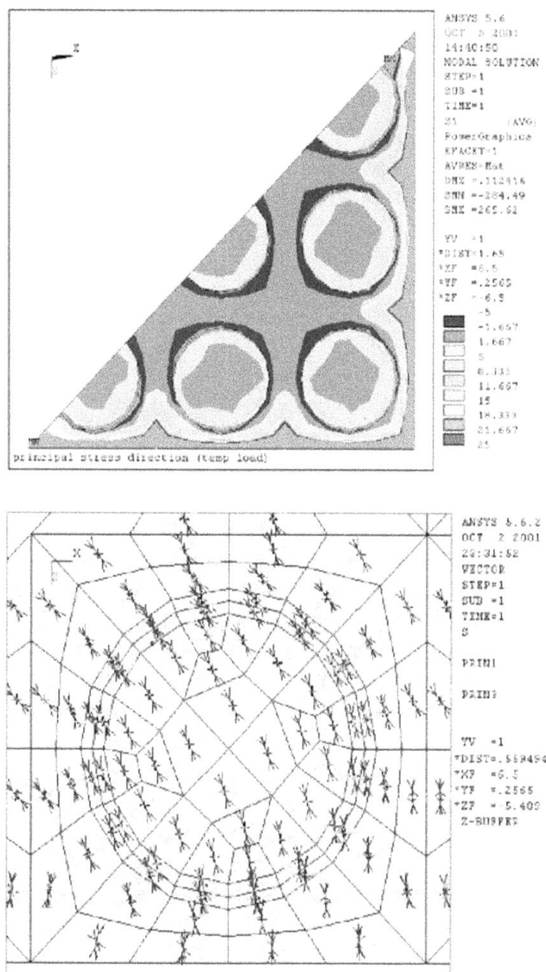

Figure 5.18 First principal stress (top) and principal stress direction (bottom) during TC.

The simulations also revealed that the stresses in the substrate and package warpage were closely related to the space between the die and the stiffener ring attached on the substrate. As shown in Fig. 5.20, the deformation on the bottom surface of the substrate varies with different spacing. As expected, the maximum warpage at the BGA side of the substrate increases as the spacing increases. However, the maximum first principal stress in the substrate decreases (Fig. 5.21) while the die-stiffener spacing increases. For this specific package, with a body size of 40 mm and a 12-mm square die, the maximum stress-die/

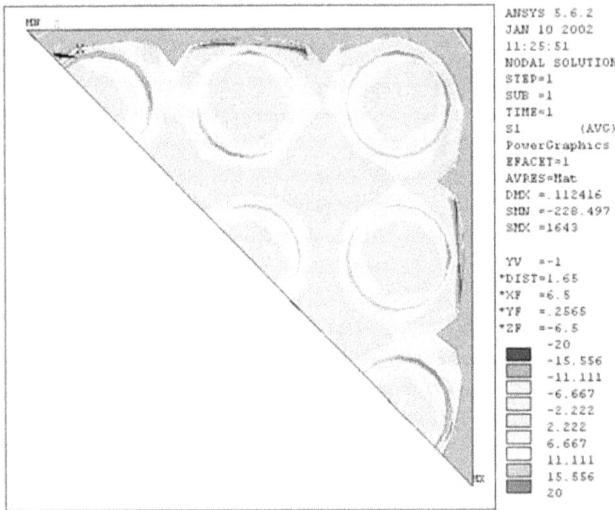

Figure 5.19 First principal stress in the dielectric layer right on the pad/SM layer during TC (bottom view).

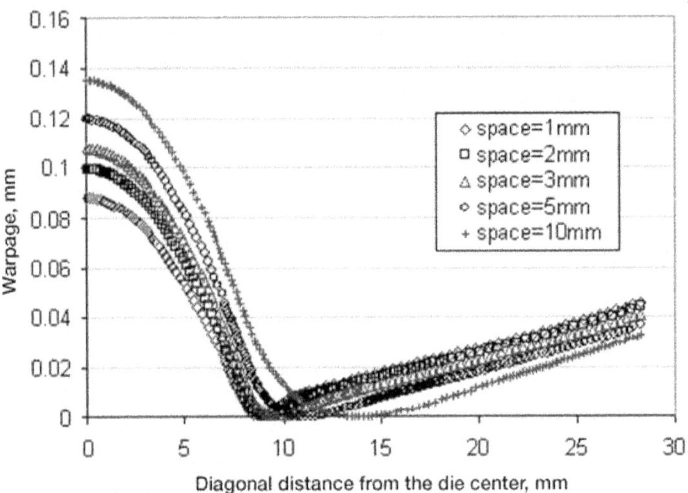

Figure 5.20 Deformation on the BGA side of the substrate varies against the die/stiffener spacing.

stiffener space curve becomes flat when the spacing is larger than 5 mm. It is obvious that the spacing is an important parameter that could be adjusted to reduce the stress level in the substrate. Of course, this must be done carefully to achieve the maximum benefit in reducing the substrate stresses without a compromise in coplanarity that

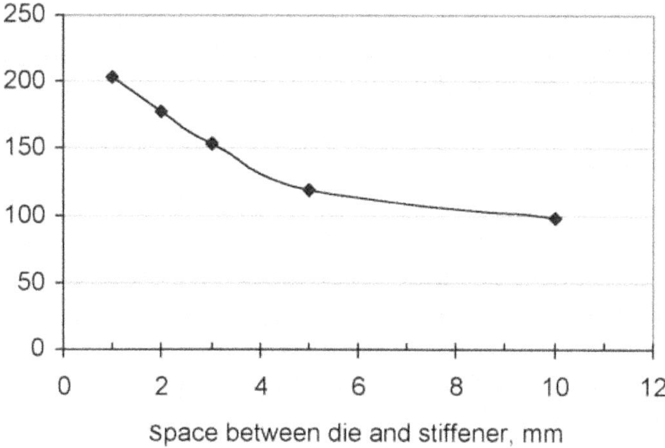

Figure 5.21 The maximum first principle stress in the substrate varies against the die/stiffener spacing.

could affect the board assembly behavior. As a final solution to address this failure mode, a combination of geometric and assembly changes were coupled with lower-stress assembly materials. The following section provides a discussion on how the failure was first isolated, plus details how the materials and geometric changes were implemented.

5.6 Case Study No. 2: Large Die Flip Chip Test Vehicle

In an effort to better understand a particular failure mode that was discovered during a prequalification evaluation (i.e., evaluation work to weed out issues prior to actual, and expensive, qualification builds) of a relatively small flip chip device on an organic substrate, a larger test vehicle, more representative of current and future product, was introduced to continue the work. This study is an extension of the work from the last section.

The first step in understanding this failure was to perform a complete failure analysis (using many of the tools that have already been described in this text) to isolate the failure mode. The large die tester, paired with the corresponding chip carrier as described in previous sections, was designed to inexpensively and more accurately isolate the failure to help mitigate this issue. What follows is a description of the prequalification evaluation vehicle, the nature of the testing conditions, a complete failure analysis, the test conditions, and results of the large die tester, plus a complete failure analysis and conclusions derived from the work.

5.6.1 Test Vehicle Description

Figure 5.7 exhibits the general cross-sectional schematic of a typical high-performance flip chip package. The following sections describe the test vehicles that were the focus of this work, which followed the form of Fig. 5.7.

5.6.1.1 Prequalification test vehicle. Although some details of this test vehicle were given in Section 5.5.1, additional details pertinent to the location of the failure are given here. The first generation of our FC reliability tester used a 40-mm BGA substrate with a 12-mm die. The purpose of the design was to test first-level (component-level) reliability, second-level (package to printed circuit board) reliability, substrate trace/via integrity, bump electromigration issues, and signal integrity. Although various substrate technologies have been evaluated with this design, a typical cross section of the one described in this case study is shown in Fig. 5.22 (refer to Fig. 5.8 for a top and bottom view of this tester).

The die-substrate interface was chosen to make use of the bump pattern on our existing FC technology qualification vehicle. There were several benefits to this approach. First, an existing die design could be modified to meet the first-level daisy chain requirements as opposed to having to develop completely new silicon (consistent with Section 5.4.4, where simple opens and shorts tester generation was

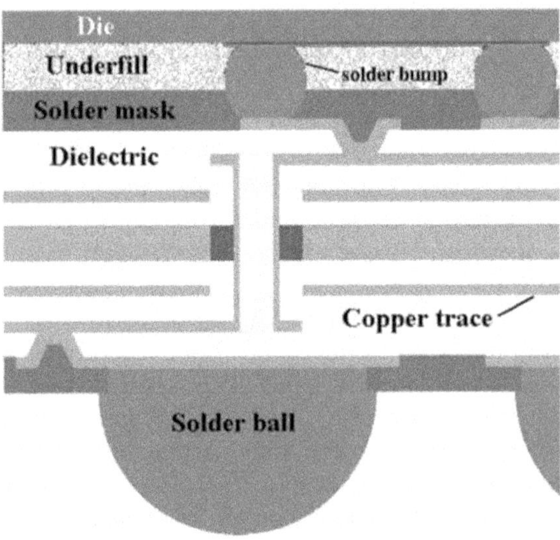

Figure 5.22 Cross-sectional schematic of the high-performance flip chip BGA substrates used in this study.

discussed). Second, since the technology proven substrate and the reliability test substrate share the same geometric characteristics (i.e., substrate type, body size, die size, and bump pattern), there was a tight linkage between the two vehicles and, subsequently, the test results.

The die fabrication process utilized a subset of a full 0.25-μm aluminum metal wafer process. The pattern consists of a 50 × 50 array of eutectic bumps (80-μm cap opening, 102-μm under-bump metallization) on a 230-μm pitch with depopulated areas. Figure 5.23 shows the die bump pattern.

First-level daisy chains. To provide a means of sorting and finding potential failure mechanisms, two first-level daisy chains were used. The first involved the "peripheral" section of bump pads and was divided into nine segments. The second used the "core" bumps and was divided into seven segments.

Second-level daisy chains. A similar approach was used on the second-level chains. Since the corner regions were of particular interest, the four chains in those areas were made unique. The remaining chains were separated into core and peripheral regions. I/O connections were reserved in the areas deemed least susceptible to substrate failures.

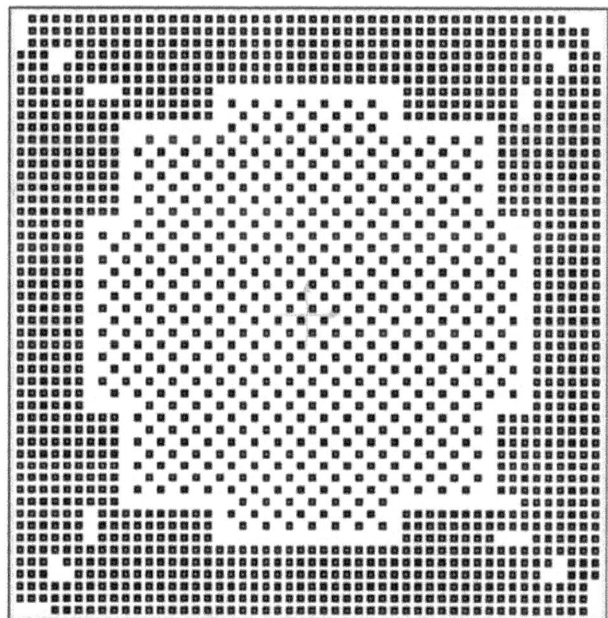

Figure 5.23 12-mm test die solder bump pattern.

Via chains. Three via chains were routed in the substrate near areas of suspected high stress. These areas were die edge, ring edge, and near-outside edge of the substrate under the ring.

Electrical characterization pairs. The location of electrical characterization pairs was specifically chosen to be in low-stress areas so that potential failures would not limit the usefulness of the tester. The tester also utilized four through (shorted) differentials pairs, and four 50-Ω terminated (on die) differential pairs to understand the signal integrity capabilities of any particular substrate/material combination.

5.6.1.2 Large die test vehicle. Based on experience gained from the 12-mm die/40-mm organic substrate (prequalification tester) program, an investment was made in a longer-term, second generation, reliability tester. The test chip was used to investigate subcontractor silicon and assembly processes as compared to our internal processes. The substrate design added large body/large die considerations and bump pattern variations for underfill study. It also refined the functionality of the prequalification tester features such that failure analysis would be easier to perform.

The die was custom designed for the tester and is approximately 21.5 × 21.5 mm square, consisting of a single-level Cu metal to reduce costs while still exhibiting (at the passivation layer) the features of a 0.13-μm wafer technology. This allows the evaluation of materials and processes for reliability prior to the introduction of 0.13-μm product to assembly, even in the presence of increasingly shorter product development spins. This design is compatible with other wafer technologies as well. The package interface is a 107 × 107 selectively depopulated array of 200-μm pitch bumps. The substrate specifics are 50-mm body with a maximum BGA array of 2401 balls. Figure 5.24 shows the die bump pattern.

This pattern, with the inclusion of large depopulated regions, transitions from populated to depopulated regions, and tight pitch to loose pitch transitions (all of which can be found in our typical product bump patterns), allowed us to characterize underfill processability in a meaningful manner. In summary, although certain features have been added to extend the capabilities of the large die tester, the technology (and layer structure, see Fig. 5.22) of the substrate remains the same as described above for the 12-mm die/40-mm test vehicle. A few of the modifications are described in the following sections. To assist in the analysis of any failures, in a similar manner to the prequalification tester, special attention was given to the arrangement of bumps as well as segregation of the first-level chains. Locating an individual bump or chain segment along edge with 107 possible sites could have

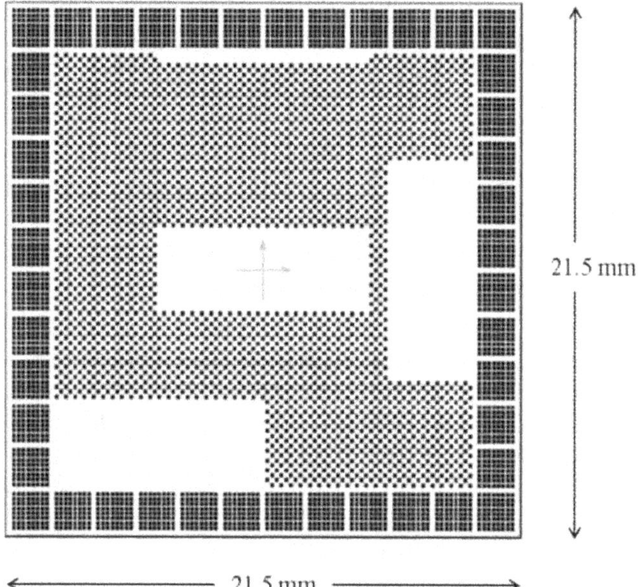

Figure 5.24 21.5-mm test die solder bump pattern. Open white areas are depopulated regions.

been tedious at best. The pattern was broken up into 8 × 8 blocks with a one-bump wide "channel" between each. When used in conjunction with an array-type (row, column) naming convention, finding specific bumps is much easier during cross-sectioning and other FA procedures. Likewise, the first-level chains were constructed so as to facilitate meaningful "binning" of failures. By breaking the chains down into small segments and using appropriate positional naming schemes, mapping the regions of concern is very simple. This allows quick isolation of failed "areas" of the substrate to help understand, at first glance, sensitive areas of the vehicle during reliability testing.

Substrate design considerations were similar to that of the prequalification tester, with certain improvements. As before, various technologies have been utilized, but the one described in this section is the same as the prequalification tester. BGA balls in low-risk areas were reserved for I/Os. To ensure connectivity to bumps, redundant traces on two different routing layers to two different BGA pins were used wherever possible.

First-level daisy chains. A total of 90 first-level daisy chains were used to isolate particular regions of interest. Suspected high-stress regions were allocated more chains of a shorter length. Each chain was routed to two unique BGA ball pairs wherever possible.

Second-level chains. A different second-level chain approach was used in the large die tester. Rather than meandering chains in specific areas of the substrate, a simple horizontal scheme was utilized. The overall pattern was defined by the second-level test board.

Via/trace chains. To expand the ability to monitor via and trace integrity over the regions most likely to exhibit problems, a series of 36 via chains were placed in those areas. The chains traversed all the layers and via types in the substrate. Although not typical of conventional routing, this approach was designed with the intent of finding potential failures.

Electrical characterization pairs. The location of electrical characterization pairs was specifically chosen to be in low-stress areas. These special traces consisted of the following: one through (shorted) differential pair (designed for 3-GHz signal integrity characterization), one four-port differential pair (designed for 6-GHz signal integrity characterization) and one microstrip/waveguide-style differential pair (designed for 10 GHz) for signal integrity characterization.

5.6.2 Reliability Testing Program

All test vehicles were subjected to a series of standard reliability tests that were run in accordance with MIL-SPEC-883, which covers accelerated life testing for electronic packages. Testing was also consistent with IPC/EIA J-STD 029. The testing regime used for this study is listed in Table 5.4. No bias testing was carried out, since all vehicles were single-level metal and contained no logic cells. Figure 5.25 shows the sequence in which the testing was conducted. In certain cases, more than one stress condition was used to provide some level of differentiation, but this work is not detailed here.

Table 5.4 Description of Reliability Tests

Reliability test	Description	Target level for "pass"
Moisture reflow test (MRT)	Precondition to JEDEC prescribed moisture level 4 (30°C/60% relative humidity [RH] for 196 hr)	Three simulated reflow @ 220°C +5/–0°C, no electrical failures from ATE, no delam from C–SAM evaluation
Temperature cycle test (TCT)	Condition "B": –55 to 125°C, air to air	1000 cycles, no electrical failures from ATE, no delam from C–SAM, post-test
Highly accelerated stress test (HAST), unbiased	130°C, 85% RH, no electrical bias	168 hours no electrical failures from ATE, no delam from C–SAM, post-test

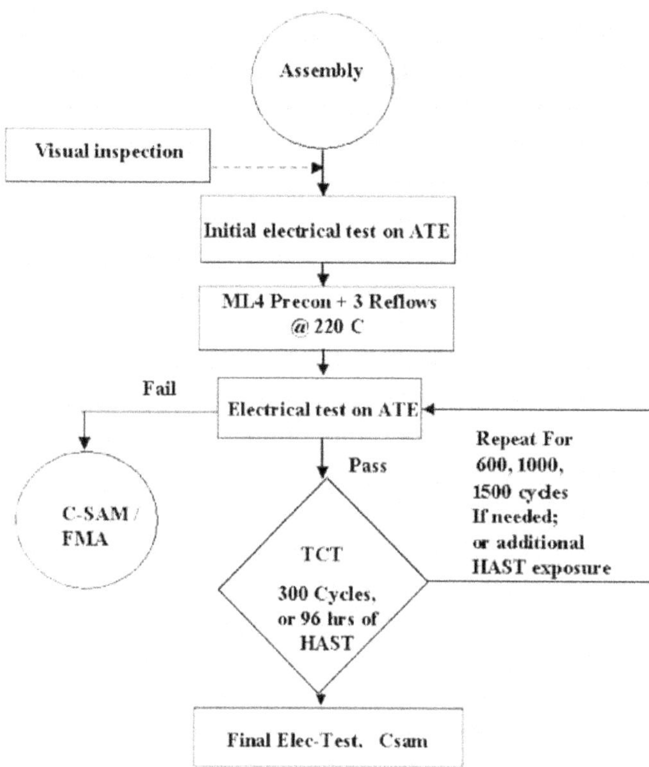

Figure 5.25 Reliability test sequence.

All electrical tests were conducted using automatic test equipment (ATE) that applied a voltage to daisy-chain structures. If damage occurred such that a trace or conduction path (e.g., a solder bump) was severed or displaced, it would be detected as an open or a short on the equipment. A confirmed open or short was then listed as a failure, which was used in the data analysis.

5.6.2.1 Material characterization. Material characterization was performed to ensure accurate, temperature-dependent material properties, which are essential to the success of FEA simulation. Package-level temperature cycling values can range from –55 to 125°C, and the FEA models require material properties over this temperature range to perform an analysis of the package. Generally, vendor-supplied material properties are only reported at ambient testing conditions. A single data point for a material property does not adequately characterize the mechanical behavior of most assembly materials across typical use temperature ranges. To further complicate the issue, reported values for a single property can differ significantly, depending on the method used to gather the property, and vendors do not always publish the accompanying testing information.

Material characterizations were performed to generate temperature-dependent material properties for use in the FEA models and to help in selection of materials for reliability evaluations. Analysis of the materials followed testing guidelines set by the American Society for Testing and Materials (ASTM).

The following is a list of the instruments employed to generate the material properties and a short description of each:

1. *Dynamic mechanical analyzer.* The TA Instruments DMA 2980 Dynamic Mechanical Analyzer was used to determine the temperature-dependent, Young's modulus values on the assembly materials. The analysis of the samples was performed using a three-point bend fixture with a 5°C/min temperature ramp from –65 to 225°C. In addition, the test was performed at a frequency of 1 Hz with a 1 newton preload force on the sample and amplitude of 25 μm.
2. *Thermal mechanical analyzer.* The TA Instruments TMA 2940 Thermal Mechanical Analyzer was used to determine the temperature-dependent coefficient of thermal expansion (CTE) and glass transition (T_g) for the assembly material samples. The analysis of the materials was performed using a quartz expansion probe with a 3°C/min temperature ramp rate from 30 to 225°C. A preload force of 0.5 g was placed on the samples.
3. *Rheometer.* The Rheometric Scientific ARES Rheometer was used to determine temperature-dependent shear modulus and coefficient of thermal expansion values on the assembly material samples. The analysis of the samples was performed using a torsion rectangular fixture with a temperature step rate of 5°C from –65 to 225°C and a soak time of five minutes per temperature step. A frequency of 1 radian/sec and a strain rate of 0.02 percent were employed in the test.
4. *RVSI.* The final tool utilized to evaluate package performance based on changing materials properties was a Robotic Visions Systems Inc. (RVSI) warpage analysis tool. This device allowed us to characterize the room temperature warpage of the package after assembly by optical means. The tool analyzes an array of predetermined sites on the device (usually BGA balls or BGA pads) and determines a "best fit" plane to describe the total package flatness. Each individual site is then assigned a coplanarity value by determining the difference, in mils or microns, between the plane of the site and the best-fit plane. This value is used to determine the 3-D

warpage of the device. A highly warped device would both fail JEDEC prescribed warpage specifications and indicate an assembly with a high stress state.

5.6.3 Design of Experiment Descriptions

5.6.3.1 Prequalification test vehicle. The goal of the 12-mm die/40-mm organic substrate evaluation plan was to identify a materials set that could be used successfully in the manufacture of products of this type. This included both processability and reliability considerations. The materials suggested were either available for commercial use or in late stages of development (the last step before commercial use) at our assembly subcontractor. Through the use of reliability tests and the reliability test sequence outlined in Section 5.6.2 above, the materials and processes were selected based on the relative performance in each of these tests. This discussion will focus only on the underfill temperature cycle portion of the evaluation. Details of how a processability evaluation was conducted are outlined in the next section.

5.6.3.2 Large die test vehicle. In the second phase of our product development cycle, which was designed to extend the die size capabilities of the same substrate technology, a more complete DoE was devised. Many of the assembly processes that were evaluated in the first phase of the work (12-mm die/40-mm organic substrate) were assumed to hold for the larger die/substrate combination. However, underfill materials were considered a crucial factor in finding a reliable assembly combination and were varied for the large die vehicle evaluation. In fact, all underfill materials for the second-phase evaluation were scrutinized using our 3-D FEA model for selection in the evaluation matrix, based in part on the findings from the 12-mm die/40-mm organic substrate test vehicle work. This process is expanded upon in more detail in the following sections.

The DoE for the 21.5-mm die/50-mm organic substrate vehicle consisted of two steps. The fist step included a processability evaluation of four candidate underfill materials, including a warpage evaluation, initial MRT response, and cursory thermal cycling evaluation. A processability evaluation was required, since all materials were in early stages of development at the assembly subcontractor. This evaluation was completed using an inexpensive mechanical version of the test substrate described in Section 5.6.1.2, in which layer-to-layer vias were eliminated to reduce costs (another cost-reduction technique that can be employed when fabricating test vehicles). Each device was X-rayed (since the lids were still in place, see Fig. 5.7) after MRT exposure to determine if solder bridging could be observed. Solder bridging will take place if delamination occurs during the MRT. Devices were then subjected to 500 and 1000 TCT B intervals. At each test point, a portion of the devices were sampled, the lids were removed, and the interfaces of interest (die-to-underfill, underfill-to-substrate) were imaged using an acoustic microscope. (Specifically, it was a Sonix C-mode scanning acoustic microscope [C-SAM] in pulse echo mode with a 50-MHz transducer. This equipment was used in other portions of the evaluations and was an important tool during FA). The copper lids had to be removed because, at the frequency used to image the device, the penetration depth was not sufficient to capture data at each interface.

The materials from the first portion of the evaluation were ranked according to lowest warpage, lowest solder bridging and delamination (from C-SAM) at each test point, lowest voiding, and filler settling from fill and cure (also determined from C-SAM). Top performers were promoted to the full-reliability DoE.

5.6.4 Failure Analysis Approach

Failures identified during the reliability testing phase were submitted to careful failure analysis using the following equipment: Tektronix Type 576 curve tracer, Tektronix TDS 8000 Time Domain Reflectometry (TDR) System, Sonix C-SAM, CR Technology CRX-1000 Real Time X-ray Imaging System, Allied High Tech Products, and Techprep Sectioning and Polishing Systems. The electrical failures captured during ATE testing were linked to root causes through a combination of hand probing, time domain reflectometry (TDR), and flat- and cross-sectioning of the devices. Once the failure mode was identified, it was categorized by the stress test generating the failure, the test point when failure was detected, and the specifics of the test vehicle containing the failure (die size, substrate type and size, materials used, and so forth). This data was coupled with the failure rates to aid in interpreting the results.

5.6.5 Finite Element Analysis Approach

A finite element analysis (FEA) model was created to study the stress state and warpage behavior of flip chip packages using ANSYS commercial software in a very similar manner as was presented in Section 5.5—this time to understand the contributions of the materials used as well. A 3-D quarter symmetry approach was used to simulate the package geometry as shown in Fig. 5.26. The similarity to the 2-D schematic shown in Fig. 5.7 is intentional.

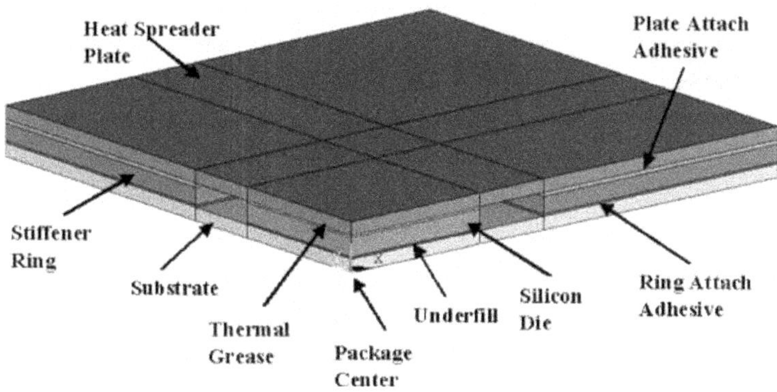

Figure 5.26 Model geometry and components.

Assuming linear elastic material behavior, the model was used to solve for the stress and displacement fields that develop at the low end of the temperature cycle. A constant mesh density was used to ensure a valid comparison of results among different geometry cases.

It is critical that proper material characterization be performed and utilized as input for the FEA model, as vendor data can vary widely due to differences in techniques and procedures. In addition, many of the package materials are polymeric and generally exhibit strong temperature-dependent behavior. Material vendor data is not typically available over the applied temperature range.

The modeling results are reported in terms of maximum first principal stress values on the BGA side of the substrate in the region of interest. Warpage values are also reported as the maximum deflection predicted on the bottom surface of the substrate.

5.6.6 Results and Discussion

5.6.6.1 Prequalification test vehicle results. All devices from the small die DoE were submitted to reliability evaluation after assembly (as per Fig. 5.25). No visual defects were found, and electrical testing was conducted according to Fig. 5.25 and Table 5.4. C-SAM imaging was conducted only on failed devices for this portion of the evaluation. No electrical failures were noted after MRT. However, failures were observed during thermal cycling. Figure 5.27 shows the results of the thermal cycling tests for this vehicle. It is clear from this figure that a real difference exists in the thermal cycling performance of this vehicle, linked solely to the underfill material used. However, to understand how material changes affected reliability results, it was important to determine the failure mode associated with each mate-

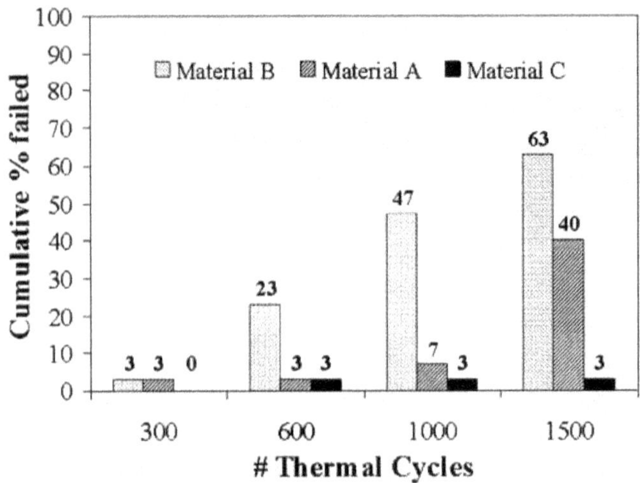

Figure 5.27 12-mm die/40-mm organic substrate temperature cycling results. (Numbers on columns are cumulative failure percentage values for each material.)

rial. Only in the presence of a common failure mode can the relative reliability performance of a particular underfill be gauged.

Failure analysis. Failed flip chip packages at 300, 600, 1000, and 1500 temperature cycles were analyzed to identify their root cause. The packages were subjected to electrical test after completing the critical test points. It was noted that the number of package failures increased with increasing temperature cycles. Electrical test data of the failed packages were analyzed and grouped in two categories—electrical opens within the substrate daisy chain nets and opens in the die chains. The specific failing BGA balls and connected net chains were identified using the die and substrate routing plots (as per Sec. 5.2).

At each test point, a visual inspection of the package BGA side showed localized cracks in the solder mask adjacent to the BGA pads for all materials, with increasing severity as the number of temperature cycles increased. On the die side, localized cracks in the underfill fillet at die corner regions were also found for material A only after 600 cycles. Figure 5.28 shows the two types of defects observed during visual inspection. Package-level electrical bench tests were performed on failed packages using a curve tracer and TDR. Both the curve tracer and TDR tests confirmed the reported electrical open failure mode on each package. The TDR performed on the failing BGA balls was an effort to localize the open failure site in the package interconnects. This technique is used to discriminate between a package- and die-related open or short. This test indicated that the failure sites

(a)

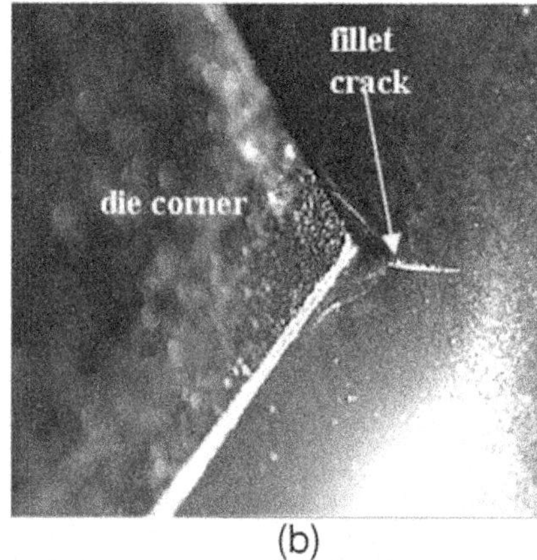

(b)

Figure 5.28 Visual defects found in (a) BGA side of substrate for all materials after 300 temperature cycles and (b) fillets near die corners in material A after 1000 cycles.

were in a substrate region close to the silicon die corners on most of the devices tested.

Radiographic inspection was performed on the failed packages using a real-time X-ray system to detect the presence of any anomalies. No anomalies contributing to the package failures were detected. Red dye was applied to the package BGA side to expose existing substrate cracks in the solder mask to determine their depth of propagation laterally and vertically in the substrate after the package is cross-sectioned.

C-SAM imaging was performed on the failed packages to document their underfill/die and underfill/substrate interfacial integrity condition prior to the start of the physical failure analysis process required to determine root cause failure mechanisms. The C-SAM indicated no anomalies on packages that did not exhibit underfill die corner cracks, whereas it indicated die/underfill delamination for packages that exhibited underfill cracks at die corner regions from the visual inspection. More clearly, the C-SAMs indicate that, for materials B and C, regardless of the level of electrical failures, no underfill/die or underfill/substrate interfacial failure was observed. However, for material A, interfacial failure begins near 1000 temperature cycles (TCs) and becomes extensive at 1500 TCs. Figure 5.29*a* through *d* show C-SAM images to support this conclusion. Figure 5.29*a* and *b* show images typical of materials B and C after 1000 and 1500 TCs, respectively, showing intact interfaces. Figure 5.29*c* and *d* show C-SAM images for material A after 1000 and 1500 TCs, respectively.

After applying a red dye to penetrate and expose any cracks, horizontal and vertical sectioning were performed on the failed packages

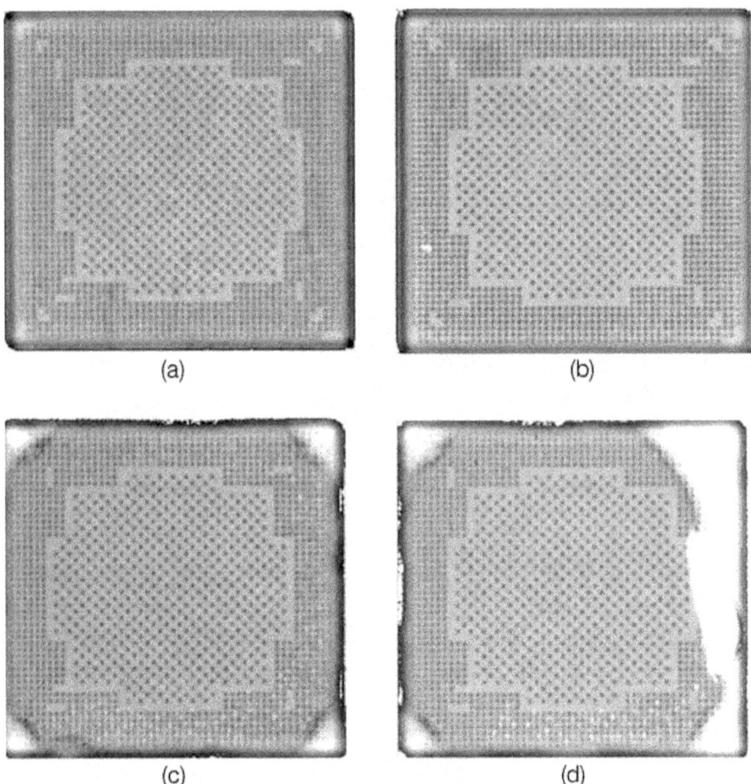

Figure 5.29 50-MHz pulse echo C-SAM images of (*a*) material C after 1000 temperature cycles (TCs), (*b*) material B after 1500 TCs, (*c*) material A after 1000 TCs, and (*d*) material A after 1500 TCs.

that completed 300, 1000, and 1500 temperature cycles. The horizontal cross sections on the BGA side show high spots at die corner regions that tend to polish first. Figure 5.30 a shows cracking in the dielectric layers of the substrate underneath the corners of the die. Final polish showed the signal traces routed through these regions to exhibit stress cracks, which caused electrical opens. Figure 5.30*b* shows the severed trace resulting from the crack seen in Fig. 5.30*a* extending through the dielectric toward the die. From the substrate routing, these signal traces were determined to be connected to the failed net chains.

The vertical cross sections show substrate cracks that originate from BGA pads localized to die corner regions. These cracks propagated through dielectric and signal traces closest to the BGA balls that are connected to the failed net chains. Figure 5.31 shows a vertical cross section of a crack similar to the one shown in Fig. 5.30, ema-

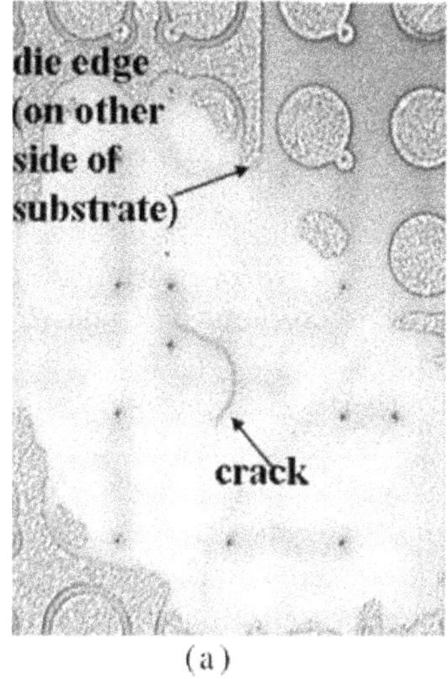

(a)

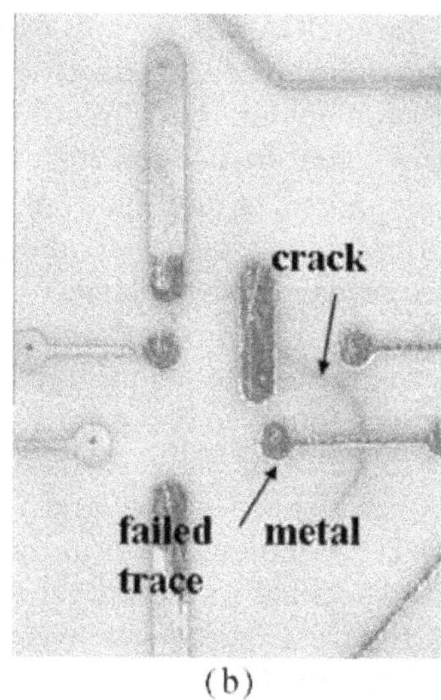

(b)

Figure 5.30 (*a*) Cracking in the dielectric under the die in the 12-mm die/40-mm organic substrate test vehicle (exposed by red dye) and (*b*) the same crack shown severing a signal trace.

Figure 5.31 Cracks from Fig. 5.30 shown in a vertical cross section, revealing propagation path through the dielectric and a signal trace, resulting in a failure.

nating from a BGA pad corner, extending through the dielectric, and severing a signal trace. These cracks were also observed to extend upward into the solder mask, resulting in nonfailure-related cracks, as seen in the visual inspection results shown in Fig. 5.28.

Failure analysis of devices from the material A cell show the type of dielectric cracking discussed above. However, an additional failure mode was revealed when devices were vertically sectioned after 1000 and 1500 TCs. The devices were sectioned along regions suggesting underfill to die surface delamination as shown in Fig. 5.29*c* and 5.29*d*. Cracking was found to begin in the fillets of the 1000-TC samples and then to extend from die corners (in the fillet region from Fig. 5.28*b*) along both edges of the die, and then underneath the die surface for the 1500-TC samples. Eventually, this crack extended to sever the bump interconnects in the delaminated/cracked region in the 1500-TC samples. Figure 5.32*a* shows a vertical cross section of a device from the material A cell after 1500 TCs, showing the cracked region near the die corners. Figure 5.32*b* shows the crack extending through a solder bump in the delaminated/cracked region under the die, resulting in an electrical failure. In summary, it is clear that two failure modes are active during the temperature cycling of the 12-mm die/40-mm organic substrate vehicle. The first is cracking of the dielectric near the corners of the die on the BGA side of the substrate (near the corners if the die outline is projected to the back side of the substrate). This mode is present to varying degrees in devices using all underfill materials. The second mode is fillet cracking, extending under the die and severing solder bumps, which is the case for material A beginning after 1000 TCs and increasing the failure rate as it approaches 1500 TCs.

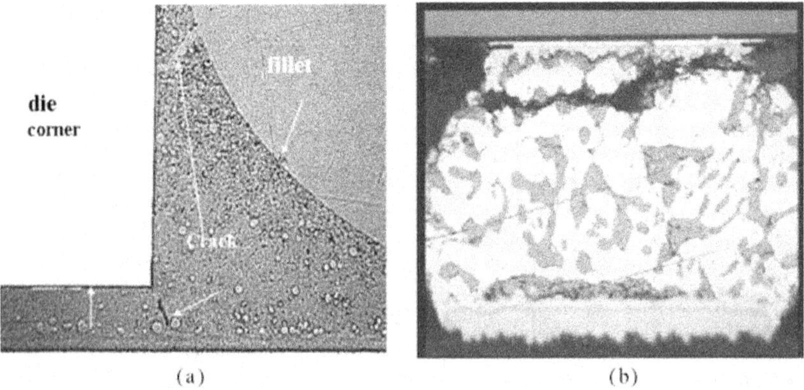

Figure 5.32 Vertical cross section of post-1500-TC failures from material A cell showing (*a*) a fillet crack extending underneath the die corner and (*b*) the same crack resulting in a severed solder bump, yielding a failure in test.

Material characterization. Clearly, from the results shown in Fig. 5.27 and the failure analysis described in the above section, the reliability results are linked to differences in the material properties of the underfills tested here. A detailed analysis of the materials used in this portion of the study was conducted, following the practices outlined in Section 5.6.2.1. Table 5.5 shows the properties determined in our lab for all critical portions of the package. Many important differences are noted in this analysis, especially in the areas of CTE, T_g, and modulus above T_g of the underfill materials. In the past, several authors have attempted to make the link between critical underfill material properties and reliability performance,[1,2] and undoubtedly this link is key to implementing a knowledge-based package development process. For this particular application, we used an FEA approach, detailed in the next section, to help understand why the reliability performance of the 12-mm die/40-mm organic substrate vehicle was affected by the properties listed in Table 5.5. With this better understanding of the underfill property/reliability performance link, the extension to large die flip chip packaging was made easier.

Finite element analysis. A detailed 3-D model of the 12-mm die/40-mm organic substrate vehicle was constructed as per Sections 5.5 and 5.6.5. From this failure analysis, cracking in the dielectric was the most prevalent failure mode. Therefore, a priority was to focus on the amount of stress and warpage (a result of CTE mismatches in the package) generated in the sensitive corner regions of the substrate as a function of underfill material. Using the RVSI equipment described in Section 5.6.2.1, it was determined the highest warpage occurred between the stiffener ring and the die. The stiffener acts to flatten the total package warpage but creates a hinge point in the substrate between the die and stiffener ring.

From Section 5.5, Fig. 5.20 shows the predicted warpage profile for this test vehicle at the highest predicted deflection, which occurs at –55°C, the lower bound temperature seen during temperature cycling (regardless of underfill material type). Figure 5.33 a shows the corresponding 3-D stress contour and shape profile for the same conditions (using a 3-mm die-to-stiffener-ring spacing), revealing the pinch point located right at the corner of the die as observed in the RVSI data. Reviewing the 2-D stress contour plot (again shown for the same conditions as Fig. 5.20) shown in Fig. 5.33*b*, observed from the BGA side vantage point, we see that the maximum first principal stress is found at the die corners. Not coincidentally, this is where we see the pinch point during thermal cycling and the presence of life limiting failures that manifest as cracks seen in FA. This agreement between the RVSI data and the FEA results provides an added level of confidence in our FEA predictions.

Table 5.5 Material Properties Determined for the Constituents for Both Test Vehicle Evaluations in this Study

Material	CTE1 (below T_g) (ppm/°C)	CTE2 (above T_g) (ppm/°C)	Elastic modulus @ –55°C (MPa)	Elastic modulus @ 25°C (MPa)	Elastic modulus @ 125°C (MPa)	T_g (°C)	Poisson's ratio
Substrate	19	N/A	19994	17865	14887	N/A	0.26*
Underfill A	3	106	7388	5580	4340	150	0.3*
Underfill B	42	134	5786	4219	3404	170	0.3*
Underfill C	39	120	8114	6370	5	56	0.3*
Underfill D	64*	154*	N/A	4400*	N/A	122*	0.3*
Underfill E	57*	130*	N/A	3600*	N/A	153*	0.3*
Underfill F	30	114	8092	6587	20	75	0.3*
Underfill G	59*	153*	N/A	5500*	N/A	153*	0.3*
Silicon die	2.6*	N/A		129500*		N/A	0.278*
Stiffener ring and heat spreader	17*	N/A		128700*		N/A	0.3435*
Thermal grease	232*	N/A	N/A	0.35*	N/A	–121*	0.45*
Ring attach adhesive	19*	N/A	14500*	N/A	N/A	220*	0.26*
Plate attach adhesive	40	108	11764	7883	291	55	0.4*

*Data provided by vendor

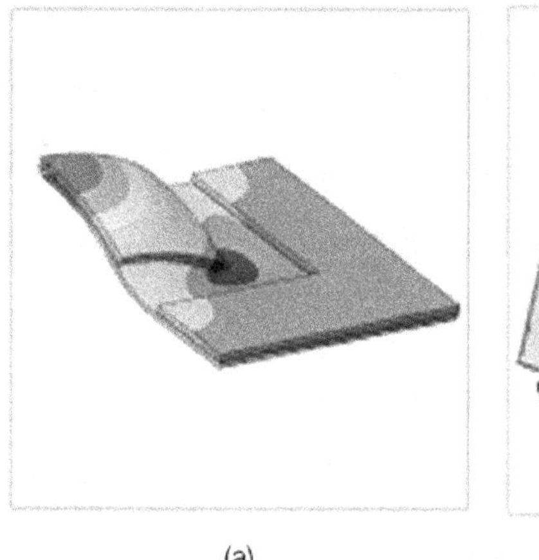

(a)

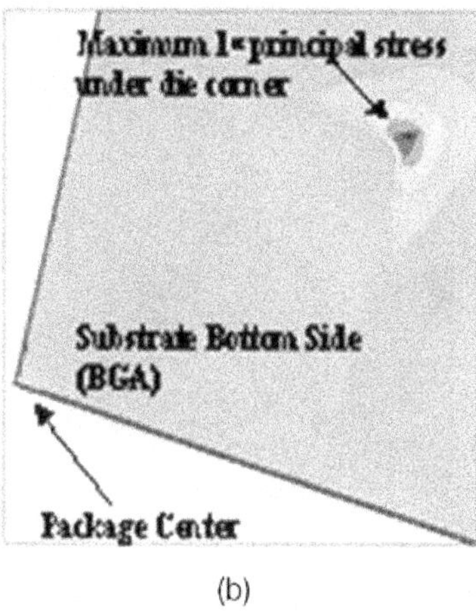

(b)

Figure 5.33 (*a*) 3-D stress contour plot of the 2-mm die/40-mm organic substrate test vehicle showing shape and pinch point at –55°C and (*b*) 2-D contour plot for the same case showing stress distribution in the substrate near the die corner.

The FEA work reveals that a stress maximum is observed in the same spot in which the failures were identified through our FA investigation. Additionally, a correlation is observed that the failures are sensitive to the properties of underfill used. To further investigate this apparent relationship, an FEA was completed to examine the maximum stress observed in the die corner region on the BGA side as a function of underfill material. Figure 5.34 shows the results of this analysis.

It is apparent from this analysis that the lower stress imparted in the substrate corners from material C provides a lower cracking propensity in the critical regions of the substrate (corners). We have chosen to examine the stress generated at –55°C since, in a linear elastic sense, this temperature represents the upper bound value of the stress in the substrate under the corners of the die. It is also the point at which all materials used in this system are the most brittle. To understand why we see differences in the relative performance of all the underfill materials, we need to examine the individual properties of the materials used in this portion of the work. Referring to Table 5.5, we can see that, for the most part, materials A and B are very similar. However, the materials differ from material C in two important ways. The glass transition temperature is much lower and, most importantly, the modulus above the glass transition temperature is very low for material C. These two features combine to yield a low stress-free temperature (the temperature which the structure becomes elastically loaded) and thus a lower stress at –55°C for material C. This lower

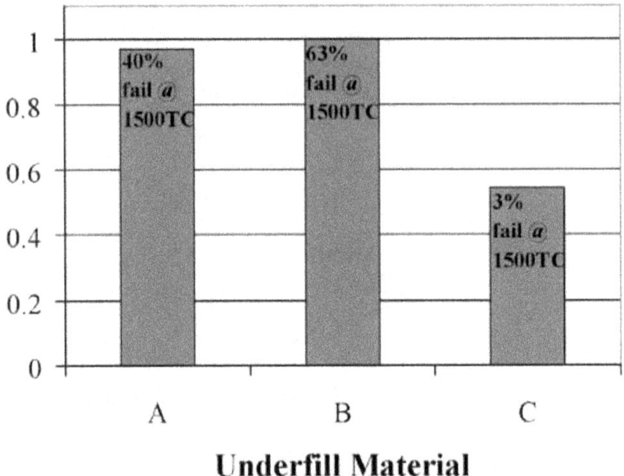

Figure 5.34 Normalized FEA-predicted package stress near die corners at −55°C for the 12-mm die/40-mm tester as a function of underfill material. Numbers on the chart are failure percentages at 1500 TCs.

stress helps alleviate the most prevalent failure mode in this test vehicle (corner cracking on the BGA side of the substrate), facilitated by simply switching to the underfill material with the best aligned material properties for the system.

The other failure mode operating in this test vehicle, underfill delamination at the die corners and subsequent bump failure, was prevalent only in test vehicles using underfill A, and only then after 1000 TCs. A comparison of FEA of the 12-mm die tester shows that the stress states and warpage associated with material A and material B are very similar (refer to Fig. 5.34 for the stress comparison). This is not surprising, since these materials have very similar material properties. The question posed is why an additional failure mode emerges for vehicles using material A. The answer almost certainly lies in the differences between the adhesion of these two materials to SiN. A previous study[1] showed that material A and material B have significantly different performances in a high-moisture, high-temperature exposure test. Material A was shown to fail due to interfacial delamination (between the underfill and SiN passivation of the test vehicle used in that study) during very early stages of test. Conversely, material B showed no evidence of delamination during the test. Although performance in this high-moisture test is not a true comparison of the adhesion values of these two materials to SiN, it does suggest a better interfacial integrity of material B to SiN than of material A to SiN. This difference does appear to be supported by the performance of 12-mm die testers assembled with materials A and B, especially for post-1000-TC results.

5.6.6.2 Large die test vehicle results. From the work presented in the previous sections, it is clear that the organic substrate technology that is the focus of this study is prone to one particular failure mode linked to the substrate alone and that another assembly-material-related mode may emerge if the wrong underfill material is chosen for the system. Additionally, since the 12-mm die/40-mm organic substrate vehicle work preceded the large die tester evaluation, we were able to modify the test vehicle design to more accurately capture potential failures linked to cracking of the substrate on the BGA side under the die corners. Section 5.6.1 detailed our design philosophy for this test vehicle. However, an important feature of this vehicle is reviewed here, especially based on the findings of the small die test vehicle work. To capture any failures linked to cracking under the die corners, a relatively high signal routing was chosen through this sensitive region. In this manner, even small failures would be noticed in the ATE very quickly. Additionally, since we could identify specifically which traces failed, we could isolate failures due to a known failure mode versus a new mode that may emerge in this new larger platform. Finally, even though the failure mode was known to potentially sever signal traces in this region, retaining them for the large die vehicle allowed us to identify assembly materials (specifically underfill) that would provide a robust assembly, even in the presence of the failure mode. More clearly, if a vehicle with routing through the corners passes stress testing with a particular material, we know this material to be part of an assembly solution for this failure mode. To this end, several materials beyond the work from the small die tester were evaluated to satisfy the expanded needs of the large die tester. The selection process for the subsequent reliability work was based on a processability evaluation in conjunction with FEA work using the candidate that most closely displayed the properties that were shown to be effective in reducing the failure mode from the 12-mm die/40-mm organic substrate tester.

Underfill processability results. The processability study was carried out using the 21.5-mm die/50-mm organic substrate tester as per Section 5.6.3.2. No material could be considered a candidate for subsequent reliability work if it was deemed unprocessable in any test vehicle. The materials considered candidates for this vehicle were chosen based on vendor recommendations and assembly subcontractor experience. The material properties of the four candidates are shown in Table 5.5. The results from the processability evaluation are presented in Table 5.6.

Table 5.6 Results of the Underfill Materials Processability Evaluation on the 21.5-mm/ 50-mm Organic Substrate Test Vehicle

Material	D	E	F	G
Avg. tray coplanarity	8.00	8.90	7.60	6.90
STDV	0.46	0.62	0.72	1.28
Post-MRT4/220°C (pass/total)	10/10	8/10	10/10	10/10
Post-TCB 500 (pass/ total)	10/10	N/A	10/10	10/10
Post-TCB 1000 (pass/ total)	10/10	N/A	10/10	10/10
Results	Retained for main DoE	Dropped for main DoE	Retained for main DoE	Retained for main DoE

All materials were observed to fill the test vehicle with minimum of voiding and filler settling. All materials passed X-ray inspection after MRT exposure except material E, which showed evidence of solder bridging after reflow, an indication of delamination of the underfill during MRT. This material was dropped from further evaluation. The results from the cursory temperature cycling using C-SAM inspection was encouraging (material E was removed prior to temperature cycling) in that no material showed evidence of delamination after 500 or 1000 TCs, which was linked to the second failure mode active in the small die tester. Figure 5.35 shows a 50-MHz pulse echo C-SAM image of material F after MRT and 1000 TCs. This image is typical of the results from all materials evaluated in the processability phase of the large die work. Additionally, the warpage behavior was somewhat sim-

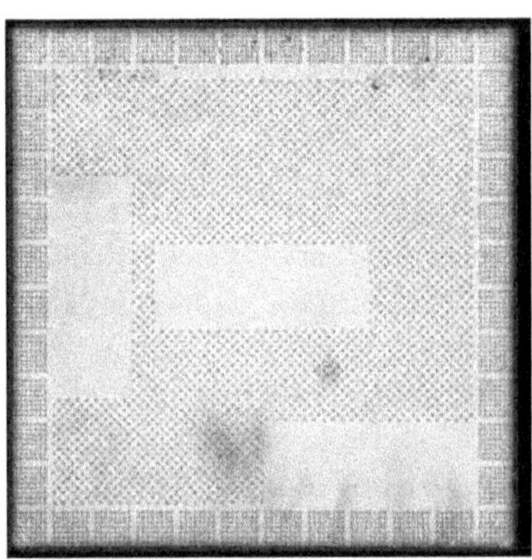

Figure 5.35 50-MHz pulse echo C-SAM image of a 21.5-mm die/50-mm organic substrate tester assembled with material F after MRT and 1000 TCs.

ilar, given the relatively small distribution in the average coplanarity data (from RVSI @ 25°C). However, some difference was observed for materials F and G. Since material E was dropped from the evaluation, all remaining materials were promoted to the main reliability phase of the large die evaluation.

Finite element analysis. Although it would appear obvious to incorporate the best-performing underfill material from our small die tester evaluation, this material was not considered for the 21.5-mm die/50-mm organic substrate tester. This decision was made based on vendor input that this material may lead to bump fatigue in a vehicle of this size. Referring to Table 5.5, material F is the most similar to material C based on the critical parameters of low T_g, relatively low CTE, and low modulus above T_g. Because of this similarity, this material was believed to be the best candidate for a reliable assembly for the large die test vehicle. A detailed FEA on this tester was completed using material F exclusively.

The detailed FEA was an important tool for understanding how the stress magnitudes and distributions, as well as warpage in the large die tester, would compare to the 12-mm die tester. Figure 5.36 shows the FEA-predicted package warpage comparison for both the small and large die tester using material F (and a 3-mm die-to-ring spacing). Also shown in Fig. 5.36 is a comparison between a vehicle with the same die size but with different body (substrate) size (21.5-mm die in

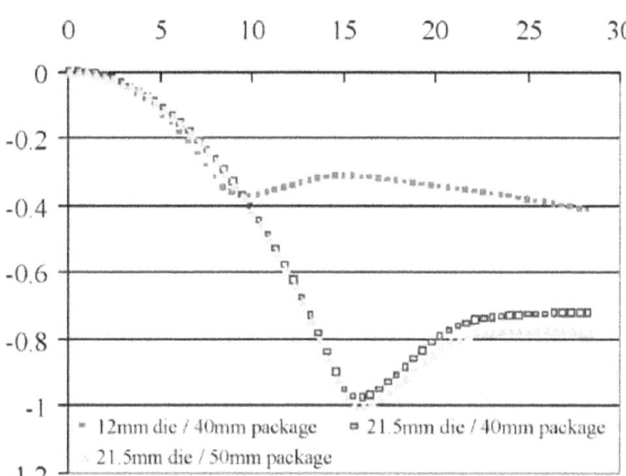

Figure 5.36 FEA-predicted package warpage for various die/test vehicle body size combinations assembled with underfill material

a 40- and 50-mm body). Clearly, body size plays less of a role on package warpage than die size. Table 5.7 shows the results of the predicted stress in the sensitive BGA corner regions under the die (based on 12-mm die/40-mm organic substrate work) at –55°C for all vehicles. This data shows that, as in the case of predicted warpage, die size has the strongest effect on predicted stress. Another important item to notice is the stress predicted in the substrate by simply increasing the die size from 12 mm to 21.5 mm. The stress increases by a factor of two in a region that has already been proven to be sensitive to stress levels, within one failure mode. This analysis also shows that changing the body size for one particular die size does not significantly impact the warpage or, in the case of Fig. 5.36, the stress under the die corners in the substrate. However, what is shown is that, even by using the underfill material most likely to provide increased reliability in this tester, a very large stress increase is predicted.

Table 5.7 Normalized FEA-Predicted Stress and Warpage Magnitudes for the Test Vehicle/Body Size Combinations from Fig. 5.17 (Using Material F)

Die size (mm)	12 × 12	21.5 × 21.5	21.5 × 21.5
Body size (mm)	40 × 40	40 × 40	50 × 50
Normalized predicted stress	1.0	2.0	1.9
Normalized predicted warpage	1.0	2.3	2.4

Reliability results. All devices from the large die DoE were submitted to reliability evaluation after assembly (as per Fig. 5.25). No visual defects were found, and electrical testing was conducted according to Fig. 5.25 and Table 5.4. No electrical failures were noted after MRT exposure. Figure 5.37 shows the reliability results from the thermal cycling portion of the evaluation. It is apparent that, even though the C-SAM evaluation from the processability evaluation showed no sign of delamination, all materials yielded massive electrical failures in this portion of the testing, although material F clearly showed a lower failure rate. C-SAM images on the failed devices were obtained and confirmed that, even on electrically failed devices, no delamination occurred. Only one material, material G, showed any sensitivity to HAST exposure, showing a 21 and 25 percent cumulative electrical failure at 96 and 168 hr, respectively.

Failure analysis. In the same manner as the 12-mm die/40-mm organic substrate test vehicles, the large die testers were submitted to failure analysis to determine the root cause of the electrical opens observed in

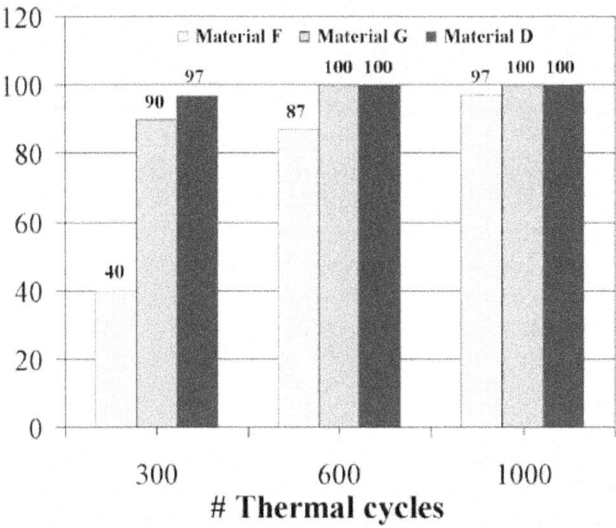

Figure 5.37 21.5-mm die/50-mm organic substrate temperature cycling results. (Numbers on columns are cumulative failure percentage values for each material.)

the ATE. The failed samples were first imaged in the C-SAM at 300 TCs and once again, although electrical failures were noted on the ATE, no delamination at the underfill/die interface was observed. This finding agreed with the observations during the processability evaluation. Successive C-SAM images for samples failing at 600 and 1000 TCs also showed little or no deterioration of the interface between the die and underfill. Also, no fillet cracking was noted on failed devices assembled with materials D, F, and G at any thermal cycling level. This indicates that the second failure mode active in the small die tester (fillet cracking followed by subsequent crack extension under the die eventually severing solder bumps) was not present in the large die tester assembled with these materials.

Due to the ability to accurately isolate failures in the large die test vehicle, the failure signatures from the ATE data were examined carefully. The data indicated that signal routing on the BGA side under the die corners had indeed failed. Flat sections of devices failing at 300, 600, and 1000 TCs were completed to determine the root cause of the failures. The flat sections revealed that the same corner cracking found in the small die tester was not only present, it was linked directly to the failures identified by the signatures from the ATE. Figure 5.38 shows an example of the flat sections completed on devices that failed during the various stages of temperature cycling. From this image, the corner cracks, enhanced with red dye, can be clearly seen to sever the traces that were intentionally placed in the corner region to determine the robustness of assembly material combinations in the presence of this known failure mode.

The device shown in Fig. 5.38 was assembled with material F, which showed the lowest failure rate of all materials tested. Failed devices from the other material cells were determined to have the same failure mode via the failure signatures observed in the ATE. It is apparent that, even though the stress level predicted (via FEA) for material F should be the lowest based on comparison of material properties of all underfills used in the assembly of the large die vehicle, it is still large enough to induce cracking in the sensitive regions of this vehicle (albeit at a lower rate). Although choosing a "low-stress" underfill material alleviates the corner cracking failure mode in the small die tester, the stress is obviously still above some critical level for crack extension when growing the die to 21.5 mm within the same substrate technology and a similarly low-stress underfill material, as predicted by FEA. Based on the 2× increase in stress predicted in the sensitive die corner regions on the BGA side of the substrate through the FEA work, the increased failure rate observed in TCB for the larger die vehicle is not unexpected.

5.6.7 Concluding Remarks

This case study outlined a detailed study on two separate flip chip BGA test vehicles aimed at examining the failure modes associated with stress testing of each vehicle assembled with various commercially available underfill materials. The goal of this work was to apply

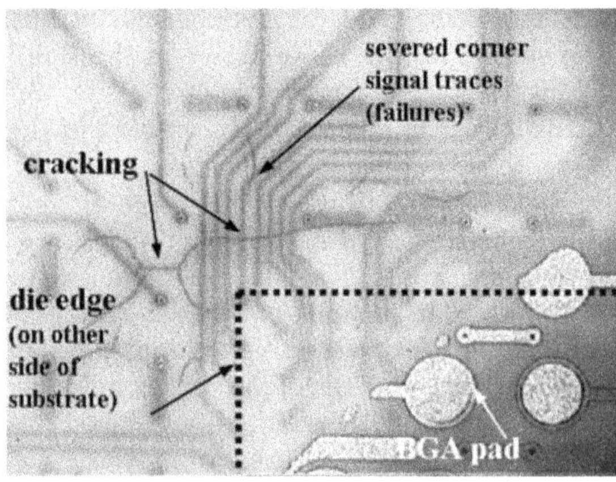

Figure 5.38 Typical horizontal cross section of the 21-mm die/50-mm organic substrate wafer TC, revealing he presence of corner cracking under die corners.

a systematic process of failure analysis on first generation test vehicles after stress testing to understand the root cause of the failure. Next, an in-depth material characterization on important assembly materials (underfill) was conducted to enable an FEA to both understand the nature of the failures in the first-generation vehicles and to influence the design of the subsequent large die test vehicle. Ultimately, the goal of both investigations was to enable our flip chip package portfolio to accommodate large die devices in a timely manner.

Two failure modes were discovered on the prequalification test vehicle, consisting of 12-mm die/40-mm organic substrate:

1. Cracking developed during temperature cycling in the dielectric on the BGA side of the substrate under the die corners, leading to severing of signal traces in this region.
2. Cracking occurred in the underfill fillet, extending under the die, severing solder bumps, beginning after 1000 TCs and increasing the failure rate while approaching 1500 TCs. (Delamination was detected at the 1000-TC point.)

The first failure mode is present in all test vehicles, whereas the second appears to be material dependent. Materials that show reduced stress (from FEA) in the sensitive BGA regions under the die corners reduce the corner cracking failure mode significantly. From our detailed materials characterization and FEA work, "low-stress" materials were identified, at least for this package geometry, to have low CTEs below T_g, low T_gs, and low moduli above T_g.

Armed with two known failure modes for this particular package/substrate type, a 21.5-mm die/50-mm organic substrate tester was designed to evaluate assembly materials and processes to enable the large die flip chip packages. This vehicle was designed to be easily analyzed during FA and during electrical testing. Signal traces were specifically designed in the sensitive corner regions to evaluate assembly solutions to at least one known failure mode. Material characterization was conducted on several commercial underfill materials, and at least one candidate was identified that showed the characteristics of a low-stress material as identified through the work on the small die tester work. Three materials were selected for a broad reliability analysis based on performance in processability evaluation (MRT, cursory TCB performance, and package warpage ranking). No evidence of failure mode no. 2 was observed from C-SAM imaging and inspection of fillet integrity. The reliability results from the main evaluation showed no failures in any device (or any material) after MRT. HAST testing showed only a low-level failure in devices built with one material, as a result of bump failure from delamination. Failure mode no. 1

was found to exist in all of the large die tester devices beginning at 300 TCs—even for the material that was identified by FEA and comparison between material properties (used in low TCB failure rate small die testers) to be the most likely to yield a reliable assembly. However, the FEA selection process and conclusions made from this work were supported in that the material chosen as "best" yielded the lowest failure rate in the tester reliability results.

Clearly, the substrate technology used in both testers in this study contains a potentially life-limiting failure mode, the rate of which scales with die size (or stress, which was shown to be a function of die size much more than body size). To extend the usefulness of this technology to large die flip chip packages, employing advanced materials does reduce the failure rate but will not be sufficient to arrive at a failure-free platform within the bounds of typical industry-wide qualification requirements. However, using a systematic combination of reliability evaluations on test vehicles, materials characterization, careful failure analysis, and finite element modeling, we have gained an in-depth knowledge of the challenges of packaging large flip chip die in one substrate technology can be gathered. In light of the failure mode discussed in this work, it would seem that a large die package in this substrate technology is not feasible. In reality, careful package design and additional steps to reduce the severity of the failure mode have proven successful in production versions of this technology with large flip chip die. However, this production capability would not have been possible without the knowledge gained through the study presented here.

5.7 Case Study No. 3: Multichip Plastic BGA

In the previous study, a failure mode was documented and well characterized prior to shifting to a package geometry that was very similar to the geometry in which the failure mode was discovered. The failure mode (substrate cracking and underfill delamination to the silicon during stress testing) was shown to be inherited by the new package geometry, and solutions were evaluated based on mitigating the known failure mode. In many cases, especially for new package technologies or for significant changes to existing package technologies, the failure mode (if any) is not known, and many of the tools discussed in this chapter and elsewhere in this text will be required to locate and engineer a solution for a particular failure. What follows is a case study where a significant change to an existing package technology introduced new failure modes that required a multitiered approach to help isolate and resolve the failures. This included proper use of inexpensive test dies, close scrutiny of the electrical test data, and use of the design files to aid the detailed FA process.

5.7.1 Device Description

Most multichip packages consist of at least two die, and in some cases several passives (devices containing passives will sometimes be referred to as multicomponent modules), arranged in a fashion that is consistent with the application and the product type. For instance, many MCPs are used in the cell phone sector because of the critical limitation on available board space. In this case, the die will usually be stacked in a way that allows for some manner of conventional interconnect process (wirebond, flip chip bumps) so that the footprint of the package is minimized. Stacked die packages are not limited to plastic, ceramic, lead frame, or substrate types, as instances of die stacking can be found in all of these package technologies. In other cases, MCPs are used to combine the functions of many pieces of silicon in one package. This solution is a higher-yielding, faster, and many times cheaper approach to system on a chip, and it provides footprint reduction as compared to many separate packages.

The device described in this case study is a 45 × 45 mm, four-die plastic overmolded MCP. Each die was identical (14.1 mm × 14.1 mm × 13 mils thick). The die were mounted in the six-layer BT (bismaleimide tiazine) laminate substrate and wirebonded with gold wire (wire size and pitch) with limited die-to-die interconnects. Table 5.8 summarizes the details of this test vehicle. This package presented a new packaging challenge in that it was the largest body size MCP on the market at the time, both the substrate supplier and assembly subcontractor had limited experience with a package of this size, and this experience preceded the product by only a few months. Figure 5.39 shows the die placement footprint (as well as wire locations) and body size for this test vehicle. Figure 5.40 shows a schematic cross section of a MCP device for reference.

Table 5.8 Device Details of the Case Study No. 3 Four-Die Plastic Overmolded BGA MCP Test Vehicle

	Device details
Device name	45 × 45 1152 I/O four-die MCP
Package type	PBGA
BGA grid	Fully pop. (except corners) 34 × 34 array, 1.27-mm pitch
Package size	45 × 45 mm
Die size (ea. × 4)	14.1 × 14.1 mm
Die attach/EMC	1 mil thick/1.17 mm thick
Substrate	6 layers; solder mask/Cu/BT, 0.61 mm thick

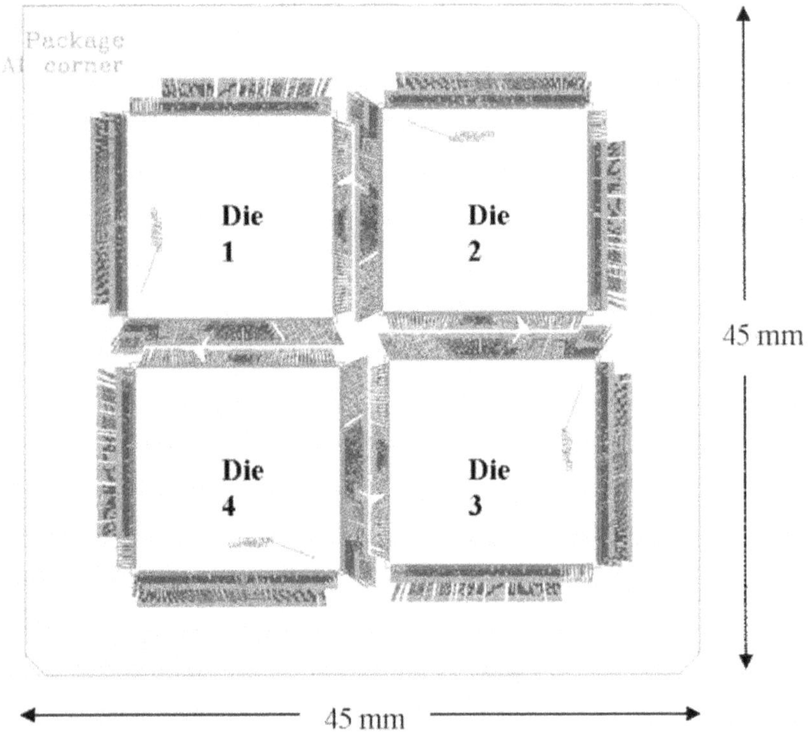

Figure 5.39 Die placement layout in four-die plastic overmolded BGA MCP.

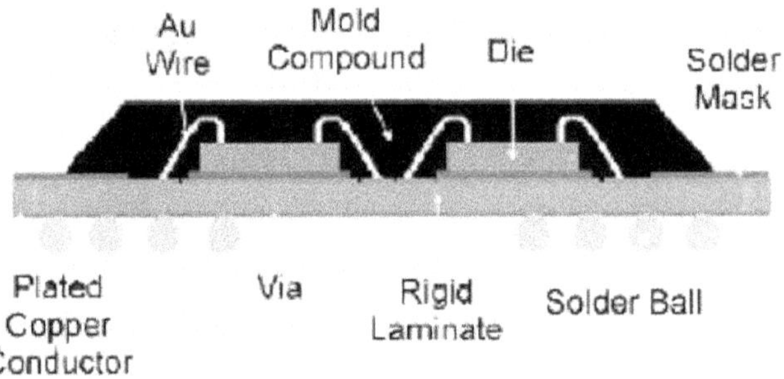

Figure 5.40 Schematic cross section of a four-die plastic overmolded BGA MCP.

Assembly of the device proceeded using conventional assembly processes and materials commonly used for single-die applications of similar body size (overmolded, PBGA). Figure 5.41 provides an image of the finished device. However, in anticipation of issues introduced by including such a large amount of silicon in a plastic package, a test ve-

Figure 5.41 Fully assembled four-die MCP described in case study no. 3.

hicle was designed to evaluate the available package material sets during qualification tests. This evaluation, coupled with finite element analysis of the package during thermomechanical stress and detailed FA, helped arrive at a reliable package available to the customer base in a short amount of time. What follows is a description of the tester used, the stress testing employed, the FEA used to help guide the FA and the materials selections, and detailed FA techniques that helped to understand the failures observed during the prequalification evaluation of the test device. All of these efforts employed tactics that are consistent with the tools outlined in this chapter in specific and this text in general.

Because the product for which this activity was initiated used very expensive die, the costs of doing package-level testing (which usually requires hundreds of devices) quickly became prohibitive. To combat this, techniques outlined in Section 5.4.4 were utilized in that a specialized die was used to create easily testable (opens and shorts only) conductive pathways in the existing substrate. In fact, the exact same structure outlined in Section 5.4.4.1 (wirebonded) was used in this case study. A single mask layer was changed in the commercial silicon mask stack, which, as described in the preceding sections, allowed for quick introduction of an easily and rapidly testable structure. For all testing completed in the following sections, this "daisy chain" tester was used to gather data and facilitate in product improvement work prior to the availability of a much more expensive product silicon and prior to costly and time-consuming qualification work.

5.7.2 Reliability Test Program

To gather as much information as possible on this new packaging technology before proceeding to full qualification, a series of prequalification evaluation tests were run on the daisy chain test vehicle. The reliability testing program is summarized in Fig. 5.42. The reliability tests are the same as described in Table 5.4. All other testing was conducting in the same manner as described in Section 5.6.2.

5.7.2.1 Results from prequalification evaluation. In the first phase of this work, a sample size of only 44 devices, representing the baseline materials and processes available for this packaging technology at the assembly subcontractor, was submitted to the reliability testing program just described, except only two test points were recorded during temperature cycling. This process acted as a screen for any potential issues. The results of this work are captured in Table 5.9. At the 1000-cycles test point, the failure level was 57 percent. The product reliabil-

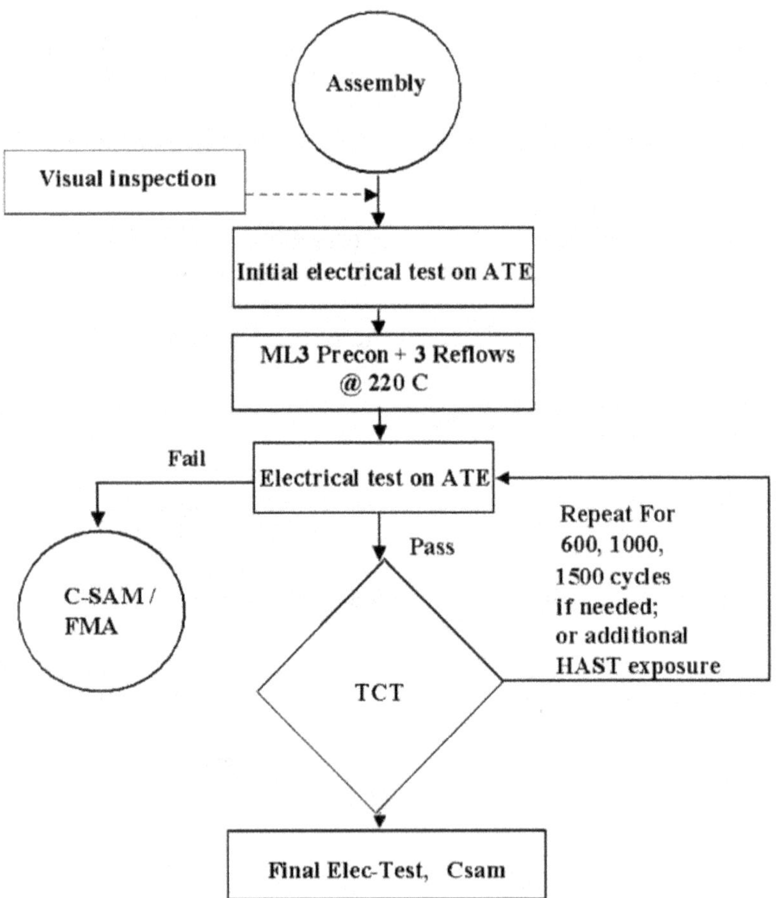

Figure 5.42 Reliability test program for case study no. 3.

Table 5.9 Results from Screening Work from Case Study No. 3

Test	Level	E-test results (failed/tested)	C–SAM observations	Comments
MRT (JEDEC level 3 [60°C/60% RH for 52 hr] reflow temp. = 225°C)	Moisture exposure + 3 reflows	0/44	No chip surface or D/A delam	No anomalies noted in optical inspection
Temperature cycle B (–55 to 125°C)	500 cycles	0/44	No chip surface or D/A delam	Some surface (solder mask) cracking noted on BGA side of substrate
Temperature cycle B (–55 to 125°C)	1000 cycles	25/44	No chip surface or D/A delam	Extensive surface (solder mask) cracking noted on BGA side of substrate

ity goal was to pass 1000 temperature cycles failure free to guarantee the required service life according to the end-use conditions for the product. Clearly, additional development work was required to improve the survivability of the device to power cycling that correspond to temperature cycling during field use.

5.7.2.2 Failure analysis. The first step in developing a more robust packaging solution for this product is to conduct a detailed failure analysis on the devices failing the initial screening evaluation. As noted in Table 5.9, the visual inspection results for the screening work revealed a large number of cracks on the BGA side of the device, some showing up at 500 temperature cycles and growing in number and severity at 1000 TCB. Figure 5.43 shows an optical micrograph of the cracks that were noted during temperature cycling. These cracks were obviously the first clue as to the potential failure mechanism operating during this evaluation. The cracking was noted to occur randomly under the die regions, in many cases emanating from the etchback holes (openings used to sever the electrical connections required during the electrolytic plating of the substrate metal layers) or from regions directly above buried vias.

Following the many techniques presented in this chapter to isolate failures in the device, the first step was to collect the error logs from the automatic testing equipment. Once completed, the failed nets could be identified, and the destructive analysis could begin with the design file used to guide the process. Unfortunately, the cracking in this case was quite widespread such that no patterns could be identi-

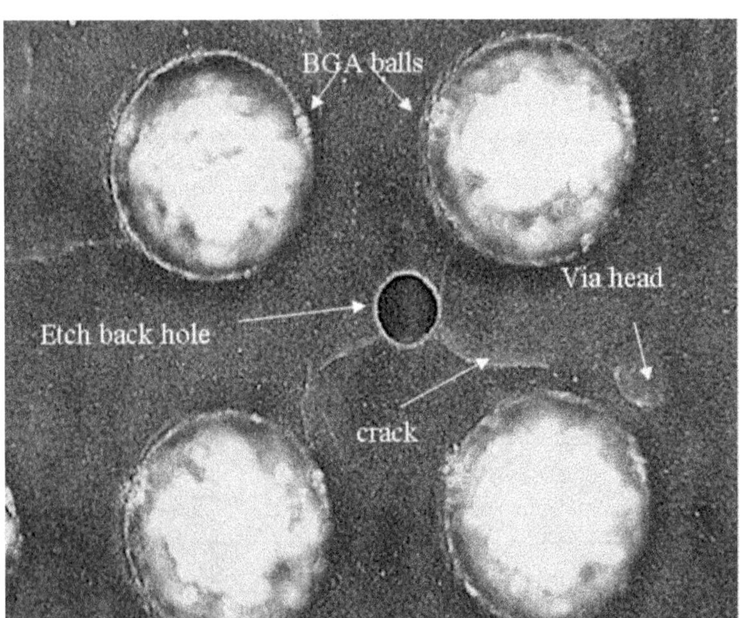

Figure 5.43 Optical micrograph of the BGA side of the four-die MCP test vehicle after 1000 TCBs during the screening evaluation.

fied as far as the general location of the failure was concerned. It was noted that failures were occurring in nets that contained routing on only top or bottom metal layers, indicating that the failure was occurring through the entire thickness of the substrate. One failed net was identified as a candidate for destructive analysis, as this trace was located centrally among many failed nets and both routing through a high-stress region (die corners) and only on one metal layer; i.e., after wedge bond connection, the trace is routed only to a via on one layer, and the via then drops down to a BGA ball. Figure 5.44 shows the second metal layer from the design file for the test vehicle. The failed net is highlighted, revealing the single metal layer routing. (No other layers are visible in this image, and no other routing on other layers was used.) Using the design file with the net highlighted in this fashion allows the failure analysis professional to immediately narrow the focus of the root cause investigation to the critical elements of this net on the proper layers (i.e., wires, die pads, wedge pads, and so on.)

Prior to destructive analysis, the first step in the root cause investigation was to obtain a C-SAM image on the device to determine if any chip surface or subsurface delamination had occurred. Figure 5.45*a* and 5.45*b* show both a 50-MHz pulse echo and through-mode image, respectively, of the device in question. No chip or subsurface damage

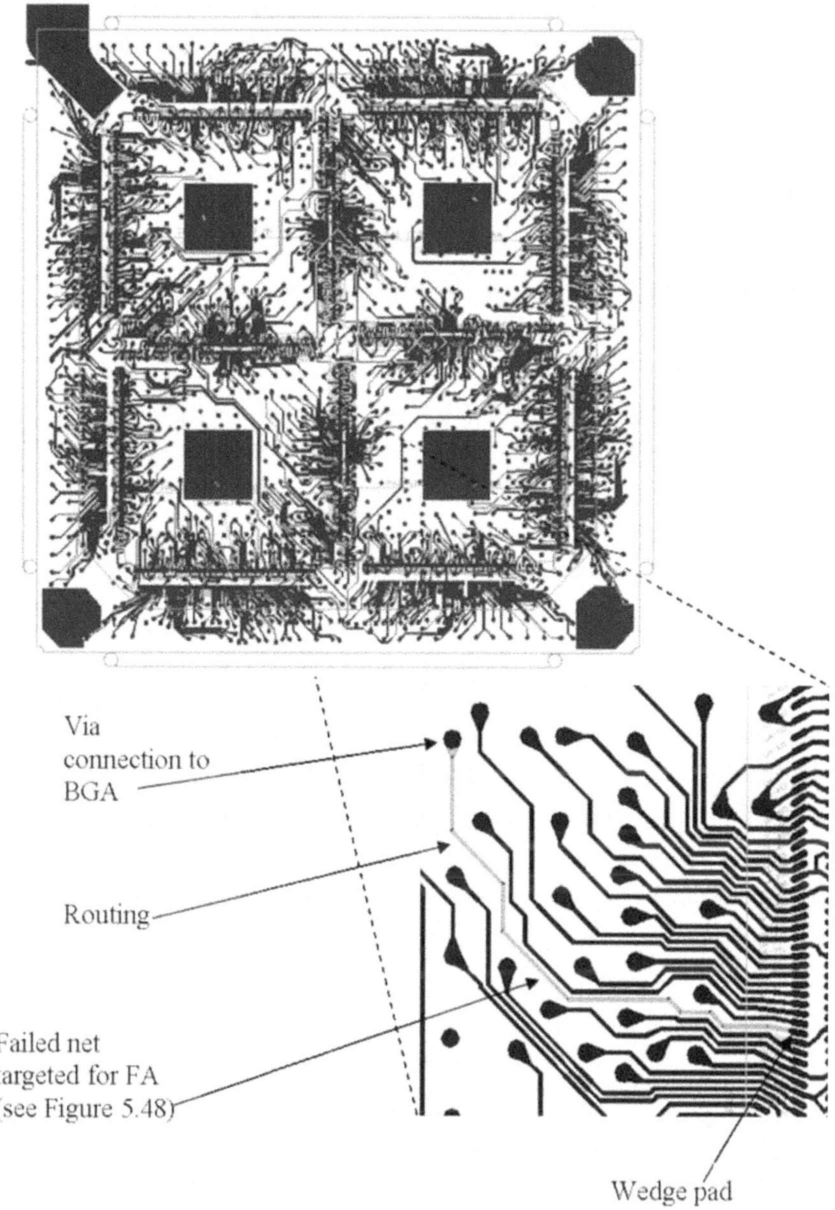

Figure 5.44 Design file for metal layer no. 2, revealing conductor structure and routing for one failure from a post-1000 TCB device from the screening evaluation of case study no. 3.

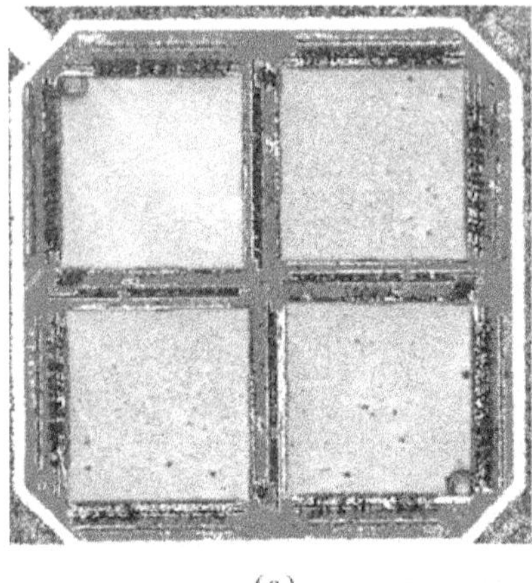

(a)

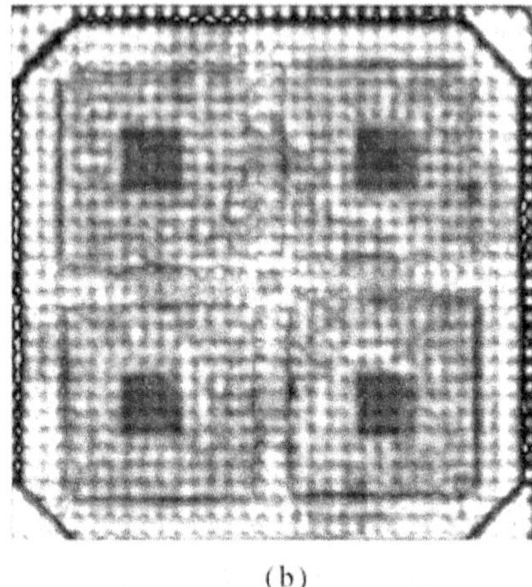

(b)

Figure 5.45 50-MHz C-SAM images after moisture preconditioning (JEDEC level 3 [60°C/60% RH for 52 hr) followed by three reflows at 225°C) and 1000 B temperature cycles) in (*a*) pulse echo mode showing intact die surfaces and (*b*) through-mode revealing intact substrate, die, and die attach layers.

was detected. Additionally, other than the BGA side cracks, no other physical defect was noted through optical inspection. This finding then leads the failure analysis process in the direction of an internal failure mode that is not detectable through a nondestructive imaging technique.

The device was then subjected to destructive analysis in a similar fashion as described in Section 5.6.4. The device was subjected to two separate processes to better understand the nature of the failure. One process was to cross-section the sample and view the region near the failed trace in an edge on manner. The other process was to horizontally section the device to expose the failing trace from the top and from the bottom to look at both the via structure (following the visible cracking from the BGA side). Figure 5.46 shows an image from a horizontal cross section of the BGA side near the BGA pin that was associated with one of the failed nets. In this case, the solder mask has been removed to reveal subsurface cracking that appears to emanate from the via land pads. This finding suggests that the cracking originates at the surface and extends below to the via and the lower metals layers, or that the cracking originates below the substrate surface (near vias) and extends to the surface and to lower layers. In either case, based on the observation (from data in Table 5.9, where cracking at the surface appears well in advance of the actual electrical failures), it is clear that the cracking in the dielectric layers occurs first, and the metal traces are fatigued to failure as a result of additional stress caused by the temperature swings. Figure 5.47 shows a close-up of the

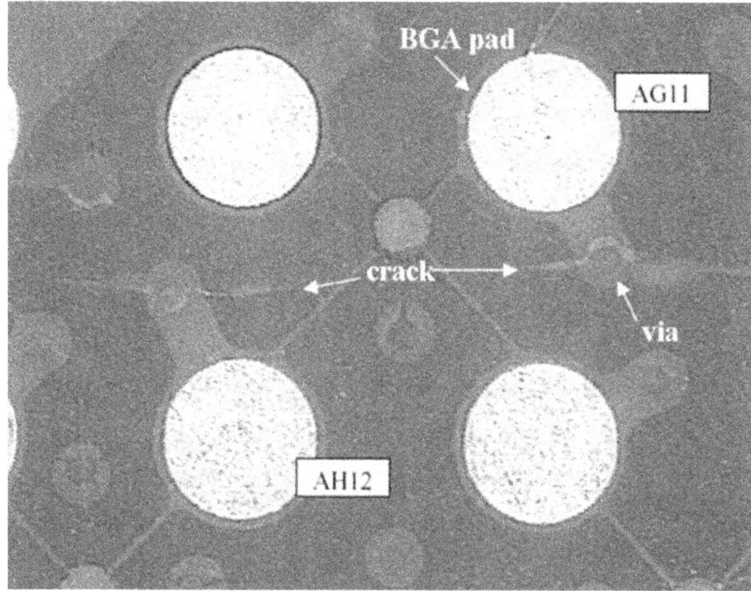

Figure 5.46 Horizontal cross section of a failed device, from the screening portion of case study no. 3. The image is taken from the BGA side with solder mask removed, revealing cracking near the failed (and intact) BGA coming from the via locations.

via in Fig. 5.46, which is part of the failed trace in question. Here we see that the crack does indeed sever the connection from the BGA pad and the via.

The sample was then polished from the other side (die side) down to the second metal layer to reveal the routing for the failed trace. Figure 5.48 shows this horizontal section with the metal routing for the trace in question (as seen from the design file from Fig. 5.44). No cracking is observed in this routing. The failure for the net in question was linked to the cracked metal via connection on the BGA side of the substrate. A traditional cross section was taken at the dotted line shown in Fig. 5.48 to evaluate the nature of the dielectric cracking observed near the via connection in this intact trace. Figure 5.49 shows the cross section of the cracked dielectric region near the via of the trace near the failed trace for this device. It is clear that the dielectric cracking near the vias is the root cause of the failures. Because the cracks in the solder mask levels (tops and bottoms) of the substrate appear to happen first right above the via locations, the hypothesis initially presented as to the cause of the cracking (an integral part of the full root cause disclosure) was that there was a local CTE mismatch of the vias and the dielectric material, causing local stress buildup during temperature cy-

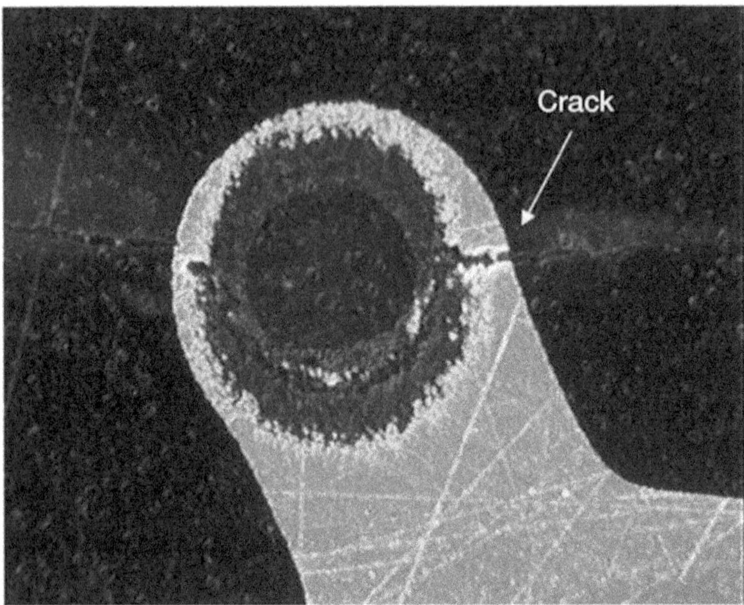

Figure 5.47 Close-up of failed BGA pad shown in Fig. 5.46.

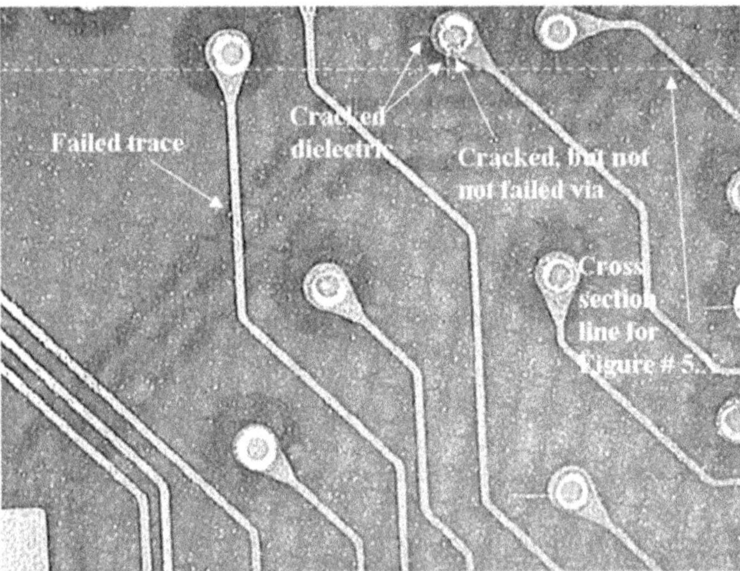

Figure 5.48 Horizontal section revealing failed metal trace from post-1000 TCB failure from screening work. Cracking in the dielectric and metal via connection in a nearby intact trace is also observed.

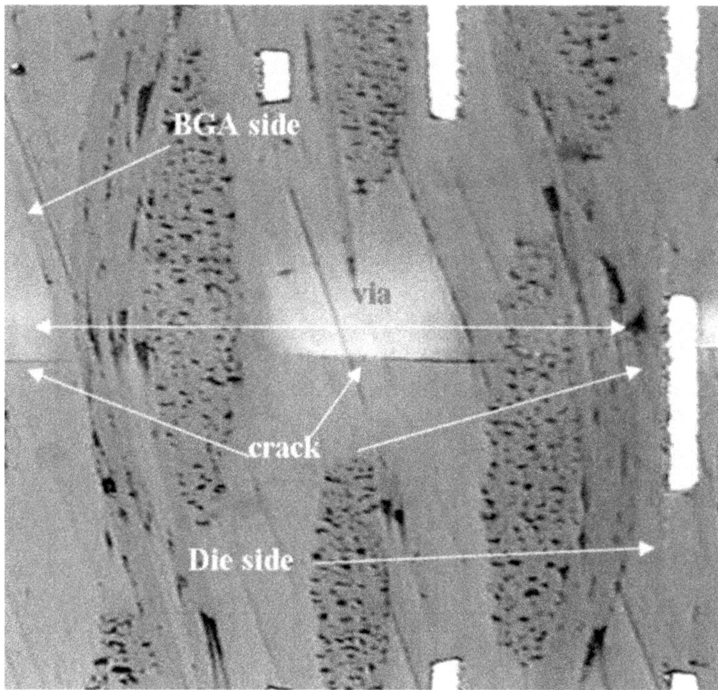

Figure 5.49 Traditional cross section taken from horizontally sectioned device in Fig. 5.8 at line indicated in the figure. Section shows cracking close to vias through the entire thickness of the substrate (in this case, not severing copper traces) in a region near the failed net from the post-1000 TCB failure of Fig. 5.48.

cling. This stress resulted in cracking, first in the dielectric and the solder mask layers and then in the metal layers near the via. To further understand how to address the local stress buildup issue, several assembly and materials solutions were proposed, building on the hypothesis that the cracking was indeed linked to the stress buildup near the vias. The following section shows how FEA can be used as a screening tool to evaluate potential corrective actions once a root cause has been established. In this case, the work was done in conjunction with a designed experiment whereby the via cracking hypothesis was tested and elements of the finite element analysis were experimentally verified (as corrective actions to the via cracking if indeed the hypothesis was correct). In some cases where a root cause is more clearly and certainly defined, FEA alone may be used as a quicker means to evaluate corrective actions. However, the value of experimentally verifying the root cause cannot be dismissed, and FEA often can be a very useful complementary tool in narrowing the choices evaluated in any experimental investigation. This is exactly how FEA was used in this case study.

5.7.3 Finite Element Analysis

To support the experimental evaluation, several potential assembly solutions were established that presumably would reduce the stress on the vias in the package technology used for this MCP. The working hypothesis for the root cause was local stress created by CTE mismatch of the materials near the via structures, resulting in fatigue cracking (since this is a late-stage failure, the fatigue element was hypothesized as well) in the metal conducting layers during temperature cycling caused by the lack of support in the metal structure resulting from the presence of the crack. To evaluate and rank the effectiveness of possible materials and assembly process changes to the standard assembly solution, a matrix of possible changes was constructed and is presented in Table 5.10. The properties of the new materials, along with the properties of the nonvaried package materials, all of which were used to construct the FEA model, are presented in Table 5.11.

Table 5.10 Process and Materials Varied in the FEA Simulation Work for Case Study No. 3

Cell	Substrate thickness (mm)	Die thickness (mils)	BT	EMC	Die attach	Evaluated in DoE
A*	**0.61***	**13***	**A***	**1***	**X***	**X***
B	0.8	13	A	1	X	
C	1	13	A	1	X	
D	0.61	8	A	1	X	X
E	0.61	13	A	2	X	
F	0.61	13	A	1	Y	X
G	0.61	13	B	1	X	X

*Baseline materials (Cell A).

A 3-D finite element analysis on the package was then completed, evaluating the stresses at both 25 and –55°C (the latter because this is the lower end of the temperature cycle and the highest stress point for a linear elastic evaluation). Because the layout of dies is nearly symmetric, only one-fourth of the package needed to be analyzed. The analysis was carried out using ANSYS® v. 7.0. using a linear hexahedral solid element meshing process to build the model. Figures 5.50 and 5.51 show the finite element meshes used and the global model and submodel, respectively (submodeling is used to obtain more accu-

Table 5.11 Material Properties of the Baseline (Bold Italic) and New Assembly Material Options Used in the FEA Work Case Study No. 3; Material Properties of the Nonvaried Packaging Materials Also Shown

Components	Type	E (GPa)	α (ppm/°C)	T_g	Poisson's ratio, ν
Silicon chip	Silicon	131	2.8	–	0.23
Solder mask	AUS-303	0.35	60/160	103	0.30
Trace plating	Copper	130	17	–	0.30
BT	***A***	***2.2***	xy/z direction ***15/60***	–	***0.30***
	B	2.4	15/45		0.30
EMC (molding compound)	***1***	***26.5***	***8/34***	***135***	***0.30***
	2	29.42	15/50	120	0.30
Die attach	***X***	***1.5 @ 25°C***	***65/200***	***60***	***0.30***
	Y	1.5 @ 25°C	70/130	31	0.30

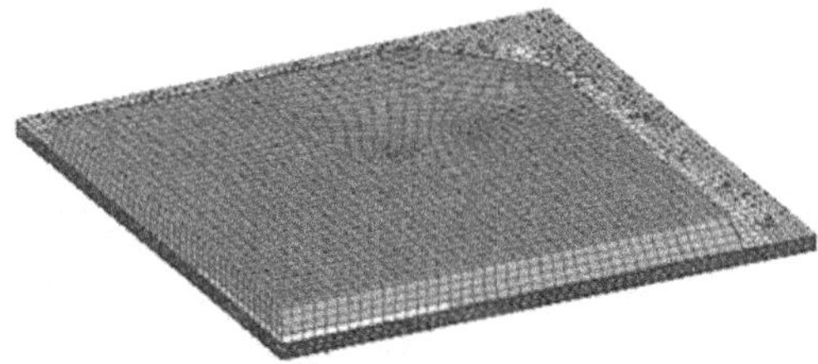

Figure 5.50 Global, quarter-symmetry FEA model used to evaluate package materials and process choices outlined in Tables 5.10 and 5.11 for the four-die MCP in case study no. 3.

racy for smaller structures, such as a via, in the global model). In these figures, different shades represent different package materials (molding compound, substrate, and so forth). Stresses were analyzed on the via at the location shown in Fig. 5.52, at the two temperatures of interest.

Several stress indices are commonly used to investigate the possibility and locations of mechanical failures. Among them, the von Mises equivalent stress, is defined by

$$\sigma_{eqv} = \sqrt{\frac{1}{2}[(\sigma_x - \sigma_y{}^2 + (\sigma_y - \sigma_z)^2 + (\sigma_x - \sigma_z)^2 + 6(\sigma_{xy}^2 + \sigma_{xz}^2 + \sigma_{yz}^2)]} \quad (5.1)$$

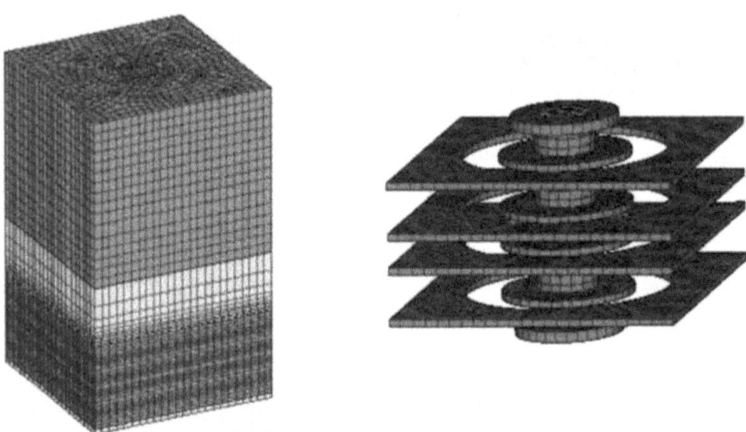

Figure 5.51 Local submodel showing the via structure used in the FEA work from case study no. 3.

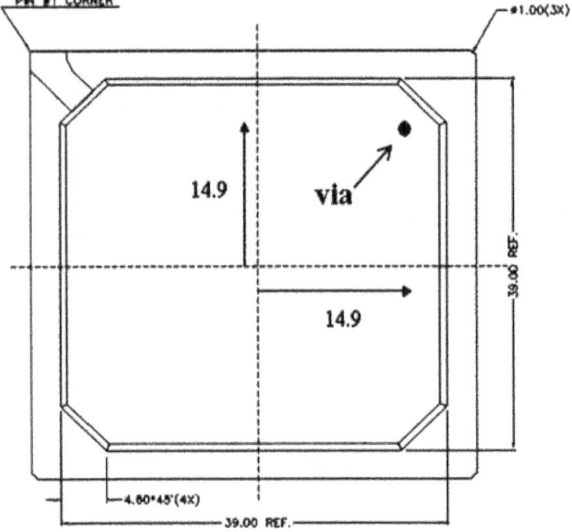

Figure 5.52 Location of the via, using the package outline drawing as a reference, for the FEA evaluation in case study no. 3.

It is closely related to material failure criteria. Without proper yielding criteria it is not possible to tell if failure actually occurs, but the equivalent stress is still a potential indicator and was used in this analysis to rank the contribution of the various elements. In Fig. 5.53, we see an example of the contour plots representing the distribution of the equivalent stress on the copper layer for different cells. The maxi-

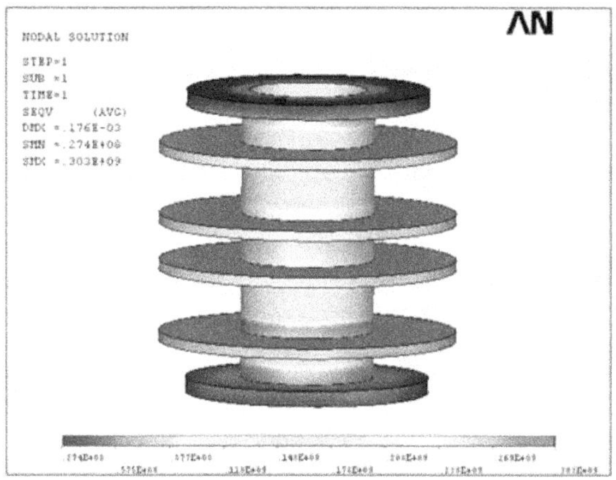

Figure 5.53 Representative contour plot showing an example of he equivalent stress distribution for the FEA work detailed in case study no. 3.

mum equivalent stresses for different cells are summarized in Fig. 5.54. It is clear from this analysis, disregarding synergistic effects, which were not evaluated, that the most effective means to reduce the stress near the vias (and specifically on the vias themselves) is to move to the BT core material "B" outline in Table 5.11.

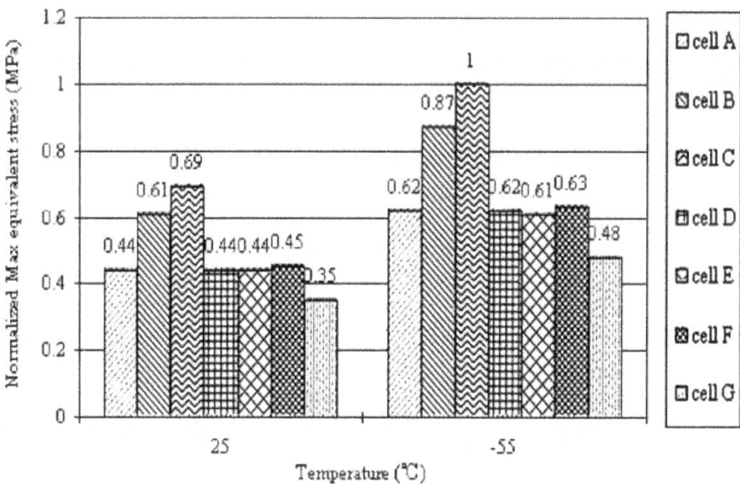

Figure 5.54 Maximum equivalent stresses predicted by FEA on substrate vias for cells A–G from Table 5.10.

Table 5.12 Reliability Results for the Verification DoE from Case Study No. 3

Temperature cycle B cumulative failure percentages								
		Temperature cycling level						
Cell	Post-MRT failures	200 TC	500 TC	700 TC	800 TC	900 TC	1000 TC
A	0	0	0	0	0	13	31
F	0	0	33	33	33	33	83
D	0	0	0	0	7.1	7.1	7.1
G	0	0	0	0	0	0	0
Screening data	0	0	0	No data	No data	No data	57

5.7.4 Design Of Experiment

To verify the results of the finite element analysis and to verify the hypothesis that the cracking in the substrate was indeed due to the expansion and contraction of the regions near the vias, an experiment was designed to test these elements. However, it was clear that certain elements contained in the finite element analysis matrix did not need to be evaluated, since the responses were predicted to make the stresses on the vias higher. To this end, only certain cells were evaluated in the experiment such that the DoE was a much less than full factorial design. This, once again, pointed out the usefulness of the FEA approach to minimize costs and effort in subsequent verification work (many interactions were ignored in both the DoE and the FEA matrix, since many combinations were not possible in actual manufacturing). Referring again to Table 5.10, the legs of the FEA matrix that were verified by actually building samples and running them through a follow-up reliability analysis, in the same manner as Section 5.7.2, are indicated in the last column.

5.7.4.1 DoE results. Table 5.12 shows the results of the reliability analysis from samples built as a result of the FEA work outlined in Section 5.7.3. From the FEA work it was concluded that moving to the lower CTE BT material in the substrate would indeed reduce the cracking near the vias resulting in the opportunity for fatigued metal and subsequent trace failure. This conclusion was borne out in the verification DoE, since no failures were observed in samples constructed according to the properties from cell G, whereas the control cell, cell A, did indeed show electrical failures (a comparison to the screening work is provided in the last column of Table 5.12). The failure rates between the control cell and the screening data do not agree and may have to do with the variation in the flaw distribution from one sample lot to another. However, the fact that no electrical failures are observed in the reduced CTE BT cell (cell G) at 1000 TCB does lend credence to the verification, since this agrees with the predictions of the FEA work (a 23 percent reduction in the via stress distribution is predicted by utilizing the lower CTE material over the standard process). Also, further credence is given, since visual inspections reveal that no cracking was observed on the BGA side of the cell G samples, where BGA side cracking was observed for the control samples, in the same fashion as detailed in Fig. 5.43. This final observation also provided strong evidence that indeed our hypothesized root cause was confirmed (local CTE mismatch of the vias and the dielectric material causing local stress buildup during temperature cycling, cracking first in the dielectric and the solder mask layers and then in the metal layers near the via). Since no cracking on the solder mask was observed in the passing cell G, it was concluded that eliminating the expansion near the via eliminates the surface and subsurface cracking, removing the possibility of subsequent fatigue failure of the copper traces, eventually causing failures. Therefore, the corrective action for this failure was to implement the new lower-CTE substrate material.

5.7.5 Concluding Remarks

This section described an important application of many of the techniques outlined in this chapter. It started from a failure noticed in an early development spin and progressed all the way to a root cause disclosure and the all-important verification stage, providing proof of the effectiveness of the corrective action as determined, in part, by a detailed finite element analysis (the verification would be essential for implementing the corrective action to the product line). The use of many of the tools from this text not only reduced the cost of the work, it also reduced the development time dramatically. Both are essential in effective product development in today's market.

5.8 Conclusion

When pushing the limits of the product development envelope in the microelectronics industry, it is inevitable that failures will result from the introduction of new packages, or combinations of new materials and/or processes, to existing packaging technologies. In the past, the market would tolerate somewhat lengthy product development intervals, allowing for detailed step-by-step, serial engineering work. However, with the increasingly short time to market demands placed on companies supplying the microelectronics sector, different tactics must be adopted to provide robust products in the reduced budgetary and time frame environments in which we find ourselves today. It is clear that we must adopt a multifaceted, parallel process using the tools outlined in this chapter to quickly identify and correct product reliability and performance issues so as to compete. This chapter has focused on the many external and internal clues that must be followed to successfully accomplish the task of locating, and at times correcting, failures. Other sections of this book (specifically, Chapter 6) will focus more on the mechanics of isolating the failures within the devices (much as was accomplished to a small degree in Sections 5.6 and 5.7) once a defect has been located.

5.9 References

1. Goodelle, J. P., J. J. Gilbert, and R. E. Fanucci, "Characterization of Underfill Materials for High Reliability Applications," *Proc. 50th Electronic Components and Technology Conference,* 2000.
2. Nguyen, L., et al., "High Performance Underfills Development—Materials, Process and Reliability," *Proc. 1997 PEP,* 1997, p. 300.

Chapter

6

Failure Analysis Techniques

Barry J. Dutt
Albert C. Seier
Anthony J. Bucha
James T. Cargo
Joseph W. Serpiello
Ronald J. Weachock
Frank A. Baiocchi
Joze E. Antol
John M. DeLucca
Kultaransingh N. Hooghan
Richard L. Shook
Huixian Wu
Weidong Xie

6.1 Imaging Techniques

The expression "a picture is worth a thousand words" is never truer than in a failure analysis laboratory. A well composed image of a defect site on a carefully prepared sample can provide the solid evidence required to ascertain the nature of the failure and, often, its root cause. In the following section, a variety of common imaging techniques based on visible light, X-rays, sound, heat, electron beams, and ion beam interaction with materials are discussed. These are the basis of important tools found in the failure analysis laboratory, such as the optical microscope, real-time X-ray imaging camera, acoustic microscope, infrared camera, scanning and transmission electron microscope (SEM and TEM), and focused ion beam (FIB) equipment. In addition, some techniques specifically used in package analysis are described, ranging from shadow moiré to image mechanical warpage, SQUID microscopy to image current flow in a powered device, dye penetrant leak detection to image physical leakage paths in hermetically sealed packages, and freeze/thaw techniques to isolate hot spots at the package level. While this is not an exhaustive discussion of every imaging technique available (e.g., the various surface scanning microscopy techniques are not discussed, except for scanning SQUID microscopy), it does cover most of the important ones for package analysis.

6.1.1 Optical Microscopy

Optical microscopy is a basic and indispensable tool in the failure analysis laboratory. Most labs possess a variety of microscope types, which are suitable for different applications. The optical microscope is used extensively throughout all phases of analysis, from initial inspection through various stages of deprocessing. In addition, optical microscopes are integrated with other instruments used in the failure analysis laboratory, such as electrical probe stations and fault isolation tools, to help locate and document specific sites on the failed device. Physical analysis should begin with a thorough optical inspection of the device as received. It is important to note the condition of the part before any further handling and during the course of analysis. During inspection, one should be mindful of the electrical failure signature as well as any other diagnostic test results (TDR, SAM, X-ray, and so forth) to aid in defect localization. Visual inspection using a low-power optical microscope can highlight any external physical anomalies such as codemark issues, bent or missing package connections (leads, balls, bumps, or landings), "popcorn" effects (i.e., internal delamination), evidence of overheating, cracks or gouges, and planarity issues. If the device fails for shorts or leakage, special attention should be paid to the material around the leads or connections on the package. Opens can be caused by misaligned connections, missing or undersize balls/bumps (in the case of BGAs), or debris on the external conductors, as well as any internal package issues. Evidence of thermal damage can appear as discoloration, cracking, or warping of the package.

Optical inspection may also include various tilt angles (typically for low magnification) to better identify and highlight any anomaly. High-power inspection (500× up to 5000× with multipliers) may also be necessary to fully examine an area of interest. More advanced optical techniques, such as confocal microscopy, contrast enhancements (i.e., differential interference contrast), polarized light/filters, and ultraviolet illumination, can provide enhanced detection capability. Water or oil immersion lens objectives also offer advantages in resolution due to a better numerical aperture for a given magnification. Immersion lenses also add the advantage of improved light transmission and the ability to view through oxides. Water immersion is preferred for its easy cleanup with just an air duster.

Another important consideration with optical microscopy is the ability to capture and enhance any image acquired in the microscope. Digital image capture will provide increased resolution over video capture (at the expense of file size), while a camera controller with enhancement capabilities (such as contrast, brightness, gamma, sharpness, color shift, and annotations) will aid in documenting any observed anomaly. Photographic film and cameras are still used in some cases where excellent resolution is required. The disadvantages of film are the need to scan the images to transmit them electronically and excessive cost with high-volume use.

Many failure mechanisms, such as delamination, lifted ball bonds, internal package shorts/opens/leakage, warpage, and solder bump anomalies, may require deprocessing of the package for further investigation. Deprocessing techniques are described in Section 6.4 and include chemical etch, dry etch, cross-sectional mechanical polish, planar polish, and mechanical separation techniques to expose the area of interest. An optical microscope inspection after each deprocessing operation is essential for proper removal of material and identification of possible anomalies. A clear, high-contrast optical image of the failed site is often all that is needed to help elucidate the failure mechanism.

6.1.2 Real-Time X-Ray Analysis

The real-time X-ray system is one of the key tools for performing nondestructive analysis of integrated circuit packages. With typical imaging resolution of about 5 μm and magnification up to 200×, the tool is well suited to the observation of structural detail within packages. Most X-ray systems have multiple axis control (x, y, z, tilt, rotate) to allow the sample to be viewed from various perspectives. In addition, beam power can be controlled (HV up to 160 kV or higher with beam power up to 10 W) to allow viewing of samples with various densities. The X-ray penetration depth is proportional to beam voltage. Higher-power systems are necessary for packages that incorporate copper heat slugs and so on. A typical system is shown in Fig. 6.1.

Real-time X-ray systems are used to inspect and evaluate basic assembly characteristics such as wire sweep and wirebond integrity, including ball and wedge bonds, solder bumps in flip-chip packages, evidence of electrical overstress (EOS) damage to wires, shorted/bent lead frames, proper bonding connections, package voids/cracks, and die attach coverage. Several examples of package failures identified by X-ray imaging are shown in Figs. 6.2 through 6.4. Gross die-to-package delamination can also be viewed in the real-time X-ray system when observed parallel to the die surface, although the scanning acoustic microscope is much better suited for this application. In the X-ray system, gross delamination would appear as a white gap line at

Figure 6.1 FeinFocus® FOX-160.25 Nanofocus X-Ray Inspection System.

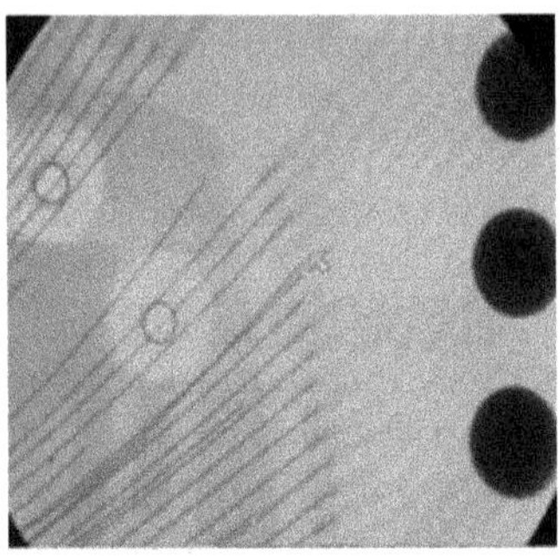

Figure 6.2 Improper wedge bond placement.

the delaminated interface. When determining wire-to-wire or wire-to-lead-frame clearance, it may be necessary to rotate the sample to ensure that bond wires do not appear to short to each other or to the lead frame. The sample may need to be oriented in an upside-down position to provide correct pin sequence on the monitor for certain X-ray systems. Digital image enhancement algorithms and image capture capability are common on most real-time X-ray systems to enhance and store the images to a local or remote database. X-ray image applica-

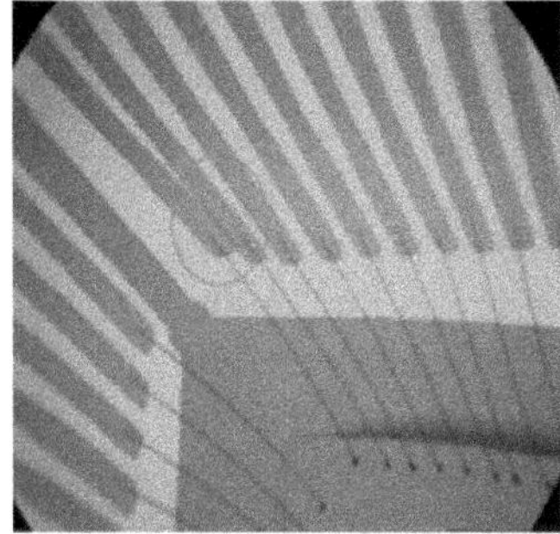

Figure 6.3 Lead frame short.

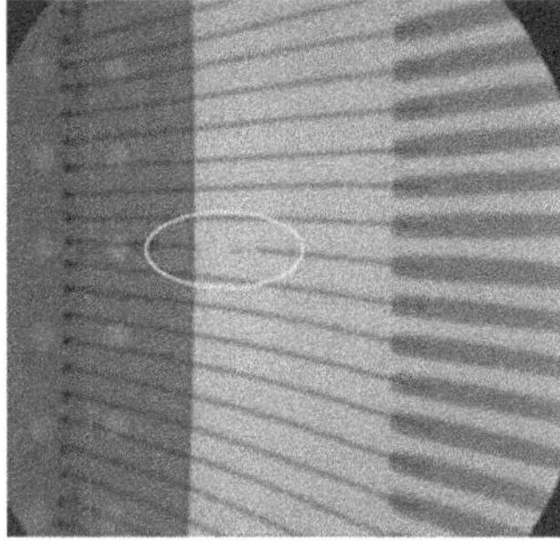

Figure 6.4 EOS-damaged wire.

tion programs are also available to analyze such parameters as percentage of die attachment voiding, wire sweep, and size and number of BGA solder balls.

6.1.3 Scanning Acoustic Microscopy

Scanning acoustic microscopy (SAM) is widely used for failure analysis (FA), reliability testing, qualification, and process control to inspect for delaminations, voids, interface degradation, cracks, and package integrity in IC packages. SAM has also been used for research to investigate new materials and characterize material properties for the package industry.

SAM uses reflected or transmitted acoustic signals to produce high-resolution images of the interior structure of a sample so as to detect and localize defects such as interfacial delamination, solder ball delamination, die cracks, die attach voiding, and so forth.

SAM is a nondestructive imaging technique. An ultrasonic wave is produced from a transducer and interacts with a sample by passing through an acoustically coupled medium of deionized water. The reflected or transmitted signals are collected by the receiver. There are three operational modes in reflective SAM: C-scan (C-SAM), an x-y planar image; B-scan, a time-of-flight projection; and A-scan, a pinpoint time-domain signal. Through-mode imaging is also possible wherein transmitted acoustic waves are collected by a receiver positioned beneath the sample under test.

C-SAM provides two-dimensional plane projection views of the internal planes of packages, which is the most common mode used to detect interfacial delaminations and cracks in IC packages. The critical interface for an overmolded IC package is the die-top/package material, where delamination and cracking at this interface can result in sheared or lifted wirebonds, passivation cracking, metallization shifting, or intermittent electrical failures. It also can provide a path for the ingress of process chemicals, leading to metallization/bond pad corrosion. Interfacial delaminations result in submicron air gaps that are easily detected by C-SAM, provided the spatial size of the delamination is equal to or greater that the resolved spot size of the acoustic signal.

There are two inspection modes for C-SAM: pulse-echo and through modes. For pulse-echo mode, ultrasound is produced by a rapid pulsing of the transducer; the generated acoustic waves are reflected from the sample and then collected by the transducer that now is alternately pulsed by the electronics to collect and receive the echoed signals. Images are generated by raster scanning the transducer over the sample in the x-y plane. Peak amplitude, phase inversion, and time of flight (TOF) information can be obtained for pulse-echo mode. The peak amplitude and phase inversion information is normally used to determine if there is die-top delamination. Normally, high-frequency transducers (>35 MHz) are used for pulse-echo mode to produce images of high spatial resolution. In through mode, ultrasound is transmitted through the sample and collected by the receiver underneath the package. Through-mode C-SAM is used to determine if there is any delamination within secondary interfacial features contained beneath the die/encapsulant interface (provided that no first reflecting interface is delaminated), such as the die attach. It is also used to verify acoustic interpretation of pulse-echo images. Figure 6.5*a* shows pulse-echo mode C-SAM images indicating there is die-top delamination (light area). Figure 6.5*b* shows the through mode C-SAM images verifying there is delamination inside the package (dark area), which resulted in lifted ball bonds as shown in Fig. 6.6. Die cracking can also be detected in IC packages as shown in Fig. 6.7, which is a pulse-echo C-SAM image.

B-scan is a time-of-flight “cross-section” mode, providing a nondestructive image of the 2-D depth profile. B-scan is used to detect die tilt, and cracks. Figure 6.8 shows the B-scan SAM images of a package with traditional “popcorn”-type package crack.

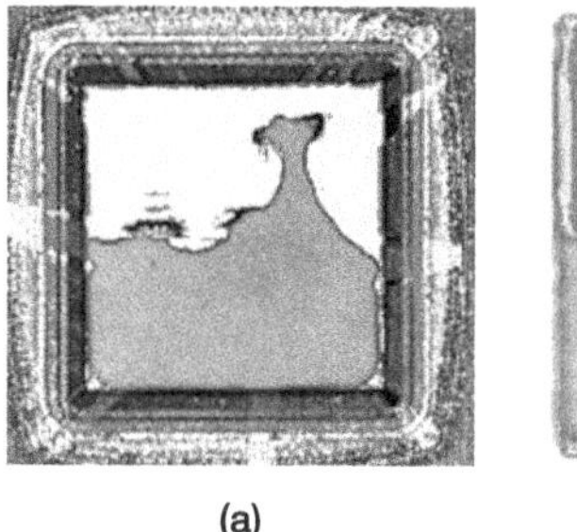

(a)

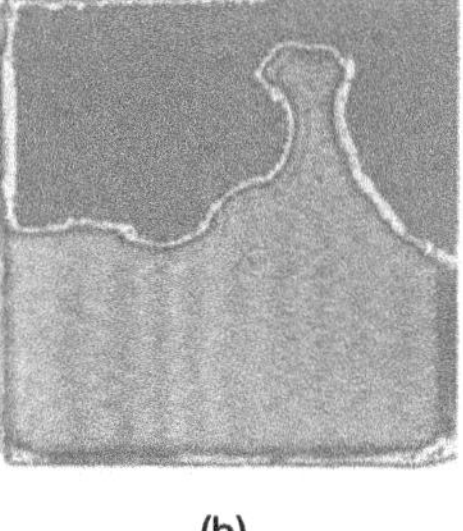

(b)

Figure 6.5 C-SAM image: (*a*) pulse-echo mode and (*b*) through mode.

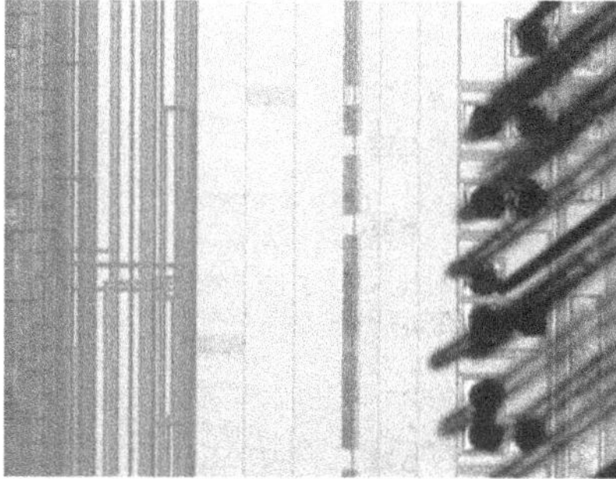

Figure 6.6 Lifted ball bonds.

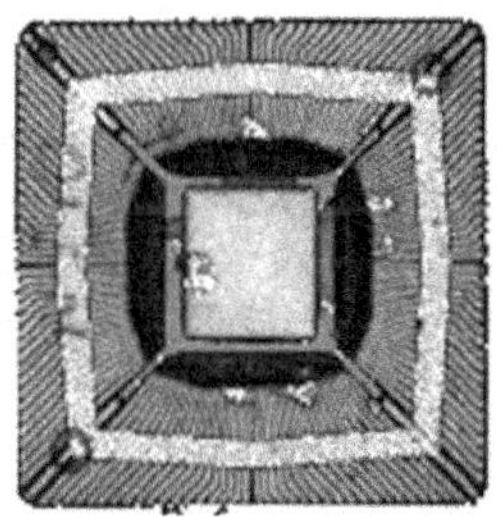

Figure 6.7 Cracked die.

A-scan can pinpoint a specific area and view all echoes, including the shape, amplitude, and polarity of the detected echoes, which can be used to interpret the interfacial bonds for the echoing interfaces. The SAM techniques discussed above are time-domain image techniques. Frequency-domain imaging can provide further information by using the frequency content of the signal filtered by means of a fast Fourier transform (FFT). This works very well for detection of small voids and cracks in flip chip packages.[1] Due to the continued miniaturization of package structures, SAM analysis using time-domain imaging is approaching resolution limits. Frequency-domain analysis may help to overcome this limitation due to frequency-dependent attenuation.

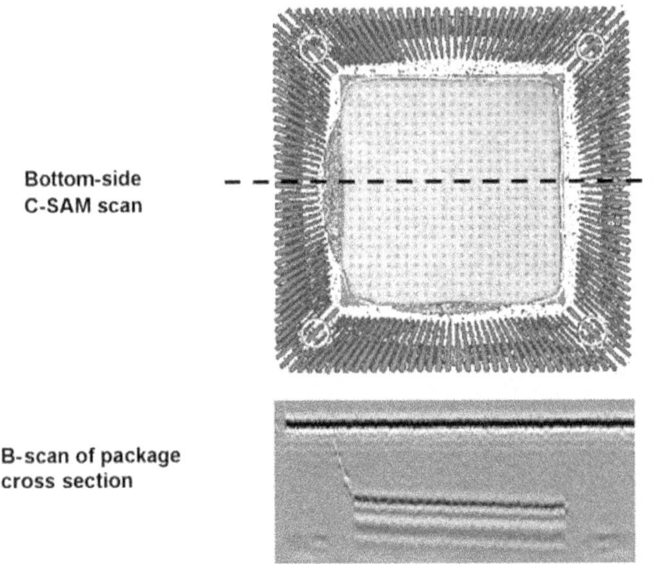

Figure 6.8 Images showing package "popcorn" cracking and paddle tilt.

A wide range of transducer frequencies, from 10 to 230 MHz, provide the capability to conduct SAM on various electronic products and components. The choice of transducers depends on the thickness of the packages, acoustic impedance of the materials, and the required resolution. Low-frequency transducers (<25 MHz) have high acoustic penetration capability and long focal lengths but lower resolution. Low-frequency transducers have been proven to work well for the thick, older generation of IC packages such as QFPs, PLCCs, and SOICs. High-frequency transducers produce ultrasound with shorter wavelengths, which improves the spatial resolution, but they have higher acoustic attenuation, which reduces the depth of the penetration into the package. High-frequency transducers are suitable for imaging thin packages, but they have limitations for multilayer packages such as stacked chip scale packages (SCSPs). Because of their multiple interfaces and the resultant convolution of the reflected signals, interpretation of the acoustic signals is challenging for SCSPs. In addition, for SCSPs, a ringing effect resulting from coincidence of multiple reflections from different dies will make it difficult to deconvolute distinguishing signals from multiple die interfaces.[2]

SAM is a critical tool for inspection of surface mount plastic IC packages. Common defects in plastic packages are delamination, lifted ball bonds, die crazing and fracture, package cracking, and paddle shifting. Plastic packages absorb moisture and then are susceptible to moisture/reflow-induced delamination and package cracking.[3] When the plastic packages go through the high-temperature reflow assembly process, the trapped moisture vaporizes and causes interfacial delamination. C-SAM becomes a powerful tool for package development. The thermal and moisture-induced damage in plastic packages can be detected by C-SAM. C-SAM can also be used to evaluate the moisture sensitivity for plastic ball grid array (PBGA) packages.[4–7] It was shown that the susceptibility to package delamination and cracking depends on the amount of absorbed moisture, die size, package material, package design, and solder reflow temperature profile.[5,8] Figures 6.9 and 6.10 show the pulse-echo and through-mode C-SAM for a PBGA package, detecting gross package delamination, which results in multiple open pins. C-SAM was also used to evaluate the effect of moisture diffusion on the moisture/reflow performance.[9]

Nondestructive SAM has been extensively used for qualification, reliability test, and quality control of electronic packages. It has also been widely used in the research area for investigation of underfill materials, solder bump integrity, solder joint integrity, and substrate evaluation for flip-chip packages. High-frequency transducers with high resolution are used for C-SAM inspection of flip chip solder bumps and underfill materials, and for analysis of flip chip packages.

The thermal expansion mismatches between the die and substrate could result in rapid fatigue failure of solder joints. Underfill material is used to buffer the coefficient of thermal expansion (CTE) mis-

Figure 6.9 Pulse-echo mode.

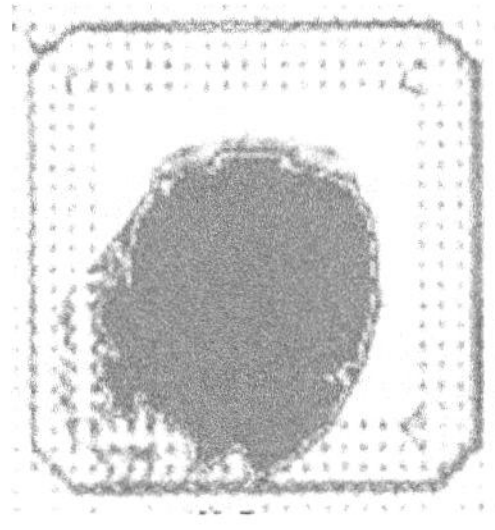

Figure 6.10 Through mode.

matches between the die and substrate. SAM has widely been used to investigate and qualify new underfill materials.

For flip chip BGA packages, solder bumping is migrating to less Pb (and eventually Pb-free), primarily as a result of legislative requirements to eliminate Pb in electronic assemblies, along with technical requirements to reduce alpha particle emission associated with Pb-containing alloys. Substrate technology is migrating from ceramic to organic technology, while die-level interconnect metallization is migrating from aluminum to copper technology. Flip chip interconnect integrity is becoming a critical performance and reliability issue.

Flip chip solder joint interconnect is a complex combination of under bump metallization (UBM) and bump metallurgy combined with the substrate pad metallurgy. Temperature cycling can aggravate the CTE mismatch stresses developed between the die and substrate, resulting in deformation and cracking of the solder bumps.[10,11] Electromigration can cause voids in solder bumps, resulting in contact failure. Elevated temperature can also accelerate the diffusion and interactions in solder alloys, creating excessive growth of intermetallics with a result of joint embrittlement and eventual failure due to cracking.[12] Nondestructive C-SAM can be used to detect the solder joint voids, delamination, and cracking. C-SAM can also be used to characterize thermally accelerated diffusion mechanisms that degrade flip chip solder interconnect integrity.[1] C-SAM results are consistent with the analytical and electrical continuity tests. Figure 6.11 shows the peak amplitude C-SAM image of a flip chip package, detecting delamination and solder bump failures in the dark regions of the image.

Die attach voiding can lead to die cracking, die attach fracture, or thermal runaway due to poor heat dissipation through the die attach, especially for power devices. C-SAM can be used to detect die attach voids. Figure 6.12 shows peak amplitude mode C-SAM images of die attach voids.

The changes/variations of material properties and boundaries can cause acoustic impedance changes, and this affects the phase or peak amplitude of the reflected or transmitted ultrasound. By measuring

Figure 6.11 C-SAM image of a flip chip package showing delamination.

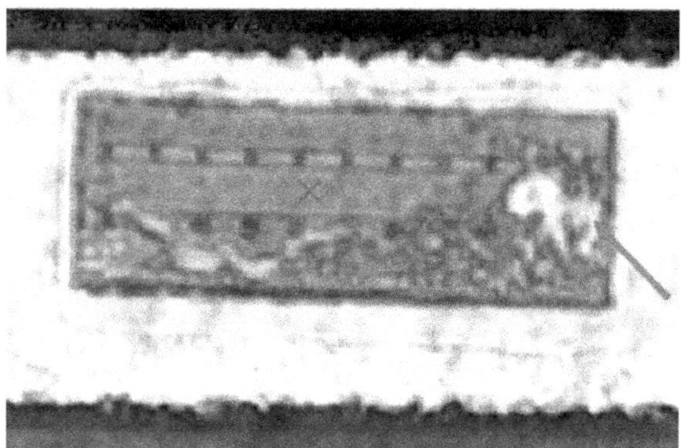

Figure 6.12 Pulse-echo mode C-SAM showing die attach voids.

the amount of sound returned as a function of the amount of sound entering the boundary, characterizations of the boundaries/materials can be made.

Scanning acoustic microscopy can also be used for thickness measurements of thin layers sandwiched between two thicker materials, based on the ultrasonic reflections at interfaces between two different materials with different acoustic impedance. The thickness of the layer can be obtained by multiplying the propagation time lag between two reflections and the velocity of ultrasound in the thin layer. This technique works well for a thin layer thickness of a few times greater than the ultrasonic wavelength, which will produce a clear separation of two reflections from the front side and backside of the thin layer. However, for layer thickness less than the ultrasonic wavelength, the time difference is not easily measured because of the signal overlap of two reflections.[14] Scanning acoustic microscopy can be also used for other material characterizations such as acoustic impedance, density, porosity, doping levels, Young's modulus, and elasticity.

Continuous wave scanning acoustic microscopes (CW-SAMs) can be used for investigations of thin film adhesion.[15] SAM can be used to study the thermal resistance degradation of alloy die attached power devices[16] and the characterization of the interfacial thermal resistance.[17] In summary, SAM is a powerful tool and technique for failure analysis, reliability, qualification, process control, material characterization, and new package development in the package industry.

6.1.4 Thermal Imaging

Thermal analysis techniques are primarily used to identify areas of a device that exhibit excessive heating associated with electrical leakages or shorts. Typically, the device is powered with the nominal operating voltage and checked for abnormal thermal sites as compared to a good reference device. High-current conditions isolated to the IC package may indicate issues with the lead frame, bond wires, or other internal multilevel metal/conductive package structures. Thermal analysis techniques are useful for isolating the problem to a specific site that can be opened for further structural characterization.

One of several thermal analysis techniques[18] can be applied, depending on the degree of current leakage, defect location, package type (BGA, TQFP, and so on) and analyst preference. These include the "freeze/thaw" method, liquid crystal, fluorescent microthermal imaging, and IR imaging (in cases where the backside of a device is polished), listed in order of ease of use. All these techniques are sample and condition dependent. Typically, high leakage currents are needed to observe package thermal emissions. Once the leakage is confirmed to be package (not die) related, the analyst can then initiate thermal analysis.

The simplest and fastest method is the "freeze/thaw" technique. This novel technique requires only a freeze can and, usually, a low-power microscope. The technique uses the condensation accumulation on a chilled package that will then thaw first in and around the area of leakage. This technique works best for gross leakage, but all the thermal techniques depend on the IR drop across a leakage path in the package. In most cases, the die can be removed from the package if it has been decapped; however, it is recommended to keep the die in the package to avoid disturbing any package structures associated with the leakage. The following thermal analysis procedures can be used as a guide:

- Set up the package with the appropriate electrical configuration required to duplicate the abnormal leakage current. A low-power optical microscope is also required to examine the package during a power-up condition.
- Apply power (DC preferred) to the device under test. The device can be either socketed or the package pins/solder bumps can be manually probed. If the die is exposed and if necessary, the bond pads can be probed.
- Next use a freeze spray to coat the device so that the whole package (or suspect leakage area) frosts up. A dust-off can may also be used upside-down for cooling action. Once the package is covered with frost, it is important to quickly power up the device and look for a thermal site before the entire package returns to room temperature.

- Carefully apply a DC voltage. This is a crucial step for the following reasons:
  - First, if too much current is applied, it may rapidly heat the device and defrost the whole package.
  - Second, too much current may cause the leakage to increase and create more severe damage, or it may burn out the leakage path, making it difficult to determine the root cause for the failure.
- When the package is covered with frost, carefully apply power so that heat from the defect will initiate a thawing effect at the suspect location. This "frost/thaw" technique can be applied several times to find the exact location. Figure 6.13 shows an example of a located short.
- Once the thaw spot is identified, carefully mark the location using a scribe or pencil.
- The defect site can now be carefully exposed using a variety of options, including chemical etching or mechanical polishing techniques. Figure 6.14 shows examples of the exposed defect.

Another convenient and quick method for thermal imaging makes use of liquid crystal analysis techniques. The method relies on detecting phase changes in the liquid crystal at a site where the local temperature exceeds a specific value known as the *transition point*. This technique is primarily used for die-level thermal analysis. However, in some cases, it can be used to detect package-related defects. The application of liquid crystal for package analysis is similar to the methodol-

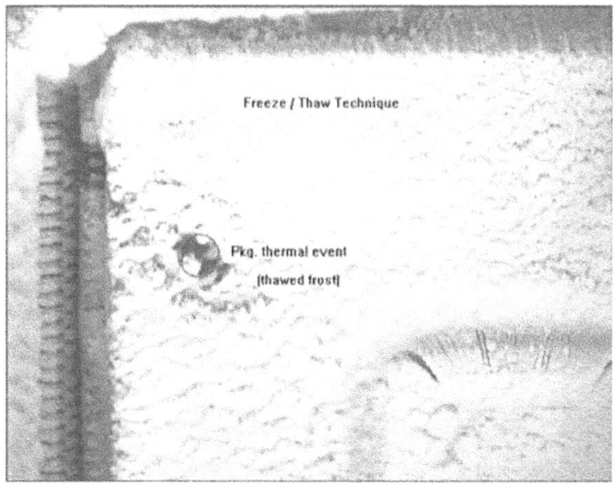

Figure 6.13 Thermal emission site observed under frosted/power-up conditions.

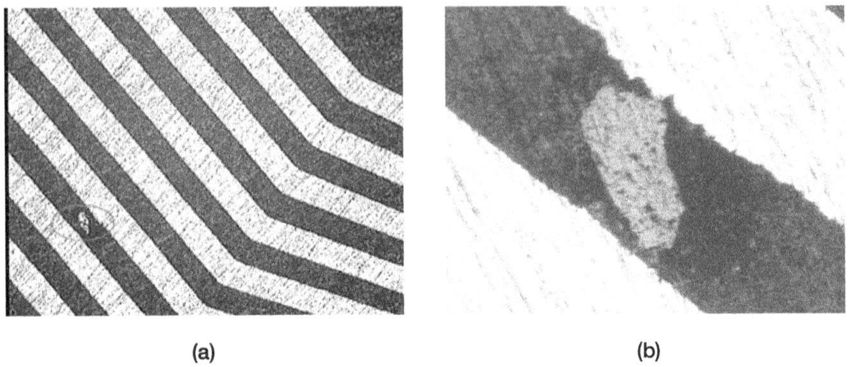

(a) (b)

Figure 6.14 Exposed defect after coarse and medium mechanical polish. (*a*) IR analysis on packages on PWB and (*b*) FMI on IC package.

ogy used for die, but abnormal thermal site emission is more difficult to detect due to poor light polarization contrast on package surfaces.

IR imaging from the backside of the semiconductor die has become increasingly important for flip chip package technologies. Since silicon is essentially transparent at IR wavelengths of 0.8 to 1.3 nm, a microscope using special IR transmitting optics, combined with an IR camera and monitor, can readily be employed to investigate failure modes in microelectronic products. This technique is used primarily for identifying emission through the back of a polished, electrically stimulated die. However, this technique can also be applied to package thermal analysis. Refer to Fig. 6.15*a* for an example.

Temperature distribution on the specimen surface can be displayed using a scanning infrared microscope and a gray scale to differentiate

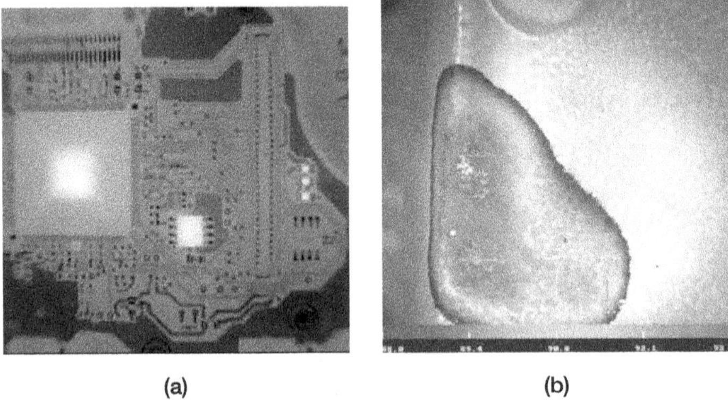

(a) (b)

Figure 6.15 (*a*) IR analysis of packages on PWB and (*b*) FMI on an IC package.

the temperatures and pinpoint hot spots.[19] This can be performed using one of several thermal imaging systems available on the market. However, the one used in our lab was exclusively designed and developed by Agere Systems (Bell Labs)—the Fluorescent Microthermal Imaging System (FMI). This system is used to detect thermal emission and/or light emission by coating a sensitive rare-earth chelate that produces fluorescence on the sample when excited by a UV source. With device under test, images can be produced with a spatial resolution of 0.1 μm and temperature resolution of 0.01°C. The FMI system is capable of front side and backside thermal imaging and photon emission imaging. Measurement accuracy depends to a considerable extent on magnification, average temperature, and emissivity. Refer to Fig. 6.15*b* for an example image.

6.1.5 Scanning Electron Microscopy

Scanning electron microscopy (SEM) is one of the most widely used imaging techniques after optical microscopy and is used in most failure analysis laboratories.[20] In essence, the SEM scans a primary electron beam across the region of interest and detects the emitted and backscattered electrons. All this occurs within a vacuum chamber, thus requiring some sample preparation in contrast to optical microscopy. The output of the detector is imaged as a function of electron beam position on the sample using a display monitor synchronized to the beam scan rate. The high resolution and almost 3-D clarity of the images make SEM one of the best tools for microstructure characterization. High-resolution imaging is made possible by the small spot size of the primary electron beam, which can be as small as 1.4 nm. This allows a magnification in the range from 100× to 1,000,000× with appropriate adjustment of the beam x-y scan lengths. Clear imaging of three-dimensional structures is made possible by the small divergence angle of the primary electron beam, which gives much greater depth of focus as compared to optical microscopy. By proper collimation of the primary electron beam, it can remain in focus over a wide range of vertical heights across the sample. Finally, the natural appearance of SEM images makes them intuitively easy to interpret. By proper choice of detector location and image orientation, the SEM image appears as if it were illuminated from the top, making it easy to determine hills and valleys in the microstructure. For this reason, SEM images are often preferred over optical images, even for lower-magnification work.

The SEM is a flexible imaging and analysis tool that can be tailored to specific sample types and problems. The analyst can adjust the probe electron beam, the sample orientation, and the detector type based on the kind of information desired. The optimal conditions likely represent a compromise between resolution, analysis time, and depth of focus.

The beam parameters that have the most impact on image quality are energy, current, diameter, and divergence.[21] The beam energy determines the depth from which sample information is obtained and can typically be adjusted from 500 eV to 30 keV. Low beam energies are used to enhance surface sensitivity. The beam current determines the signal-to-noise ratio for a fixed image acquisition time. High beam currents improve the ability to detect small changes in contrast on the sample but may damage the sample faster. The beam diameter determines the resolution. Stronger focus or smaller apertures can reduce the beam diameter, giving higher resolution. The beam divergence determines the depth of focus. Collimating apertures and longer working distances tend to reduce the divergence angle, which keeps the beam focused over a longer distance. Unfortunately, the beam parameters are interrelated so that the highest current, smallest diameter, and smallest divergence cannot be achieved simultaneously. The parameters must be optimized based on the sample information required. For semiconductor package applications, the highest resolution is often not required, and it is more important to have good depth of focus and image contrast. Hence, a larger probe diameter can be tolerated to achieve higher currents or lower divergence angles.

The type of electron source used in a particular instrument affects the available range in beam parameters. The tungsten (W) filament, lanthanum hexaboride (LaB_6) filament, and field emission (FE) sources are the commonly available types, with the FE source having the highest brightness and resolution. Instruments based on the field emission source are commonly referred to as FE-SEMs to distinguish them from models with the older types of electron sources. FE-SEMs are generally easier to use when scanning large areas for defects, since the higher signal-to-noise ratio allows faster scans. However, good quality images can be obtained from any instrument type with proper choice of beam parameters.

The SEM stage allows the sample to be tilted, rotated, and translated in x, y, and z directions. The ability to view the sample at different orientations with respect to the incident beam is often helpful in elucidating microstructure or enhancing contrast. The specimen holder attaches to the SEM stage and maintains the sample at ground potential. It is critical to prevent the sample from accumulating charge, since any surface potential on the sample will alter the incoming and emitted electron trajectories, causing serious degradation of focus and resolution. Insulating samples are difficult to image properly and must be grounded in some way to prevent charge buildup. In practice, conductive tapes and paints can be applied between the sample and holder to make good electrical contact. In addition, insulating samples are often sputter coated with a thin (<50 Å) conductive film to prevent charge buildup. For situations in which the majority of samples are insulating, it is worth considering the purchase of an environmental SEM, which is a relatively new innovation that maintains a higher gas pressure in the sample chamber. Water vapor can be mixed with the gas flow, which virtually eliminates charging problems on uncoated insulating samples.

As the primary electron beam interacts with the sample, a wide variety of scattering processes can occur over an average depth in the sample that ranges from 0.1 to 5 μm, depending on beam energy and element composition.[21] As a result of these interactions, photons and electrons are emitted from the sample from an area within and around the probe footprint. Emitted signals can originate from an area larger than the probe diameter because of scattering events in the sample that divert some electrons in lateral directions. For imaging and analysis purposes, X-ray photons, low-energy secondary electrons, and high-energy backscattered electrons are the most useful signals. Measurement of the emitted X-rays is the basis of X-ray microanalysis or energy dispersive spectroscopy (EDS) in the SEM and is discussed in a separate section.

The imaging of secondary and backscattered electrons is the primary source of structural information in the SEM.[21,22] Secondary electrons are relatively low in energy (<50 eV), and only those formed relatively close to the surface escape from the sample and are detectable. The yield of secondary electrons is affected by changes in surface topography, material type, and surface electric and magnetic fields. Contrast in images formed from secondary electrons will reflect these changes across the surface of the sample, with a primary emphasis for package analysis on topography. Secondary electron contrast due to changes in electric field across the surface of a semiconductor die may be related to open vias or other electrically floating structures, and this specific application is often referred to as *voltage contrast.*[21]

The other main category of emitted electrons is the backscattered component. These are incident electrons that elastically scatter from the atoms in the sample and retain between 60 percent to 90 percent of the initial beam energy. The backscattered electron energy is related to the atomic number, with higher energy corresponding to scattering from heavier elements. Because backscattered electrons have much higher energy than secondary electrons, even those originating deeper in the sample are detected. The yield of backscattered electrons is mainly affected by changes in surface topography and atomic number. Contrast in images formed from backscattered electrons shows changes in topography and element composition. At high beam energies, backscattered electron images can provide information from structures as deep as 1 μm below the sample surface.

One or more electron detectors are arranged around the specimen stage to allow flexibility in imaging the secondary and backscattered electrons.[21] The Everhart-Thornley (E-T) detector is the most common detector and is sensitive to both types of electrons. A grid placed in front of this detector can be biased to screen out the lower-energy secondary electrons, if desired. The passive scintillator, or Robinson detector, is also fairly common and is sensitive only to high-energy backscattered electrons. Finally, some instruments place an E-T detector above the final lens so that there is no direct line of sight to the sample. This is referred to as a *through the lens* (TTL) or *upper* detector, depending on the manufacturer. This allows only the lower-energy secondary electrons to be deflected into the detector. By proper choice of detector, the analyst can optimize the image contrast to emphasize topography, composition, or surface morphology.

Some examples of SEM images obtained from various package samples are shown in Figs. 6.16 through 6.20. Samples can be examined as received, decapped, etched, or cross-sectioned and polished to focus on specific regions.

6.1.6 Transmission Electron Microscopy

Transmission electron microscopy (TEM)[23,24] has become an integral and valuable tool in the practice of failure analysis. This fact has become particularly evident with the continual reduction in scale of fea-

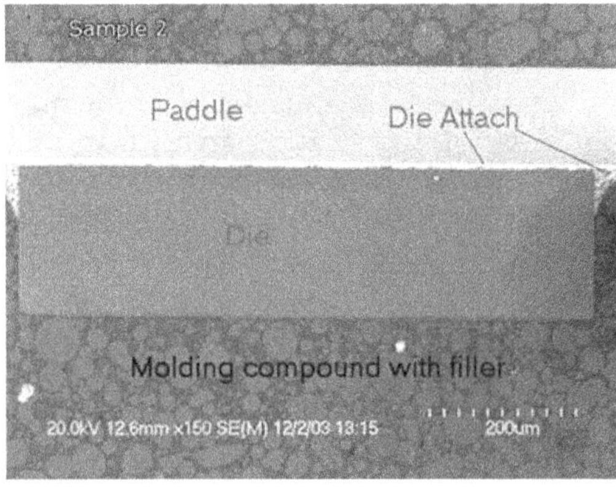

Figure 6.16 Polished cross section of package showing die, lead frame, and package molding compound.

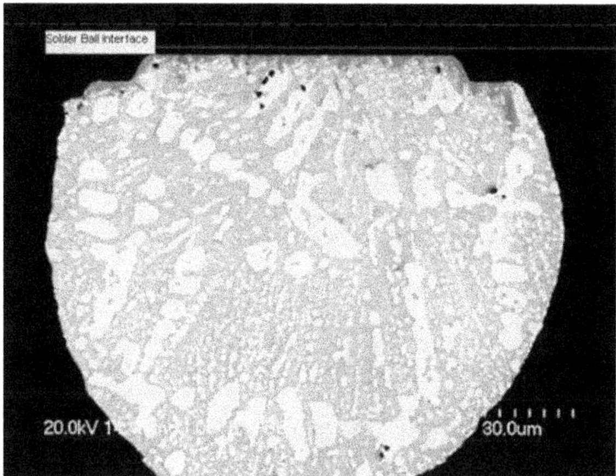

Figure 6.17 Image of solder ball cross section using backscatter detector to show element contrast. The brighter areas correspond to lead regions in surrounding tin areas. The darker areas at the solder ball interface indicate a lighter element diffusing into solder.

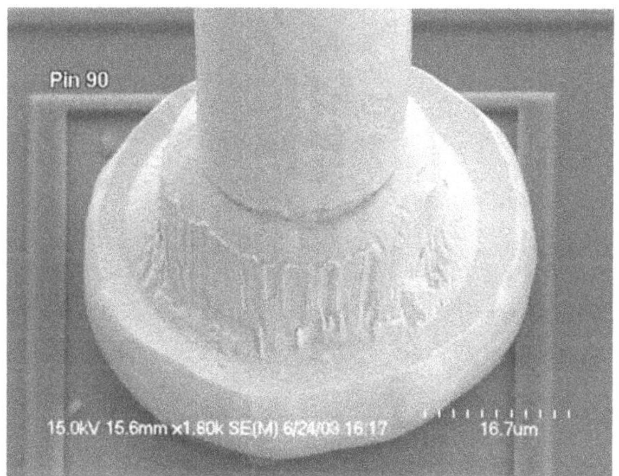

Figure 6.18 Tilted SEM image showing excellent depth of focus on exposed ball bond.

ture sizes in the semiconductor industry over the past several decades. Two developments have largely facilitated this integration. The first is new sample preparation techniques, such as use of the focused ion beam (FIB), which allows more timely preparation of samples from highly specific sites. The second, as with scanning electron microscopy (SEM), is the evolution to a computer-based interface with the microscope. The latter development has facilitated the addition of digital

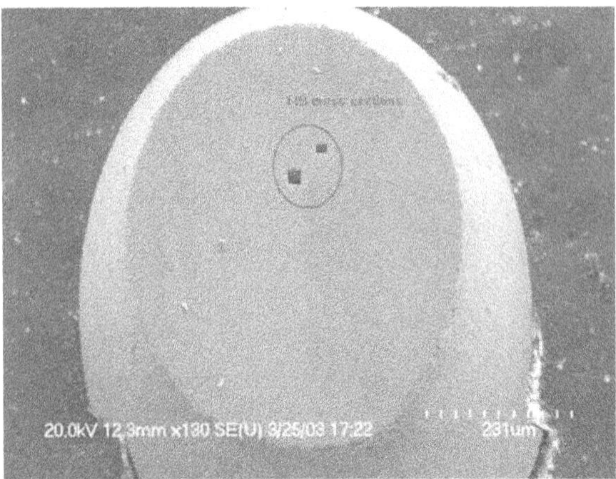

Figure 6.19 SEM image of solder ball with FIB cross section through connection tab.

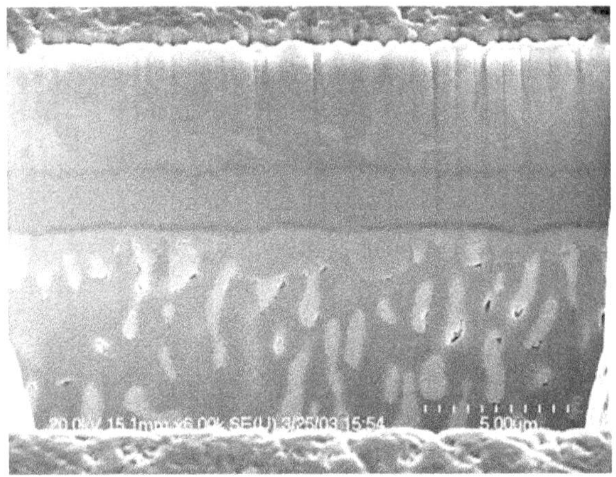

Figure 6.20 SEM image showing detail of interface in FIB cross section from previous figure, viewed at a 45° angle into the FIB cavity.

image capture media to the TEM, such as charge coupled device (CCD) cameras and has enhanced scanning transmission electron microscopy (STEM) capabilities. These developments have reduced the total cycle time (i.e., specimen preparation to analysis and distribution of images) from several days to less than one day.

TEM requires the preparation of a very thin sample, ~100 nm or less, although useful information can be obtained from samples more than twice as thick. An electron beam, typically of 100 to 300 keV, is generated by the microscope and passed through the sample. Depending on the microscope settings, this transmitted beam may provide various forms of imaging contrast that reveal microsctructural information concerning the specimen. Such information may range from the simple presence of a void to a variation in composition within a layer, to the distribution of grain size, to the presence of an atomic scale defects. The image is magnified using electromagnetic lenses (EMs) and projected onto a viewing screen or other detector, with the available magnification ranging from ~100× up to several million times. Refer to Fig. 6.21 for an example of a TEM sample image.[29]

In addition to the image, a wealth of other information may be obtained from the interaction of the TEM electron beam with the sample. Typically, a TEM is outfitted with an *energy dispersive spectrometry* (EDS) detector, allowing the identification of elemental constituents within the sample. An *electron energy loss detector* is often added to the TEM as well, and it enables elemental identification as well the analysis of bonding states. Finally, the TEM can be quickly switched from imaging to electron-diffraction mode, allowing the identification of crystalline materials and their orientations with respect to each other.

The use of STEM has become increasingly important because of the contrast mechanisms now available with the multiple detector

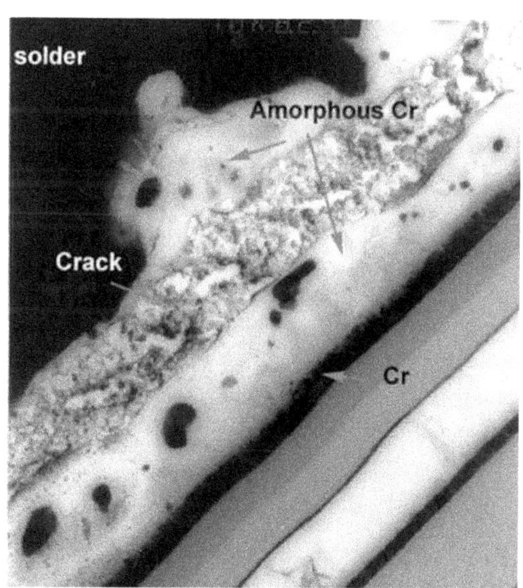

Figure 6.21 TEM image of solder bump interface from FIB prepared sample shown in FIB section.

systems. The convenience of the system is also important, particularly as related to EDS analysis. STEM operation is very similar to that of a SEM. The beam is condensed to a fine spot using EM lenses and apertures, and the beam is then rastered across a defined area of the sample. However, instead of collecting secondary or backscattered electrons as in SEM, electrons from the primary transmitted beam are collected and used to produce the image. The very thin nature of the specimen, combined with the higher accelerating voltages employed by TEM, allows for a much smaller interaction volume of the beam with the sample. These factors and others provide improved imaging and EDS spatial resolution over that available on the SEM.

With the progression of materials and products into the realm of nanotechnology, one can only imagine that the role of TEM will be an ever-expanding one. This role will expand not only for research and development purposes but also as an enabling technique in the science of failure analysis.

6.1.7 Focused Ion Beam

In the past decade and a half, focused ion beam (FIB) systems have become among the most important tools in the failure analysis, characterization, and analytical laboratories. The basic operation of a FIB is similar to that of a scanning electron microscope (SEM). Whereas the SEM uses electrons to image the surface, the FIB uses gallium (Ga+) ions to do the same. In addition, because the Ga+ ions are heavier than the electrons, more surface damage, or sputtering, is obtained, allowing the tool to be used for sample modification.

The operating principle of the two instruments is very similar. The FIB system employs a gallium liquid metal ion source (LMIS) to produce the primary ion beam. The ions are extracted from the source, collimated, focused, and rastered on the sample surface. Acceleration voltages for the LMIS vary from 25 to 50 kV, depending on the specific FIB system. Electrostatic lenses are used to steer the ion beam, as the traditional magnetic lenses used to steer electrons in the SEM cannot apply enough force to deflect the much heavier gallium ions.[25] A comparative schematic is given in Fig. 6.22.

Rastering the beam on a sample surface gives rise to a variety of emitted secondary particles. Secondary electrons or secondary ions may be used for imaging in FIB systems. An appropriate detector collects the secondary ions or electrons and, after signal amplification, its output is registered on a computer screen as an image. The different types of detectors used are as follows:

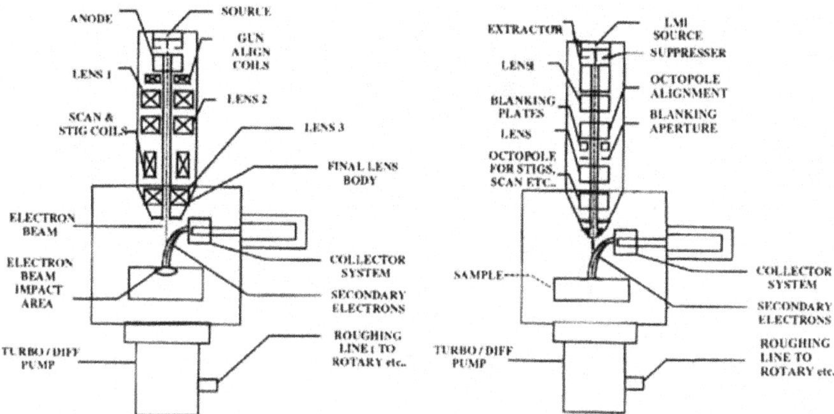

Figure 6.22 Comparative schematic for SEM/FIB. (*Courtesy of Pete Carleson, FEI Company.*)

1. Multichannel plate (MCP)—for ions and electrons
2. Channel electron multiplier (CEM)—for ions and electrons
3. Scintillator—for electrons[26]

Imaging mechanisms on FIB systems are as follows:

1. Topographic contrast (depending on surface topography, similar to a SEM)
2. Materials contrast (depending on the different materials)
3. Channeling contrast (depending on the channeling in different crystal orientations in a material)
4. Passive voltage contrast (due to variation in surface charge resulting from differences in floating versus grounded structures on the sample)[26]

Figure 6.23 shows an example of the imaging capabilities of a FIB system.

Specialized software present on a FIB system allows precise control for drawing or rastering the beam in a desired pattern across the sample so as to carry out mechanical milling via physical sputtering of the surface using the ion beam. The beam spot is defined through a set of real or virtual apertures to produce a known beam current. Different beam currents correspond to different spot sizes, which are optimized either for bulk milling or fine polishing. The sputtering process can be aided by the introduction of gases that enhance the milling process by

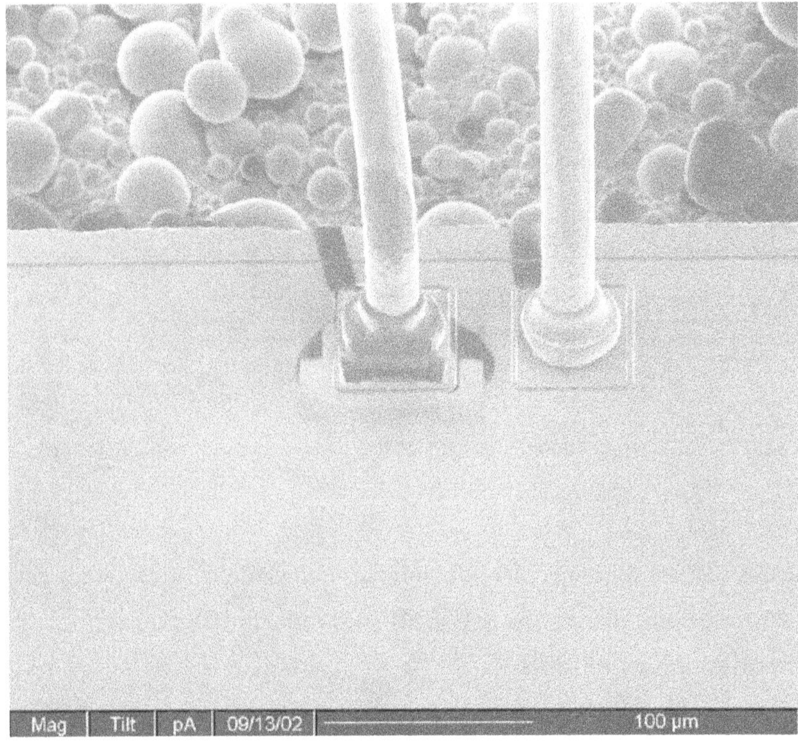

Figure 6.23 Bond pad cross section overview (800×, 45° tilt).

converting the sputtered material into a volatile gas that is pumped away by the vacuum system. One can also deposit metals (W and Pt) using ion beam assisted CVD.[26]

FIB systems are used for both device repair (typically in the die) and analysis. For device repair, prototype circuits can be modified so as to fix design flaws prior to mask changes.[27] For analytical work, cross-sectional samples for SEM and transmission electron microscope (TEM) imaging can be prepared in precise locations on the sample. These are the major uses of FIB systems, and the versatility of the system allows changing from one mode to the next as easily as changing samples.

The advantages of using a FIB system for analytical work are as follows:

1. *Repeatability.* One can select a specific feature on all lots going through a process and monitor it very accurately.
2. *Site specificity.* One can section targets in micron and submicron dimensions.

3. *Stress-free use.* There is no stress imposed on the sample while it is being prepared for SEM or TEM analysis.
4. One can image the sections for SEM using the ion beam, and the FIB image is often sufficient, with no need for further SEM imaging.

Figures 6.23 and 6.24 are examples of a FIB cross section through a particular bond pad that was used for in-process monitoring.

An interesting point here is that TEM samples can be lifted out from any sample that can be fitted inside the chamber and can tolerate the vacuum.[28] Also, one can lift out samples from potted samples (which can be hand polished), identify defective sites using an SEM, and then prepare a site specific sample for detailed TEM analysis.[29,30] Figures 6.25 and 6.26 show FIB images of a solder bump that has been FIB prepared for liftout and TEM analysis.

In summary, FIB systems are extremely versatile tools that have become a necessity in FA labs. These tools have made site-specific sample preparation convenient, rapid, and highly reliable.

6.1.8 Moiré Interferometry

Moiré interferometry is an optical technique used to analyze the in-plane displacements on a microelectronic package under thermal load-

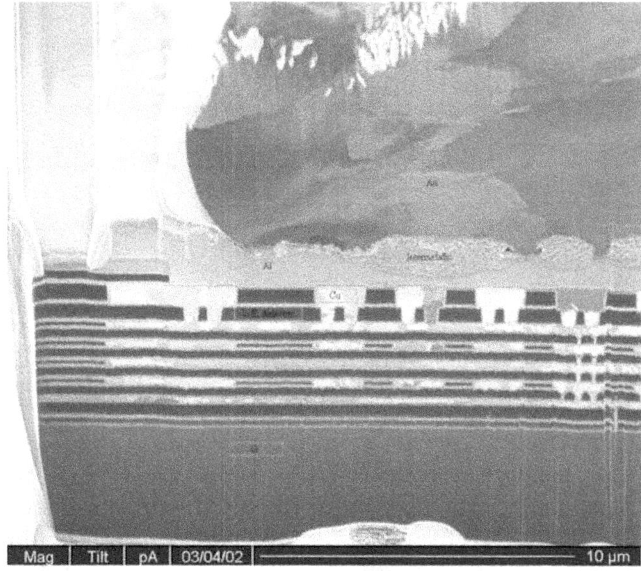

Figure 6.24 Imaging capabilities of a FIB system—bond pad cross section (15,000×, 45° tilt).

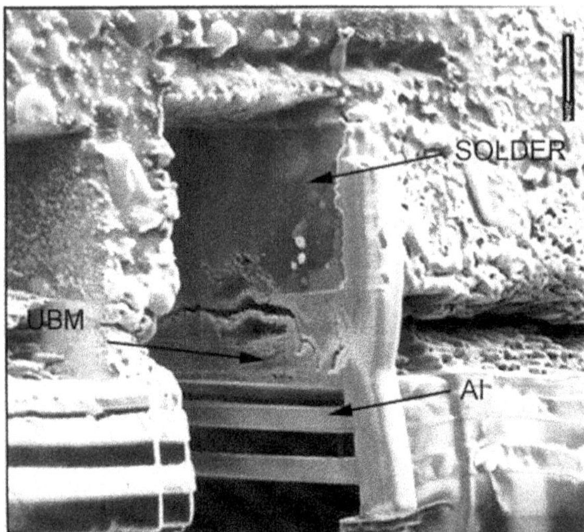

Figure 6.25 FIB image of TEM section lift out on solder bump.

Figure 6.26 FIB image after final cut, ready for lift out and TEM analysis.

ing. Figure 6.27 schematically shows the optical concepts of moiré interferometry. A high-frequency diffraction grating must first be transferred onto the surface of the specimen. To map the displacement in the x direction, U, and y direction, V, an orthogonal grating, should be used that covers the area of interest and deforms along with the specimen. A virtual reference grating is produced by interfering two parallel laser beams as shown by B1 and B2 in Fig. 6.27.[31] The virtual

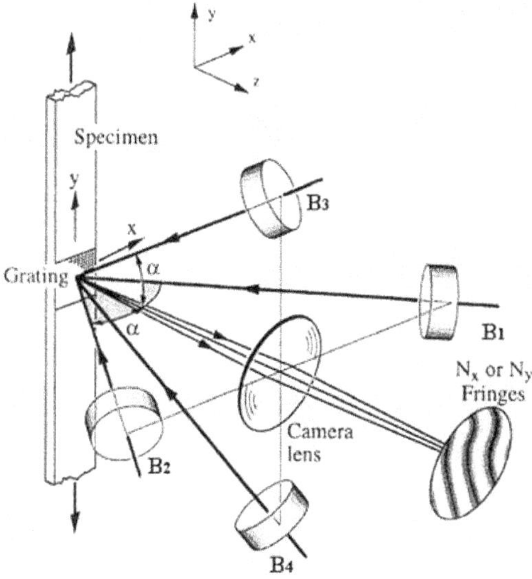

Figure 6.27 Schematic illustration of four-beam Moiré interferometry. *(Figure partly or wholly extracted from Ref. 31. Copyright © Springer-Verlag, reprinted with permission.)*

reference grating is superimposed with the grating on the specimen surface. When the specimen deforms, the virtual grating interacts with the deformed specimen grating to form the moiré fringes. These moiré patterns represent contour maps of the U displacement field corresponding to x direction. Similarly, laser beams B3 and B4 are arranged to generate the moiré patterns of the V displacement field corresponding to y direction.

Figure 6.28 shows a moiré interferometer, the PEMI II, developed by ESM Technology, Inc. This compact system has a very high sensitivity and can be applied to the analysis of the displacements/strains on small samples such as microelectronic packages. Figure 6.29 shows a typical example.[32] As a newly developed experimental technique, moiré interferometry has been successfully applied to the assessment of various types of electronic packaging.[33–43]

6.1.9 Shadow Moiré

Shadow moiré is another type of optical method for capturing out-of-plane deformations, such as the warpage of a microelectronic package under thermal loading.[44,45] Measurements are made by placing a reference grating of constant pitch, usually a sheet of low-expansion quartz glass etched with equally spaced parallel lines, above the spec-

Figure 6.28 A Moiré interferometer, PEMI II. (Courtesy of ESM Technology, Inc.)

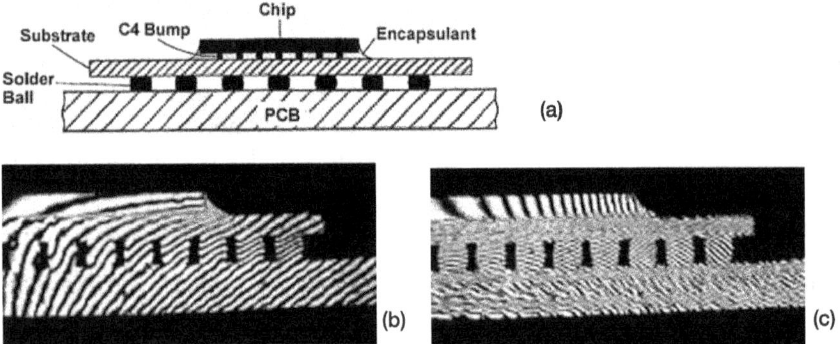

Figure 6.29 An example of Moiré interferometry application on microelectronic packaging. (a) Flip chip BGA package assembly, (b) U displacement field, and (c) V displacement field. *(Courtesy of EMI Technology, Inc.)*

imen as schematically shown in Fig. 6.30. No moiré fringe pattern would be observed if the specimen was flat and parallel to the reference grating. When the specimen deforms, such as a package that has sustained a temperature change, the shadow grating on the warped surface of the specimen interacts with the reference grating to generate moiré fringes. The resulting moiré pattern is the contour map of the specimen surface shape. The gap between the specimen surface and the reference grating, w, is calculated in the following equation:

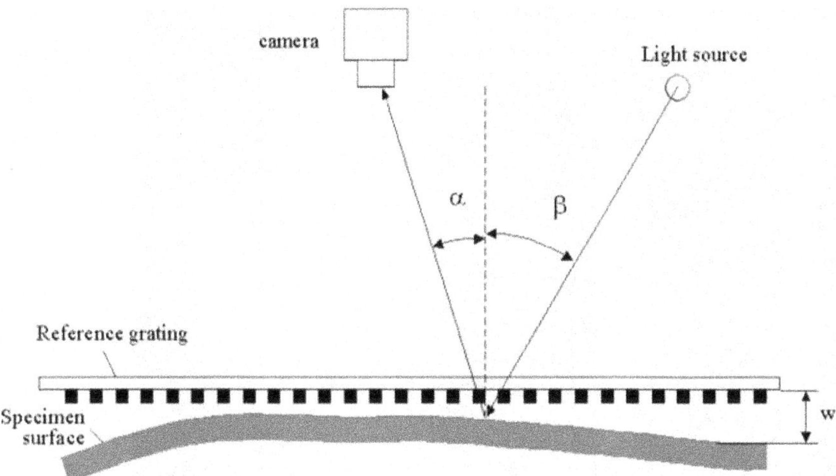

Figure 6.30 Schematic illustration of shadow Moiré method.

$$w = \frac{g \cdot N}{\tan\alpha + \tan\beta}$$

where g = reference grating pitch
N = number of fringes in resulted moiré fringe pattern
α, β = incident angles of the camera and the light source, respectively

The out-of-plane displacement can be estimated by comparing the moiré fringe patterns before and after the specimen is deformed.

Normally, a commercial shadow moiré system equips a CCD camera for capturing moiré fringes, software for data analysis, and an environmental chamber to capture a complete history of a specimen's behavior for a given time-temperature profile. Figure 6.31 shows such a shadow moiré system, the TherMoiré PS400 from AkroMetrix LLC, which is capable of measuring warpage of a sample up to 400 mm square in a thermal range between –55 to 350°C (300°C sustained). Figure 6.32 shows a typical contour plot output from this system. Figure 6.33 is an example of the warpage response of a PBGA package for a specific reflow profile.

6.1.10 Scanning Superconducting Quantum Interference Device Microscopy

The superconducting quantum interference device (SQUID) has been used since the 1970s as the basis of a sensitive magnetometer for mea-

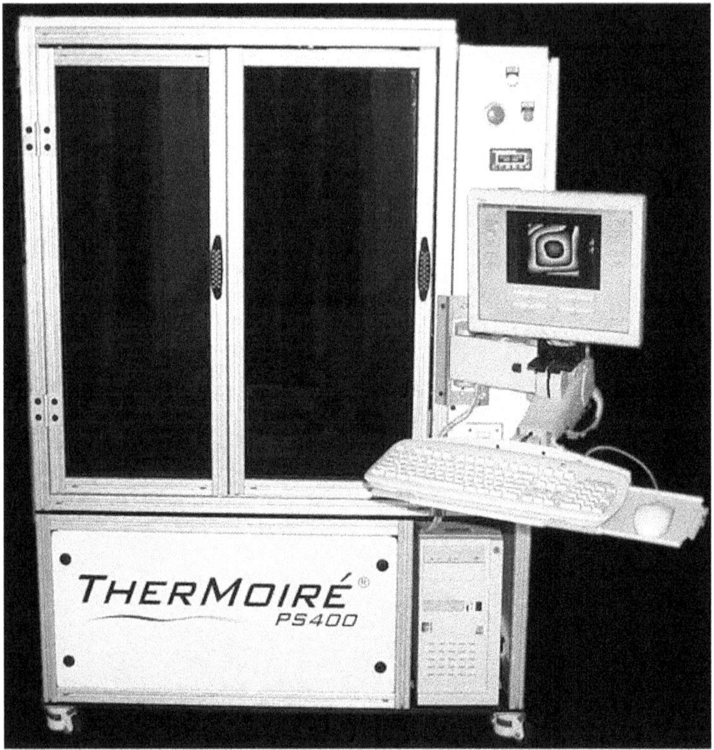

Figure 6.31 A shadow Moiré system, the TherMoiré PS88. *(Courtesy of AkroMetrix, LLC, http://www.warpfinder.com.)*

surement of weak magnetic fields. The device makes use of the quantum mechanical current tunneling across a Josephson junction and its variation with magnetic field strength.[46] The development of high-temperature superconductors has made possible SQUID magnetometers that can operate at the relatively more convenient temperatures achieved with liquid nitrogen cooling. When coupled with the instrumentation developed for scanning probe microscopes, a scanning SQUID microscope (SSM) has recently become available that allows magnetic field imaging of semiconductor devices.[47] Figures 6.34 and 6.35 show examples of the Model Magma-C10 manufactured by Neocera, Inc. The sensitivity of the SQUID sensor allows it to detect magnetic fields two million times smaller than the Earth's magnetic field and allows it to detect current flow within a semiconductor device as low as 10 nA.

SSM is a nondestructive, noncontacting technique that can be applied to traditional BGA and TQFP packages as well as newer flip-chip technology. As it scans across a powered device, the magnetic field

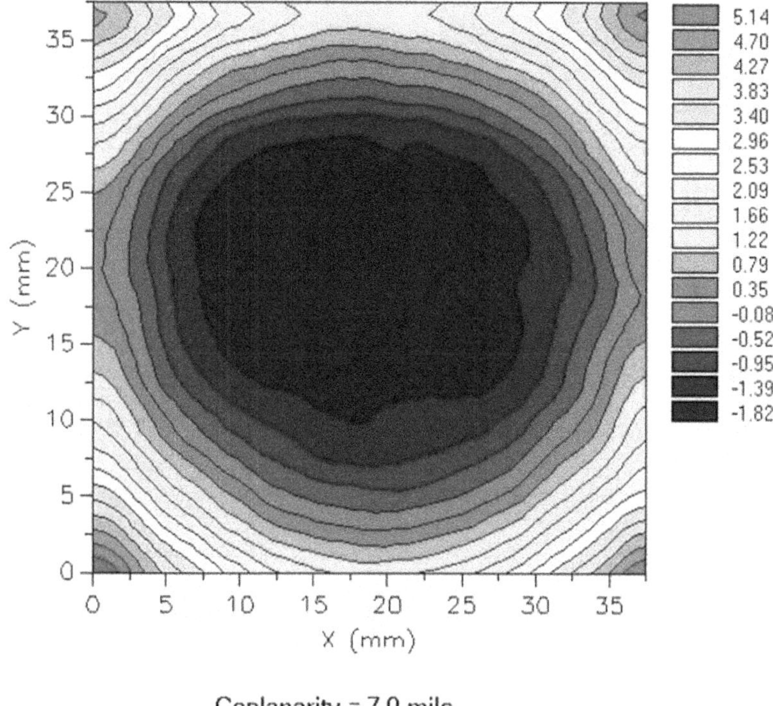

Figure 6.32 Contour plot representing the warpage of a PBGA package at 220°C.

measurements are used to generate a direct image of the current paths in the device. Imaging from the back or front of the device is possible, as the interference from plastic package materials is minimal. Because only the SQUID probe is cooled, the device remains at room temperature, allowing easy connection for various electrical tests during current imaging.

The SSM provides complementary information to other fault isolation techniques available to the failure analyst, such as time domain reflectometry (TDR). By directly mapping current flow, it can be used to isolate faults in the package versus the die. Its main application has been in determining the location of shorts and other current leakage paths, but it is expected to find application in any situation where current flow in the failed device deviates from the reference device.[48] The main disadvantage of the technique is its spatial resolution, which is approximately 10 to 50 μm, depending on working distance, current magnitude, and magnetic field noise. However, this resolution is suitable for the wiring dimensions in most device packages, and SSM is expected to become a more widely used fault-isolation technique for

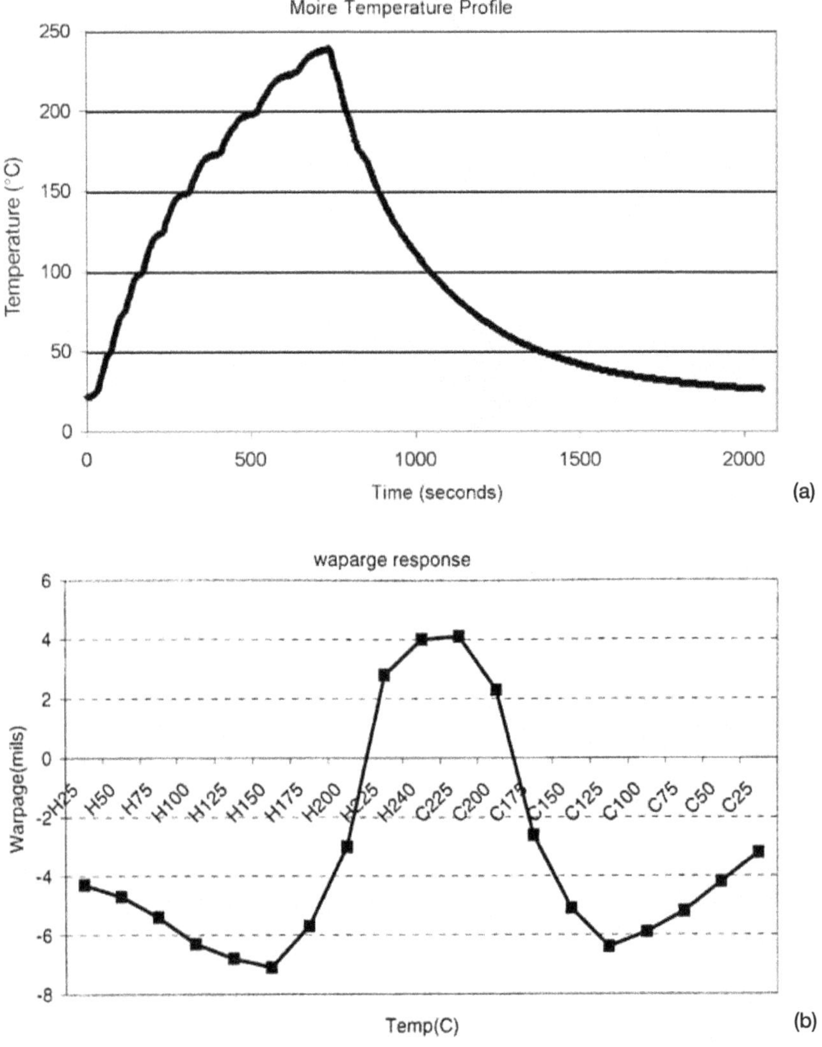

Figure 6.33 Package warpage measurement using shadow Moiré. (*a*) Temperature profile and (*b*) warpage response of a PBGA package (H = heating, C = cooling).

package failure analysis. Further improvement in the spatial resolution will extend its use in chip-level failure analysis.[49]

6.1.11 Dye Penetrant Leak Detection

An advantage of hermetically sealed ceramic packages over plastic packages is that they are not subject to moisture absorption. These

Figure 6.34 Scanning SQUID microscopy. *(Courtesy of Neocera.)*

packages use a metal or ceramic lid to enclose the chip in the package cavity. The metal to substrate seal should prevent any leak from the external package environment.

A hermeticity test using leak or seal testing on these package types confirms seal integrity. Leak testing is generally applied to devices that have designed cavities, as opposed to plastic or epoxy devices, which are fully enclosed or encapsulated. This is done in two steps.

1. Fine leak, which looks at leak rates in the 5 x 10^{-8} cc/sec range
2. Gross leak, which looks for devices with gross seal defects (see MIL-STD-883, Method 1014)[50]

When hermeticity test failures occur, the analyst must identify where the leak occurred in the packaged device without compromising the results. This leak can be extremely difficult and cumbersome to identify unless the proper equipment and techniques are used for investigation. A dye penetrant is often used to help identify leaky areas of the package seal.

Usually, one of the first places to look is at any seal of the package, particularly at the junction of dissimilar materials with different coefficients of thermal expansion. This includes the package pins/solder

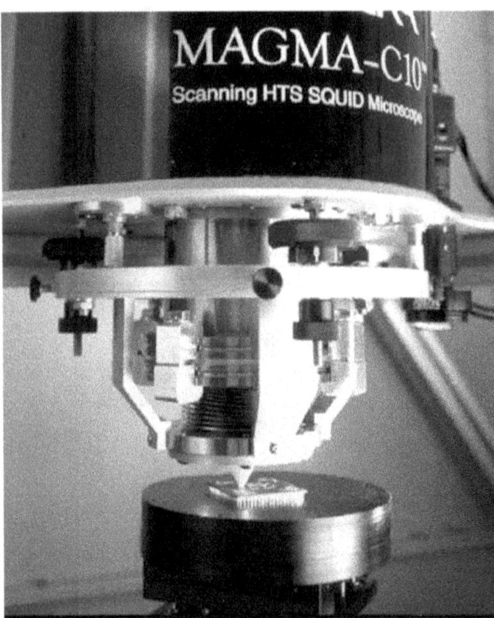

Figure 6.35 Detail of SQUID sample probe station. *(Courtesy of Neocera.)*

bumps to package interface and the lid seal to the ceramic package. The lid seal can be one of several types of material, some of which include glass, and others of gold or lead/tin. Initial inspection is typically performed by both low- and high-power optical inspection using either white light or confocal viewing with differential interference contrast or polarization. Subsequently, the analyst may choose to perform SEM analysis. This may be appropriate to view the interface seals, but usually it will not confirm the void causing the leak unless the leak is a major one. Most of the time, optical inspection is fruitless for leak detection unless a die penetrant is used.

Several types of die penetrants can be used. One example is Zyglo. This die can be used with stability under pressure to assure permeability. The following equipment and procedures are required for accurate leak analysis (see Fig. 6.36):

Equipment

- Zyglo die
- Cotton swabs
- Autoclave (with gauge)
- Clock

Figure 6.36 An autoclave, 10 ml of Zyglo, and a cotton swab used under a fume hood.

- DI or other pure rinse water
- Low- and high-power optical microscopes
- SEM (may be necessary)
- UV high-power optical microscope with 510-nm wavelength filter
- Heat lamp
- Low-pressure air

Procedures

- Prepare the autoclave with a small amount of water and metal tray (on which to place devices above the water level).
- Be aware of electrostatic discharge (ESD) precautions
- Pour about 10 ml of Zyglo in a beaker.
- Using a cotton swab, thoroughly coat the entire package with the Zyglo, making sure it reaches all exposed area of the package.
- Place the device on a tray (so as not to allow the water to touch the package) in the autoclave and adjust the temperature so that approximately 5 to 10 lb of pressure is attained.
- Keep this pressure constant for 30 min.
- After 30 min, turn off the autoclave and slowly release the pressure (using the release value on lid) and remove the device.
- Cautiously rinse all Zyglo die off the package with DI water.

- Carefully remove the package lid. Be certain to know which part of the lid was impacted by the instrument for lid removal in case damage occurred at that spot.

Observe the cavity under a high-power white light then a UV light microscope. The fluorescent Zyglo die that penetrated the porous area will be observed with a trail leading to the void.

Once the void area has been observed and confirmed, SEM analysis may be performed to produce a high-resolution image.

6.2 Analytical Techniques

Once a failure is identified, a deeper understanding of the failure mechanism often requires analysis of the elemental or chemical composition of the package materials and how they interact with each other. The analysis is most successful when the starting materials used in the package fabrication are initially well characterized so that deviations in a failed device can be noted with confidence. This section describes some of the more common techniques available to the failure analyst. A wide variety of techniques have evolved hand in hand with the continued development of semiconductor technology, and it is impossible to discuss every available method. In general, many types of probe (e.g., photons, electrons, ions, heat, electricity) have been used to interact with the material of interest, and the resulting interaction products (e.g., emitted electrons, ions, photons) or changes in the material (e.g., weight, temperature, conductivity) are recorded. A physical or chemical model of the interaction is used to relate the observable products or material changes to the composition of the material. In this section, two common element analysis techniques using the electron probe in the SEM are described: energy dispersive spectroscopy and auger electron spectroscopy. Because many package materials consist of insulating, organic, or polymer-based components, the X-ray photoelectron spectroscopy and Fourier transform infrared spectroscopy methods are described, which can provide useful chemical information on these materials. Finally, the wetting balance solderability test is an analytical procedure of particular interest in packaging technology, providing useful information on solder contacts. Many other analytical techniques that are not discussed here may be of use in particular applications. Techniques such as secondary ion mass spectroscopy (SIMS), Rutherford backscattering spectrometry (RBS), atomic force microscopy (AFM), ion chromatography, and inductively coupled plasma mass spectrometry (ICP-MS) are available at several analytical service laboratories. Such laboratories can be contacted for more information.

6.2.1 Energy Dispersive Spectroscopy

One of the most useful elemental analysis techniques available to the failure analyst is the measurement of the X-ray spectrum generated by the electron beam/sample interaction in a scanning electron microscope (SEM).[51] The majority of incident electrons lose their energy within the top 1 to 2 μm of the sample by a variety of inelastic scattering processes, including atomic ionization. When the atoms return to their normal energy states, X-ray photons are emitted with specific energies characteristic of the electronic energy states of the particular atom. If an appropriate detector is used to collect the emitted X-rays, a spectrum is generated that can be used to identify the distinct elements present in the sample. In addition, the relative areas of the X-ray peaks provide a measure of the relative concentration of the various elements. For typical SEM beam intensities and spectrum collection times, distinct X-ray peaks can be observed for elements present at concentration levels of 0.5 percent or higher. This is a powerful tool that allows element analysis to be carried out in conjunction with SEM imaging. For example, an X-ray spectrum can be acquired with the electron beam focused at a specific point of interest on the sample to help identify the composition of a particle or contaminant residue. Alternatively, X-ray energies corresponding to specific elements can be monitored during acquisition of a SEM image to generate composition maps.

There are two types of detectors that can be added to a SEM to collect X-rays. The most common is the energy dispersive X-ray detector, which uses a lithium-drifted silicon (SiLi) diode, reverse biased so as to form a large depleted region. When an X-ray photon impinges on the crystal, the total energy of the photon is transferred in the depleted region, generating a number of electron-hole pairs. Essentially, the electron-hole pairs add up to a current pulse that is proportional to the energy of the X-ray photon. An electronic pulse counting system displays the count intensity as a function of pulse height. The pulse height is correlated to the X-ray energy by calibration with standard samples. A complete X-ray spectrum can be acquired in minutes. The convenience and relatively compact size of the energy dispersive detector makes it the most widely used detector for X-ray analysis, giving the technique its common name of energy dispersive spectrometry (EDS). Although extremely useful, the energy dispersive detector has two limitations. First, the resolution is limited to ~135 eV by the properties of the silicon crystal, which can make element identification more difficult in certain cases involving overlapping peaks. Second, light elements (atomic number < carbon) with X-ray energies less than 0.5 keV are not measurable because of energy loss in the detector material in front of the active region of the crystal. To overcome these limitations, the wavelength dispersive detector is sometimes used. This detector consists of a crystal monochrometer that spatially separates the X-rays based on their wavelength. This detector has much higher resolution and can measure low energy X-rays with an appropriate crystal. However, it takes much longer and requires changing to different monochrometer crystals to collect the entire spectrum. Hence, wavelength dispersive spectrometry (WDS) is typically used for specific applications such as, for example, quantitative analysis of boron.

The majority of X-ray emissions from most elements (atomic number greater than carbon) fall within an energy range of 0.5 to 10 keV. The SEM beam energy must be at least as high as the desired peak to be measured. The typical rule of thumb is to use a beam energy that is 1.5 times the energy of the highest X-ray peak of interest. The depth of the sample that is probed ranges from 0.5 to 2 μm and is proportional to the beam energy. In addition, even though the electron beam can be focused to a very small diameter (~0.02 μm), the actual lateral interaction with the sample is greater as a result of elastic scattering collisions that can divert the electron path laterally with respect to the incident direction. At typical beam energies of 15 to 20 keV, X-rays are generated from a roughly spherical volume in the sample approximately 1 μm deep and 1 μm wide with respect to the incident beam. The analyst must always consider the actual analyzed volume when interpreting EDS spectra, especially when examining particles or features on the sample that are smaller than 1 μm. The X-ray spectrum from the surrounding area must be known to properly interpret the spectrum from the feature of interest.

Qualitative analysis to identify the elements in a sample is the most widely used application of the SEM/EDS technique. Each element has a unique X-ray emission signature corresponding to the allowed electron transitions between atomic energy levels. X-ray peaks are labeled with the letters K, L, or M, which refer to electron transitions to the $n = 1$, 2, or 3 atomic energy levels, respectively. It is always best to look for all major peaks for a given element to be sure of its identity. Peak energies for each element are conveniently tabulated in the software packages that come with most commercially available EDS detectors. The software also tabulates various X-ray artifacts, such as sum and escape peaks, that are part of an X-ray spectrum. The artifact peaks must all be accounted for to avoid false element assignments. Elements present at the spot being probed at concentrations greater than ~0.5 percent can quickly be identified from the measured spectrum. Many software packages have the option to automatically identify every peak in the spectrum and list the elements. This should be used only as a guide. The analyst should carefully check for artifact peaks and overlapping peaks to ensure a valid qualitative analysis. Some knowledge of the sample and its processing history always helps in this regard.

Figure 6.37 shows an example of X-ray analysis of a foreign particle observed on a semiconductor chip sample. Two spectra are overlaid in the figure—one taken with the SEM beam focused on the unknown particle and one taken with the SEM beam focused on the area surrounding the particle. Comparison of the spectra shows the presence of iron, chromium, and nickel in the particle, which are the common components of stainless steel. Figure 6.38 shows another example of element analysis at a solder-ball-to-package interface. The X-ray spectrum was obtained with the electron beam focused at the solder to package interface. The lead and tin components of the solder are observed as well as nickel and copper from the contact pad on the package.

Once the elements are identified by collecting complete X-ray spectra at one or more spots on the sample, the distribution of specific elements throughout the sample may be of interest. The software packages available with most EDS systems allow element maps to be collected so that element distributions can be compared directly with the SEM image. In element mapping, specific X-ray energies corre-

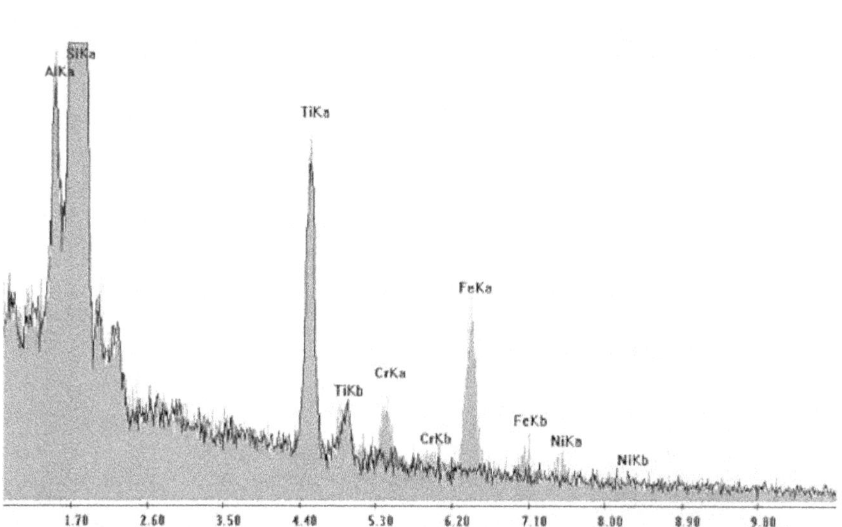

Figure 6.37 Example of qualitative analysis of a stainless steel particle.

EDS spectrum from solder ball interface

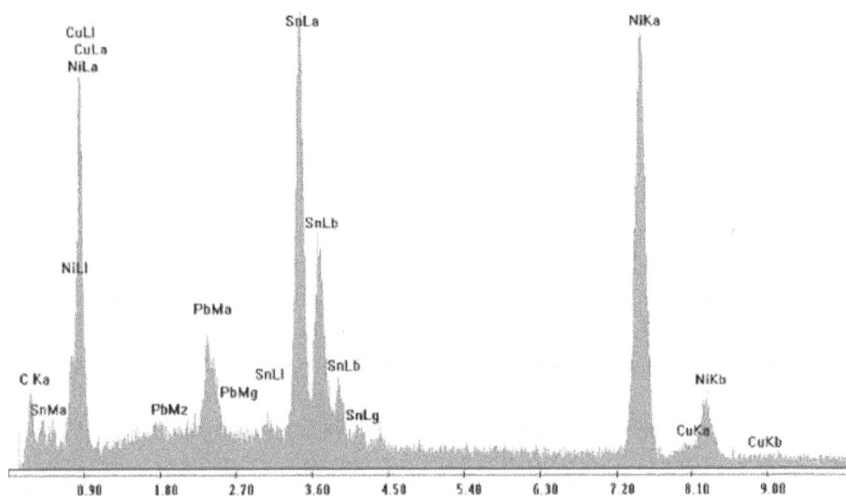

Figure 6.38 Example of element analysis at solder bump interface.

sponding to one or more elements of interest are monitored as the SEM image is collected. Figure 6.39 shows an example in which the SEM image of a solder ball to semiconductor chip interface is overlaid with the copper element map. The copper is observed to diffuse into the lead-tin solder. The copper element map can be obtained as a function of reflow temperature and time to study the metallurgical interactions. Figure 6.40 shows another example, in which the composition of an embedded particle is determined. The optical image on the left side of the figure shows the presence of a foreign particle in between two metal runners after polishing into the package. The EDS element map on the right identifies carbon as the main component of the particle.

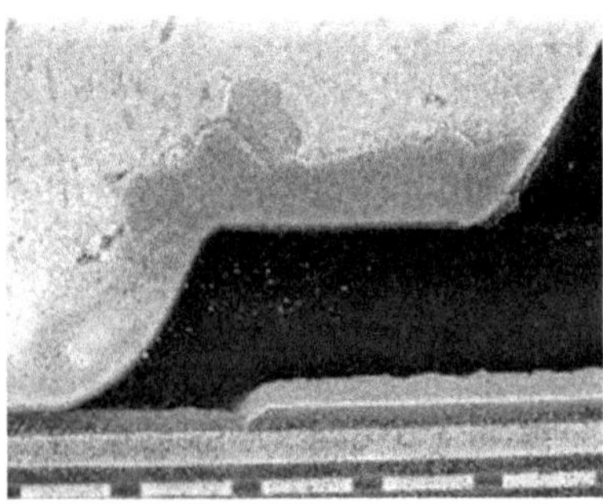

Figure 6.39 Image of solder-ball-to-chip interface with overlay of copper element map.

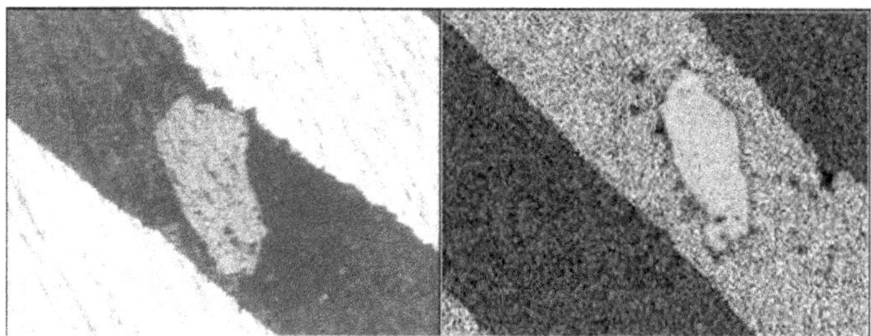

Figure 6.40 Analysis of embedded particle. Optical image on left shows a foreign particle between two metal runners in lead frame. EDS element map image on right shows that the particle consists mostly of carbon.

Finally, quantitative compositional analysis is possible with the EDS technique, since the X-ray peak area is proportional to the concentration of the associated element. However, several correction factors related to the matrix and depth of origin of the emitted X-ray must be accounted for to obtain accurate results. The software included with most EDS systems contains algorithms for standardless quantitative analysis, but these should be used with caution. If accurate quantitative analysis is required, results should be checked carefully against known standards.

6.2.2 Auger Electron Spectroscopy

Auger electron spectroscopy (AES) is an element analysis method with the unique advantages of high spatial resolution combined with surface sensitivity.[52] The technique probes the sample using a high-energy (typically 10 keV) electron beam identical to that used in a secondary electron microscope (SEM). The electron beam–sample interaction consists of a variety of possible inelastic scattering processes, including atomic ionization through ejection of an inner shell electron. Individual atoms can return to their normal energy states by the transfer of an outer shell electron to the empty inner shell, with an accompanying emission of energy. This energy emission can take the form of an X-ray photon (which is the basis for X-ray microanalysis in the SEM) or an ejected electron (which is known as an Auger electron and is the basis for AES). The probability of emitting an Auger electron versus an X-ray photon increases with decreasing atomic number, thus making AES a very good method for analysis of lighter elements (atomic number < 35). In addition, in contrast to X-rays, electrons can travel only a short distance through the sample before losing energy in a collision. Thus, only the Auger electrons emitted in the top 30 Å or less of the sample will be detected, giving the technique its surface sensitivity. This also allows the highest possible spatial resolution, because only Auger electrons formed within the surface diameter of the probe beam will be detected.

Auger electron energies are typically in the range of 100 to 1000 eV and are characteristic of the atomic energy level differences in a particular atom. An electrostatic energy analyzer is used to detect and measure the energy of Auger electrons as the probe beam impinges on the sample. Auger spectra consist of a series of sharp peaks on top of a broad background formed by other secondary electron emission events. The derivative of the spectrum is normally displayed to reduce the background and highlight the sharp Auger peaks. Spectral libraries of Auger electron energies for every element are available. As in X-ray analysis, the measured peak energies can be used for qualitative element identification, while the relative peak heights can be used for quantitative analysis. The technique is capable of detecting elements present at concentrations of about 0.5 to 1 percent.

Just as in a SEM, the electron probe beam in the Auger spectrometer can be focused at a specific point of interest or rastered across a fixed area. The resulting Auger spectra can be used to survey the surface composition at specific sites or across the imaged area. Element maps can also be obtained by monitoring Auger electron energies related to specific elements while the probe beam rasters across the imaged area. This particular method is known as scanning auger microscopy (SAM). Finally, an argon ion sputter source is usually added to the Auger spectrometer to add capability for surface cleaning and depth profiling. Auger peaks for one or more elements can be monitored during sputter removal of material from the sample to obtain concentration depth profiles.

Because of its unique capabilities, proper sample preparation and handling is required to get the most out of the technique. Because AES depends on measuring precise electron energies, insulating samples are extremely difficult to analyze. Charge buildup on insulating surfaces from the primary electron beam makes it almost impossible to measure auger peak energies reliably. In these cases, X-ray photoelectron spectroscopy (XPS) may provide an alternative for composition analysis, although spatial resolution is sacrificed. Furthermore, the high surface sensitivity of AES requires an ultra-high vacuum environment to minimize adsorption of background gases on the sample surface, and this increases the cost of instrumentation. In addition, care must be taken during sample preparation to avoid adding surface contamination that will interfere with the desired analysis. A few seconds of argon ion sputter cleaning after the sample is placed in the instrument is typically done to remove superficially absorbed carbon and oxygen contamination on the surface.

AES is a powerful analysis tool that can be used to study semiconductor package structures during development, qualification, and manufacturing phases. It is particularly useful for characterization of surface contamination, metal bonding and other interfaces, metallurgical grain boundaries, and other instances where microstructural composition analysis is required. For example, delamination and bonding failures can be characterized by mechanically prying the failed interface apart and using AES to study each side of the failed interface layers. The technique is well suited through its depth profiling capability to follow interdiffusion of dissimilar layers as a function of time and temperature, to look for unusually thick oxide contaminant layers at metal interfaces, and to find traces of chemical process residue. As one of the few analytical tools with both high spatial resolution and surface sensitivity, AES is useful for studying failures related to surface and interface problems in micron-sized structures.

6.2.3 X-Ray Photoelectron Spectroscopy

X-ray photoelectron spectroscopy (XPS) is also referred to as electron spectroscopy for chemical analysis (ESCA). It is a surface-sensitive element-analysis technique[53] that also provides information about the chemical environment of the measured element. It is a nondestructive technique in the sense that the sample is preserved for further analysis. However, the sample likely requires some modification to analyze the section of interest, as it must be placed in an ultra-high-vacuum environment. The specialized instrument required to obtain XPS data is generally not found in most FA laboratories, but the technique is readily available at various analytical service laboratories. This technique would mainly be used to study mechanisms of failure involving chemical changes of specific package components over time, such as corrosion effects or other electrochemical processes. As such, the technique would be used in special circumstances to study unique problems related to chemical change in package materials that lead to reliability problems.

The technique uses a monochromatic X-ray beam (typically Al $K\alpha$ at 1.49 keV) focused to a spot size of ~200 μm to probe the sample. The incident X-rays interact with the atoms in the sample to eject core electrons, whose energy is defined as the difference between the incident X-ray energy and the electron binding energy. Electrons ejected from atoms that are deeper than 30 Å below the surface do not have sufficient energy to escape. Hence, the technique provides surface information only, unless argon ion sputtering is used to systematically "sandblast" material away as XPS data is collected. However, sputter depth profiling is generally not done if chemical information is desired, as the ion sputtering itself will alter the chemical bonding in the sample. The emitted photoelectrons are energy analyzed to yield a spectrum of electron intensity versus binding energy. Each element has characteristic peaks, typically in the range of 100 to 1000 eV, corresponding to its unique electronic energy levels. Figure 6.41 shows a sample XPS survey spectrum from an adhesive tape used in a package application. In addition, peaks for a specific atom will shift in energy, depending on the nature of the chemical bonds to its surrounding atomic neighbors. The magnitude of the chemical shift depends on the degree to which the atom's core electrons are affected by the surrounding chemical bonds. The energy analyzer can resolve changes in peak energies less than 1 eV, so it is possible to detect small changes in chemical environment for a specific element. Several closely spaced peaks may be observed that correspond to different bonding states for the same element. The relative areas of the peaks can be used to calculate the relative amount of that element in each particular chemical state. Changes can then be monitored with time and temperature. An example of the chemical shifts observed for the carbon peak in a particular tape sample is shown in Fig. 6.42.

6.2.4 Fourier Transform Infrared Spectroscopy

Fourier transform infrared spectroscopy (FTIR) is a nondestructive analysis method[54] that is useful for characterizing materials used in semiconductor package fabrication such as molding compounds, epoxies, adhesive tapes, solder fluxes, and so on. These materials contain covalently bond compounds, such as polymers or organic molecules, which can absorb infrared radiation through changes in bond stretching, bending, or rocking vibrational energies. The allowed transitions

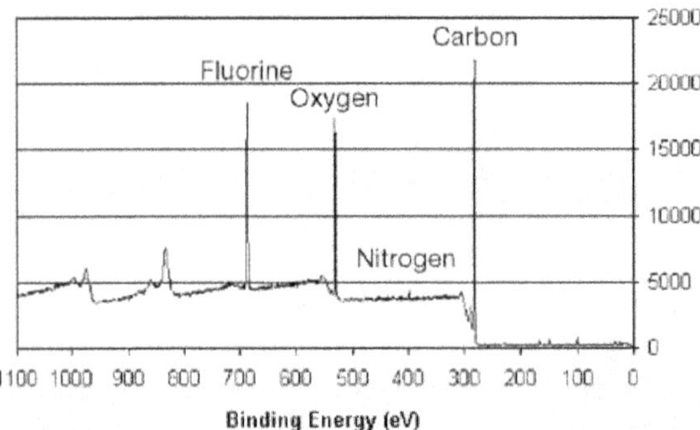

Figure 6.41 Example of XPS spectrum from a polymer tape sample.

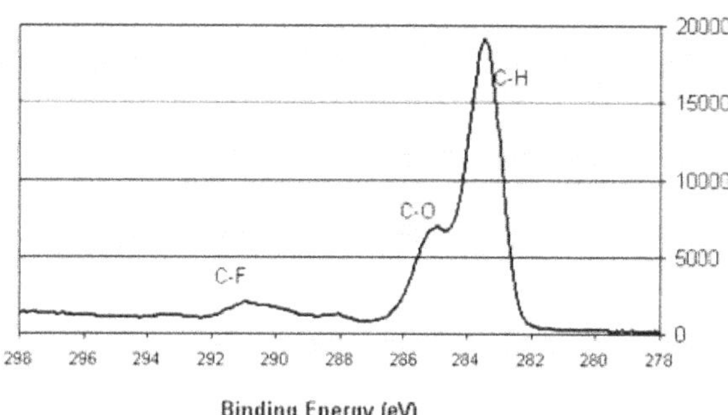

Figure 6.42 Detail of carbon peak in previous XPS spectrum from a polymer tape sample.

between vibrational energy levels follow specific rules such that unique absorption frequencies are observed. Infrared spectrometry is the general technique of irradiating the sample and measuring its absorptions in the wavelength range of 2 to 16 μm. IR spectra are typically displayed showing either transmittance or absorbance as a function of frequency measured in wave numbers (a wavelength range of 2 to 16 μm corresponds to a wave number range of 5000 to 625 cm^{-1}, respectively). In general, IR frequencies are suggestive of specific functional groups within the molecule. For example, carbonyl (–CO) groups show characteristic bond stretching frequencies in the range of 1600 to 1900 cm^{-1}, regardless of the rest of the molecule. Characteristic frequencies for a wide variety of functional groups have been tabulated in the literature. Extensive spectral libraries are also available for many individual compounds. Figure 6.43 shows an example of an IR spectrum from a polyimide tape, which plots absorption versus wave number.

Fourier transform infrared spectroscopy refers to a specific method for acquiring the IR spectra. It uses an interferometric technique to generate an interference pattern that is the fourier transform of the complete IR absorption spectrum. The FTIR approach allows rapid, repetitive spectrum acquisition giving a high signal-to-noise ratio. Modern FTIR instrumentation has been miniaturized to the point at which it can be added to an optical microscope at relatively low cost. Hence, the failure analyst can examine the sample optically and then choose to obtain an FTIR spectrum from a specific area of interest on the sample. Special lenses are included on the optical microscope to allow coupling of the infrared radiation to the sample.[55] Because most samples of interest to the failure analyst are solid, an attenuated total reflection (ATR) method is used to couple the IR energy to the sample. In this method, an infrared transparent lens is gently placed in con-

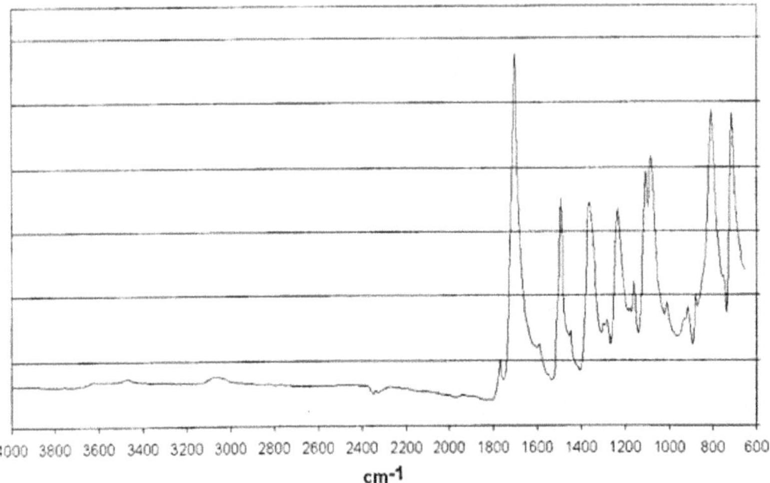

Figure 6.43 Sample FTIR spectrum from a polyimide tape.

tact with the sample. The infrared radiation essentially interacts with the top 2 to 4 μm of the sample as it reflects from the lens–sample interface. The resulting spectrum represents the IR absorptions observed from the sample in a volume that is ~10 to 100 μm in diameter and ~ 4 μm in depth from the surface.

In terms of failure analysis applications, FTIR can be used for analyzing contamination residues, tapes and adhesives, and various other components of the package that consist of covalently bonded molecules. The infrared spectra of these compounds are generally very complex. This makes conclusive interpretation of the spectrum from an unknown material very difficult. On the other hand, since each compound has a unique infrared spectral "fingerprint," it can be a useful way of monitoring changes to known materials in the package or the presence of organic contaminants (such as flux residues). The technique is most useful if the starting materials used in package fabrication are characterized initially to establish base line information. The reference spectra, along with standard spectral libraries of known compounds, can then be used when analyzing spectra taken from a failed device package. Comparison between spectra from good and bad examples of the same package also helps to eliminate uncertainty in the analysis. The convenience of the optical microscope based FTIR technique allows rapid collection of infrared spectra from multiple samples or multiple spots on one sample. Because the technique is nondestructive, further analysis or testing can be done on the sample after obtaining FTIR spectra. This will likely become an increasingly popular addition to the failure analysis laboratory.

6.2.5 Wetting Balance Solderability Test

The wetting balance solderability test for component leads and terminations is a solder force measuring device that provides dynamic quantitative information of the wettability of solderable surfaces of components.[56,57] The instrument consists of a sensitive linear variable differential transducer for measuring the solder wetting forces that act upon the solderable leads of a component as it is immersed into a molten solder bath (generally consisting of 63 percent Sn/37 percent Pb for Sn/Pb soldering or a Sn/Ag/Cu alloy for Pb-free soldering).[58] A computer controls the immersion rate, depth, and dwell time and records the solder wetting forces as a function of immersion time. A general characteristic wetting balance curve for a wettable sample is shown in Fig. 6.44. The curve consists of four characteristic regions.

A. An initial nonwetting segment produced by the sample buoyancy
B. A rapid ascending region generated as the solder begins to metallurgically wet and alloy the metal surfaces
C. A static meniscus segment produced as a result of equilibrium wetting forces
D. The maximum force region generated as the sample is lifted out of the solder bath

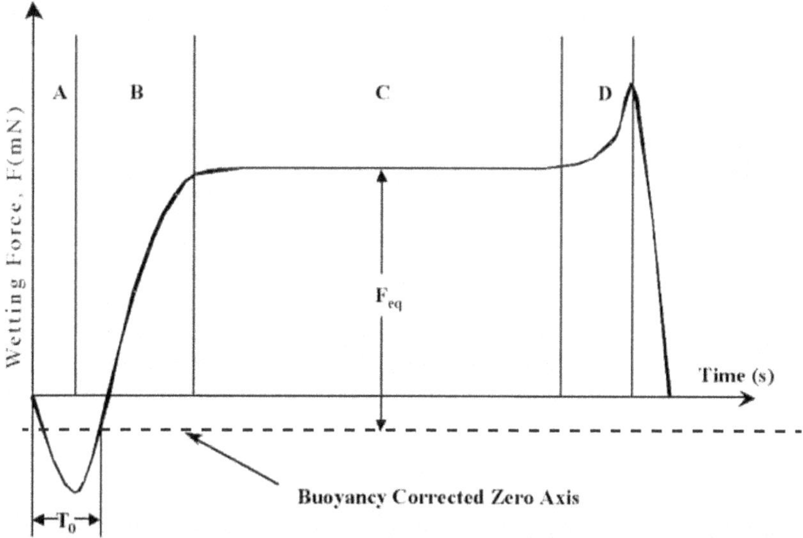

Figure 6.44 Characteristic wetting force–time curve.

As noted in Fig. 6.44, a buoyancy-corrected zero axis is applied to the wetting force curve to normalize the curve so as to adjust for sample size differences. The buoyancy force correction, $F_{b,}$ is simply a function of the total volume of the of the immersed sample and its leads as related by

$$F_b = d \times g_n \times V$$

where d = density of the solder at the test temperature
g_n = acceleration of gravity
V = immersed volume of the sample

Good solderability will be related to several effects, which include the time to cross the buoyancy corrected zero axis (T_0), the wetting equilibrium force (F_{eq}), and the total integrated area of the force curve. Coupled with a visual inspection of the soldered leads after completion of the test, an assessment of the degree of solderability performance for the device under test can be made.[59] Pass/fail criteria may be determined through criteria established in industry specifications such as IPC/EIA/JEDEC J-STD-002B.[60]

6.3 Electrical Techniques

6.3.1 DC Electrical Characterization

The primary reason for performing DC electrical characterization is to determine if the product conforms to its parametric specifications. Automatic test equipment (ATE) is initially used to test for this conformance. The ATE produces a datalog of reported failures, including open pins, shorted pins, and pins exhibiting leakage current that exceeds the specified current limits. Test equipment in the analysis lab is then used to confirm the reported failures as compared to a known good reference.

Figure 6.45 is a simplified schematic diagram of electrostatic discharge (ESD) protection circuitry typically found on the chip adjacent to the bond pads on each signal pin. The protection diodes are designed to absorb positive (upper diode) and negative (lower diode) transients that may be applied to the pin. Of course, there are limitations to the amount of energy the protection circuitry can handle, and electrical overstress due to misapplication or mishandling will cause permanent damage to the circuit. From a packaging standpoint, these protection diodes provide a point of reference for assessing potential package issues related to the associated protection circuit. The absence of current flow in the protection circuit is indicative of an open somewhere in the path. This is the first indicator of a possible open in

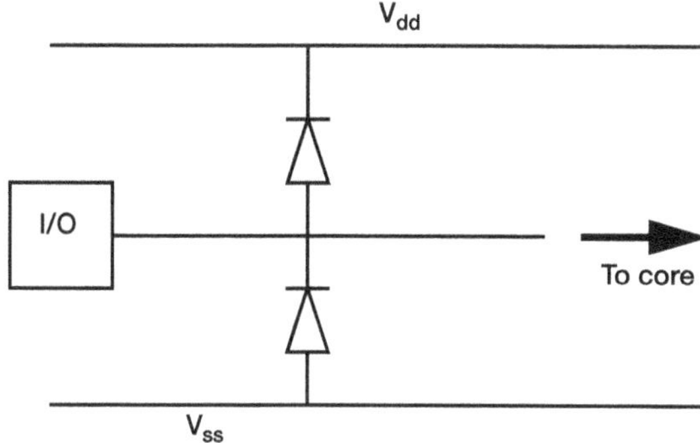

Figure 6.45 Simplified schematic diagram of ESD protection circuitry.

the package. The majority of package defects manifest themselves as opens.

A curve tracer may be used to stimulate the protection circuit in the lab, and it will display a waveform such as the one shown in Fig. 6.46. Trace 1 is a normal voltage (x-axis) versus current (y-axis) diode curve. Current flows once the PN junction forward bias voltage has been reached. Trace 2 shows no current flow and thus an open in the current path. It should also be noted that not all pins exhibit the same characteristic when electrically stimulated. For this reason, it is al-

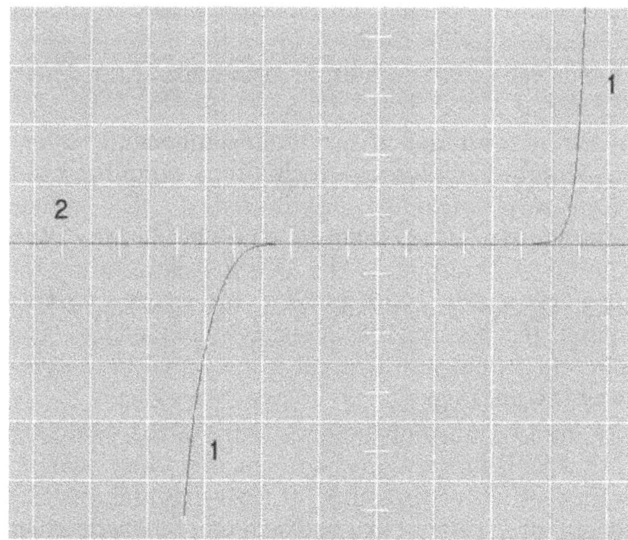

Figure 6.46 Voltage vs. current curve.

ways advisable to compare failing parts to a reference part. Once failed pins have been confirmed in the lab using this method, time domain reflectometry (TDR) (refer to TDR analysis section) may be used to determine if the cause of the failure is related to the package, the die, or the interconnections between the two. Further analysis, such as via electrical package probing, is then performed to confirm the electrical analysis. Once the defect is isolated to the package, parallel lapping or cross-sectional analysis may be used to physically locate the package defect.

6.3.2 Time Domain Reflectometry

Time domain reflectometry (TDR) is a method of detecting loading effects along electrical transmission paths. TDR has been used for many years to detect defects in signal cable paths. More recently, the same technology has been applied to failure analysis of microelectronic packages. TDR is a transient analysis technique that allows the user to closely inspect waveform perturbations that may relate to a deficient electrical path. This is made possible through the use of software tools that aid in the analysis of the reflected signal. The reflected wave holds the details of the impedance profile of the signal path being interrogated.

A typical application involves the injection of a high-frequency signal into a suspected bad pin of a package. The reflected signal is then compared to signals taken from known good substrates/modules. Analysis of the signals leads to conclusions about the presence and location of the defect. Simulations can be done, given the material properties along the transmission path to aid in the resolution of the defect site.

A typical TDR system is composed of two main components: a test fixture and a digital sampling oscilloscope, including a sampling head (refer to Fig. 6.47). Accessories include device under test (DUT) guides and DUT pin location templates (alignment tools) tailored to the BGA package type. They are designed to allow the user to locate the x-y coordinates of the package pin to be tested. An alternative to this method is an automated test fixture that identifies the package type and lead or ball pitch and automatically applies the signal to the preprogrammed points on the package.

As normal practice, and because of the sensitivity of the sampling head, it is essential that ESD precautions be taken, including wearing a garment woven with conductive thread, ESD footwear, and an ESD wrist strap. Prior to applying a signal to a package pin, the oxide must be removed from the areas of the package leads that the probe of the fixture contacts. This is accomplished by rubbing the device on an or-

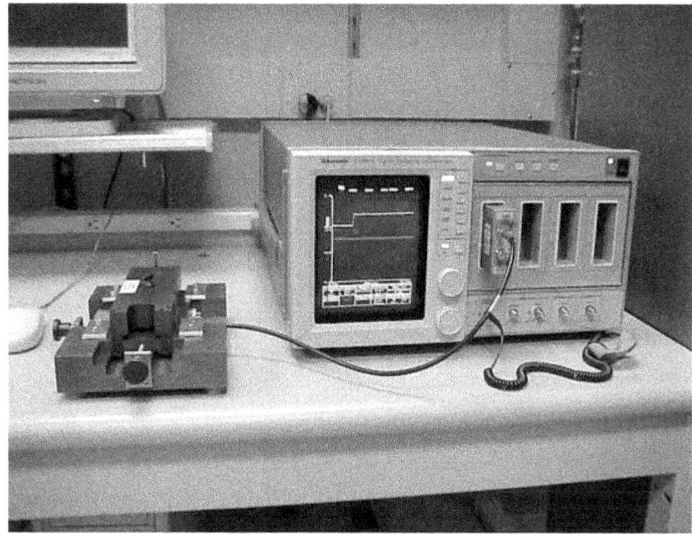

Figure 6.47 Altair Microwave T-1126 text fixture and Tektronix 11801C digital sampling oscilloscope.

dinary piece of copy paper while an air ionizer is being directed toward the sample.

6.3.2.1 Data interpretation. The following are several examples of typical TDR analysis results including both package and die related issues. The predominant failure mode in BGA packages is package substrate opens. Signal variations occur in proportion to impedance mismatches that the signal encounters on its path. In simplified terms, signal deviations (from the reference signal) in the positive direction indicate an open, and signal deviations (from the reference signal) in the negative direction indicate a short. The traces shown here were taken from various package types and dies, which explains the variations in reference wave shapes from example to example.

Trace 1 in Fig. 6.48 represents an open fixture with no device installed. Trace 2 represents the signal from the DUT. Trace 3 represents the known good signal from the reference device. Figure 6.48 is a classic example of a package-related open. Note the close proximity of the rising edge of the reference trace to the rising edge of the DUT trace. This represents a small differential in time between the two reflected signals. In this case, the open is clearly in the package substrate, close to the package ball.

In Figure 6.49, trace 1 represents an open fixture with no device installed. Trace 2 represents the reference signal, and trace 3 is the

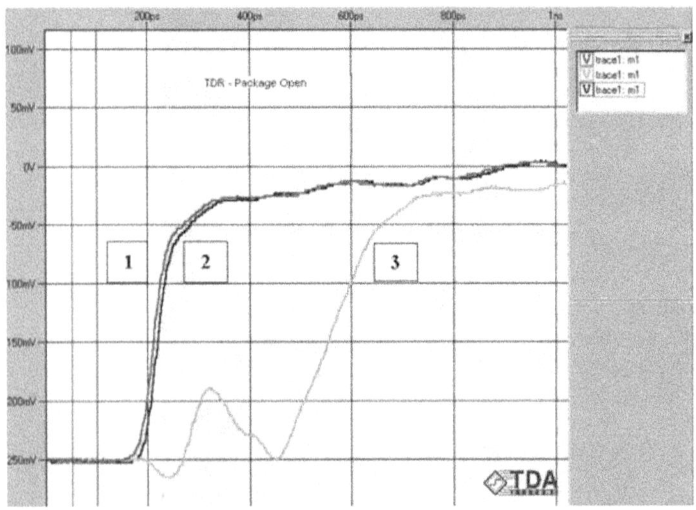

Figure 6.48 Package open.

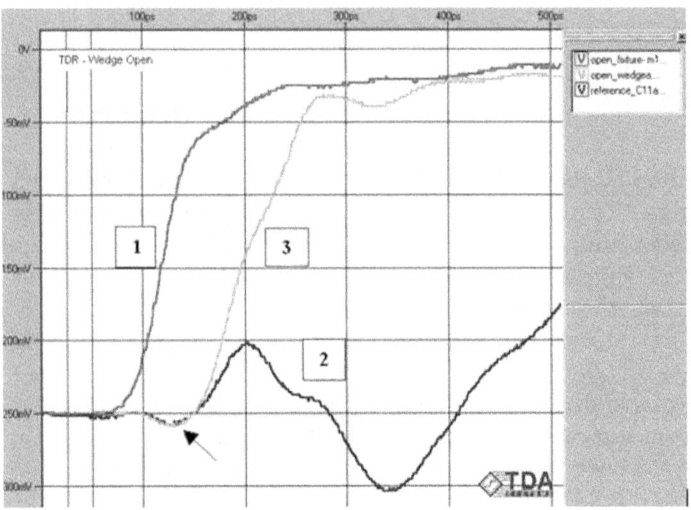

Figure 6.49 Wedge open.

DUT signal. The time scale was changed from 200 ps/div in Fig. 6.48 to 100 ps/div in Fig. 6.49 to enhance the resolution at the points of interest. In Fig. 6.49, the wire between the wedge bond and the ball bond was intentionally severed at the wedge to produce the curve. The trace segment between the arrow and the peak of trace 2 represents the impedance of the gold bond wire.

The peak of trace 2, marked with an arrow in Fig. 6.50, represents the ball-bond-to-bond-pad interface. The DUT signal (3) deviates from the reference signal (2) at the bond pad. In this case, the ball bond was lifted as a result of die top delamination. Trace 1 on this image represents an open fixture with no device installed.

In Figure 6.51, the initial peak of trace 2, indicated with an arrow, represents the ball bond/bond pad area. The signal beyond this peak represents reflections from the silicon. The negative deviation of the DUT trace (3) from the reference trace (2) indicates a short in the die. Trace 1 on this image represents an open fixture with no device installed.

6.3.2.2 Summary. Electrical probing of a package is performed after decapsulation to confirm package related opens or, in rare occasions, package substrate shorts found during TDR analysis. This adds an additional level of confidence in the TDR findings. Once a package open or short is confirmed through probing, package FA may be pursued. Package FA involves lapping and/or cross-sectioning of package with the aid of substrate drawings and any site localization data that may be gleaned from TDR results or data gathered through other site localization techniques. This process could be tedious and time consuming.

All packages analyzed in this overview were of the "wired" variety (non-flip chip). Flip chip analysis offers another level of complexity, as the chip is not electrically connected to the package with wires. Therefore, there is less line impedance at those transition points on which to

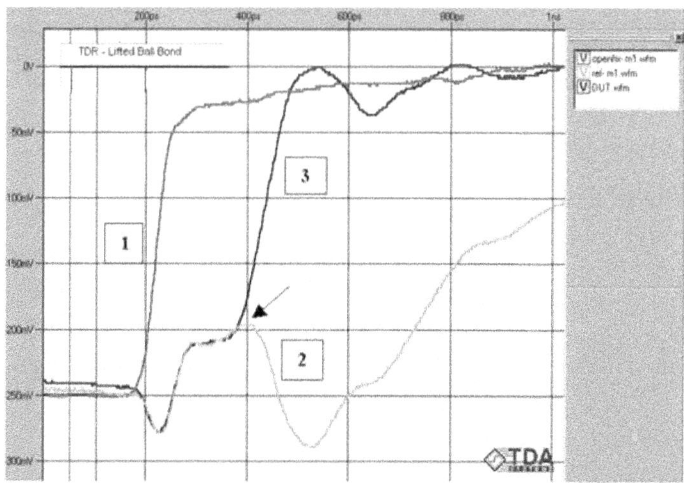

Figure 6.50 Lifted ball bond.

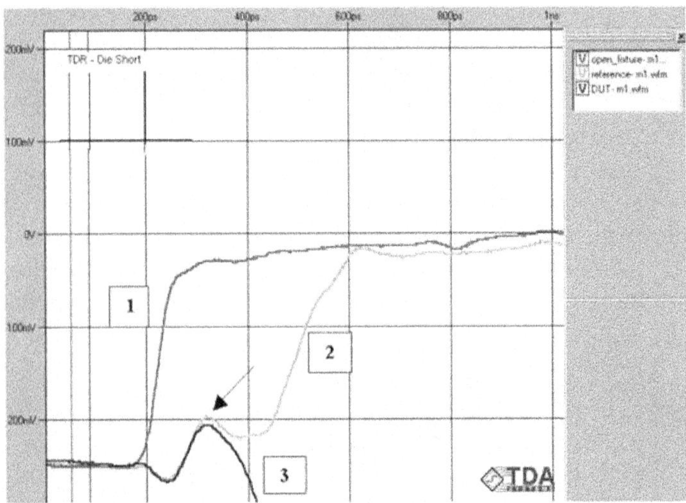

Figure 6.51 Die short.

key. When analyzing flip chips, it is beneficial to have known good substrates with and without die for use as references. Multichip modules (MCMs) also present a unique challenge, because a number of chips in the same package may share a common package pin.

6.4 Destructive Techniques

6.4.1 Package Decapsulation Techniques

Different techniques are used to dismantle a package, depending the intent of the investigation or the expected failure mode. The techniques can be grouped into two categories: chemical and mechanical.

1. Chemical
   - Automated decapsulation machine
   - Hot plate and acid drop method
   - Acid bath
2. Mechanical
   - Mechanical polish—cross-sectional, planar polish, or other orientation
   - Mechanical separation

The mechanical polish technique is usually preferred when subtle anomalies are expected, as the chemical technique can etch away important evidence. Failure modes such as cracks, delamination, opens, shorts, and leakage are best identified with a mechanical polish technique. The selected mechanical approach will depend on the area of interest, electrical failure, package type, expected failure, and physical attributes. Chemical techniques may require experimentation with various chemical compositions and mixtures, as some package materials can be very difficult to etch successfully without damaging features that will need to be analyzed.

6.4.1.1 Chemical techniques

Automated decap machine. Automated decap machines work well for clean, consistent decap while maintaining the electrical integrity of the device. Practice samples are typically used to develop recipes for particular packages to improve results on real samples for analysis. Different chemical recipes are used depending on the composition material of the package. Most recipes include fuming nitric acid, sulfuric acid, or a mixture of the two. The automated machine will heat the package and acid to a specified temperature (85°C, for example) and flow the acid over the package to etch away the molding compound. A type of gasket is employed to confine the etch to a specified area for each package type. This decap method can be used to expose the complete die in the package for wirebond, ball bond inspection (refer to Fig. 6.52). Site specific decap can also be performed to expose areas of paddle, lead frame, and wedge bonds.

Figure 6.52 Typical package decap to expose die, ball bonds, and part of the bond wires.

Hot plate and acid drop method. The hot plate and acid drop decap method is similar to the automated method but requires only a hot plate and a dropper to apply acid. Obviously, this technique is not as well controlled as the automated technique, but it can accomplish similar results with a very low investment. Again, fuming nitric acid, sulfuric acid, or a mixture of the two can be used with a hot plate set to approximately 85°C. The typical procedure would be as follows:

1. Set hot plate to desired temperature and allow to preheat
2. Place device on hot plate
3. Drop acid on the device in the area of interest
4. Allow the acid to react (foam up)
5. Rinse device in acetone
6. Repeat until desired area is exposed
7. Ultrasonic clean in solvent

Improved results can be realized be providing some way of restricting the acid to a particular area. Thicker packages can be milled to target etch areas and to speed up the etch time. Other techniques, such as surface roughing, can allow the acid to "bead" in the target area and restrict etch spread.

Acid bath. If the die is the only area of interest, another technique for molding compound removal is to simply place the device in an acid bath, which will remove all the molding compound material and most metals from the package, leaving just the die. This technique is rather crude and destructive, but may accomplish the task at hand.

6.4.1.2 Mechanical techniques

Mechanical polish. Mechanical polish techniques, either planar or cross-sectional, typically employ an apparatus that will hold the sample while a wheel with a pad provides for material removal. A sample may be "potted" first, which means that it is encased in a hard epoxy to maintain mechanical integrity during the polish process. The sample may also be cut first with a diamond saw to get near the target area. As material is removed and the target area gets closer, the polishing material is replaced with successively finer grit to provide a smooth, mirror-like finish suitable for high-magnification optical or SEM inspection. If a heat slug exists in the package, the slug can be polished off or removed before further analysis (such as acoustic microscopy or X-ray analysis) so as to obtain an unrestricted view of the target features. Refer to the "Mechanical Polish Technique" section for a more detailed procedure.

Mechanical separation. This technique works well for solder ball attached packages where the intent is to separate the package from a board or header without allowing slurry or cutting fluid to contaminate the sample (refer to Fig. 6.53). Also, the mechanical strength of each interface is tested while the parts are separated, allowing evaluation of the materials separated. This technique can be accomplished by prying the two sections apart with a prying tool.

6.4.2 Mechanical Polish Techniques

This section covers deprocessing techniques that allow analysis into the semiconductor package at specific locations or interfaces, depending on the sample orientation during the polishing process. A sample may be cross-sectioned or polished in any orientation but typically is done either by cross-sectional or planar polish (refer to Figs. 6.54 and 6.55). Cross-sectional analysis may provide a clear view of interfaces through a multilayer package, whereas a planar polish may provide a large area view of a specific layer. Either technique is frequently used to characterize interfaces in a package or to identify anomalies such as cracks, particles, voids, and electrical and/or mechanical issues. Once an anomaly is isolated and identified by these polishing techniques, further analysis and characterization can be performed by optical microscopy, SEM/EDS, or any other useful technique to identify the source.

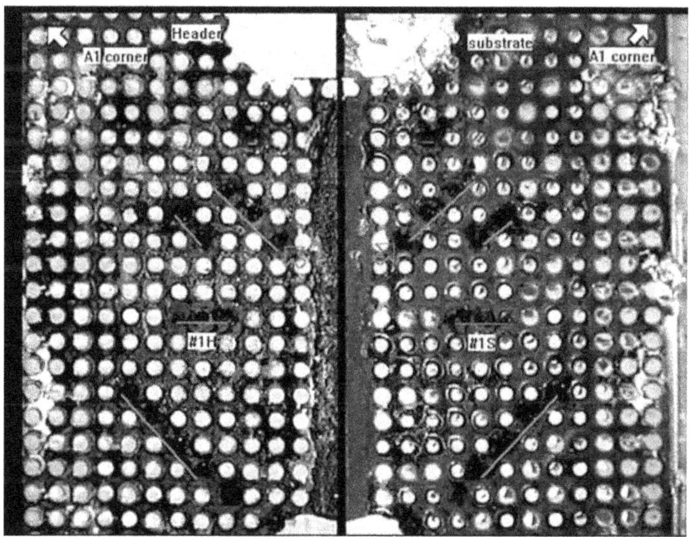

Figure 6.53 Image of two sides after mechanical pry apart, showing anomalies at solder balls.

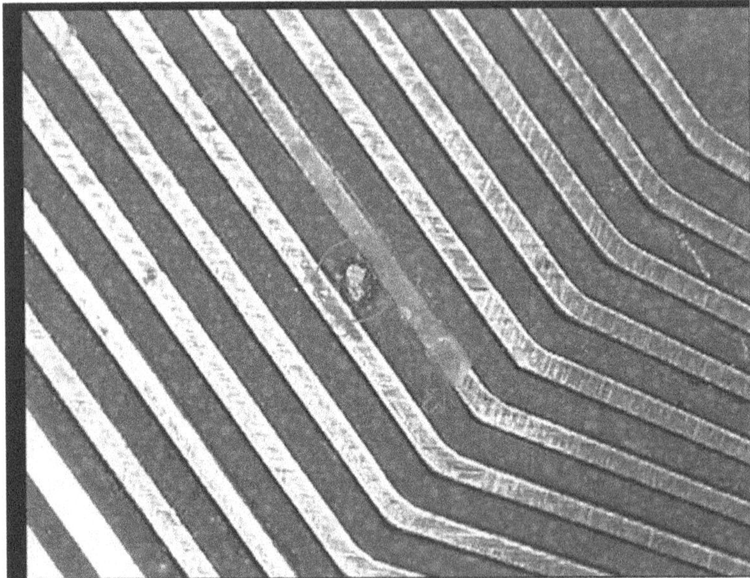

Figure 6.54 Optical image of planar polish through particle shorting adjacent lead frame fingers in package.

Figure 6.55 Optical image of same particle in Fig. 6.54 but shown here after lead frame fingers were removed and cross-sectional polish was performed.

6.4.2.1 General procedure for sample preparation and analysis. The following paragraphs are designed to give the reader a general understanding of the procedures and methodology used to examine internal package interfaces, materials, and interconnect features. Most of the procedures here will be specific to cross-sectional orientation, but similar techniques can be used for planar polish as well. There are two common methods used for cross-sectioning analysis: the manual or conventional method and the automated method. Both rely on several factors that have a definite impact on the quality of the polish and the final results. Parameters that influence the final result and the repeatability of the process are rotational speed of the abrasive material, direction of specimen grind, force applied to the specimen, type of abrasive, and amount of lubricant applied during the process. Regardless of the method adopted, it should produce exactly the same results for the same material every time the process is used.

The manual or conventional cross-sectioning method involves encapsulating the specimen in an epoxy and subjecting the sample to both a manual, hand-held grinding process and a similar manual polishing operation. This method is not as popular as the automated procedure, because its key parameters are controlled by hand pressure, and the process is basically experience driven. Unless this process is practiced routinely, it is difficult to achieve the same results each time.

The second method is an automated procedure. This method is widely used among analysts because of the ability to adjust and control the process parameters automatically. The sample is supported in a fixture that is attached to a calibrated head. Predetermined amounts of material can be removed by adjusting the process parameters as the fixture comes into contact with the abrasive material. The automated grinding and polishing sequence uses mostly diamond films and, unlike the conventional x-sectioning method, encapsulating the sample is optional.

6.4.2.2 Fact gathering/sample history. Knowledge of the package structure, material interfaces, previous exposure to environmental stresses, physical handling, electrical stress, failure signature, and other appropriate information concerning the sample is very important and should be gathered before any sample preparation or encapsulation. It is important to benchmark the integrity of your sample by knowing the internal physical makeup and interface integrity, if possible. Benchmarking your sample will assist in devising a step-by-step plan for your x-section analysis. Also, knowing how your material will perform under grinding, polishing, and exposure to various consumables and cleans will help prevent or minimize the introduction of sample preparation artifacts.

If both the manual and automatic cross-sectioning methods are options, choosing a suitable method of cross-sectioning may depend on a variety of factors. Some issues to consider are listed below:

- Number of samples to be analyzed
- Post-analysis inspection technique to be used (optical or SEM) (be aware of SEM chamber size limitations and potential contaminates)
- Support needed for sample stability during the grind and polish process, edge protection
- Precision of analysis, smallest feature to analyze
- Speed of analysis (encapsulated cold mount cures in a few hours, long-cure epoxy takes up to 24 hr)
- Sample size (fixturing restrictions)
- Direction and precision of cut (fixturing)
- Option of encapsulating (encapsulants make SEM analysis more difficult)
- Vacuum impregnator (backfill and support unstable interfaces, sample surface cracks, or remove air pockets from the sample)
- Dye penetrants or crack delineators (epoxy dye helps identify external cracks)
- Choice of grinding papers or diamond films (manual process = low-cost SiC paper, automatic process = diamond films)
- Consumables (diamond slurries or paste)

6.4.2.3 Sample preparation/encapsulation procedure

Sample resizing. Prepare the sample for encapsulation if necessary. X-section analysis using the conventional manual grinding will require encapsulation. Automated mechanical grinders/polishers use fixtures that will accommodate either encapsulated or unencapsulated specimens. Depending on the size or shape of the sample, it may be necessary to resize the sample using a diamond saw as long as the area of interest is removed intact. Resizing the sample will minimize the amount of grinding necessary. The sample must be thoroughly cleaned to remove any contamination and allow good adhesion of the epoxy to the sample.

Mounting. Select a suitable mold cup for mounting your sample. Use a mold release agent to coat the entire inside of the mold cup prior to inserting the sample. Use a sample clip (uniclip) to stabilize the sample inside the mold cup. Position the sample in the mold cup with the surface containing the structure for analysis face down. Pour the encapsulant (epoxy) slowly over the sample from one side to minimize air pockets from forming around the sample. Allow the sample to cure according to the manufacturer's specifications. Once the encapsulant has cured, the sample is ready for the grinding operation and can be removed from the mold cup.

Grinding. The actual grinding material [SiC papers (120 to 2000 grit), diamond films (45 to 0.1 μm), or diamond grinding discs] and the range of abrasives used to remove sample material may vary with each sample. The initial stage of rough grinding is performed to bring the vertical plane and all the interfaces of the sample to the same plane. Next, consider the initial material type that will be contacted and whether the sample requires a close cut to the area of interest. The grind sequence may start with a less aggressive (fine grind) grit paper or abrasive sequence in cases where less material is to be removed. For the same initial contact material, a long cut start may require a coarse (rough grind) grit as a starting abrasive because of the excessive material to be removed. For both a close cut and a long cut start, the grind sequence at the target plane is generally terminated with an abrasive paper of 1200 grit or 15 μm diamond film. In general, the grinding sequence consists of mechanically removing material by applying successively finer grain sizes, regardless of the starting grit size. During each step of the polishing process, it is important to use an abrasive that will maintain a flat surface and cause the smallest amount of damage, and that can be easily removed by the next process step in the shortest amount of time.

Intermediate polish. An intermediate or rough polish can be performed using a similar sequence of steps using successively finer abrasive particles. Typically, an intermediate polish will range from 9 to 3 μm on a napless cloth, depending on the amount of deformation or damage that was left from the previous step. For diamond polishing, a lubricant must be used. The choice of diamond particle size and paste or slurry will depend on the material to be polished and the analyst's experience.

Final polish. Final polishing requires a very fine (1 to 0.05 μm) diamond particle paste or slurry on a napless cloth. Polish time will vary according to the desired results.

The cross-sectional sample prep procedures and process steps outlined provide a general guideline for performing cross-sectional analysis. Procedures and use of materials may vary with each sample, depending on fact gathering and uniqueness of the failures. It is not possible to define a specific process that will cover all electronic package samples. Regardless of the choice of grinding material and the analyst's choice of grinding method, the basic steps for material removal remain constant. Ultimately, methods of sample preparation and choice of materials will depend on one's own personal experience.

6.4.3 Dry Etching for Package FA

Dry etching is another sample preparation technique available to the failure analyst that can complement wet chemical etch methods. Its main applications to package FA are to remove contamination from package contact pins and to decapsulate or otherwise expose structures within the package. Dry etching has been applied mostly to deprocessing of the semiconductor die within the package, but it is expected to become more important for packages as their complexity increases and as new materials are introduced.[61] The advantage of dry etching for package FA is the possibility for precise control of etch rate and selectivity, which can be tailored to specific materials. For example, dry etching may be preferred in a situation where internal metal leads may be attacked by decapsulation using acids. It is also the preferred method for removing organic contamination from metal contact pins. The disadvantages of dry etching are the lack of standard recipes tailored to package applications and long decapsulation times as compared to acid jet etchers. Newer tools employing high-density plasma sources can improve the etch rates substantially, but it will take time to develop optimal etch recipes. In this section, the basic concepts of dry etching will be discussed, along with a brief description of the main parameters that affect the etch and common equipment types.

When radio frequency (RF) power is coupled to a low-pressure (0.001 to 1 torr) gas or mixture of gases, a small fraction of the gas species become ionized, resulting in a specific type of plasma referred to as a *glow discharge*. An equilibrium distribution of neutrals, energetic atomic and molecular species, ions, and electrons is maintained in the plasma until the RF power is shut off, at which point the gas species rapidly recombine or relax to their normal state. Hence, a relatively inert, easy to handle gas mixture can be turned into a highly reactive substance by the application of RF power for a fixed period of time. This has become a powerful and flexible tool for etching, especially in the semiconductor industry.[62]

The reactive species in the plasma consist mostly of neutral atoms and free radicals (which are unstable molecular fragments), and these are the species that do most of the etching.[63] Oxygen and fluorine atoms are the most useful etch species for package applications, since they can form volatile reaction products with the carbon compounds and polymers used in package technology. These reactive components are usually introduced using the stable gases oxygen (O_2) and Freon 14 (CF_4).

A much smaller concentration of positive ions, with a balancing number of negative ions and electrons, exist in the plasma, and it is the positive ions that can add directionality to dry etching by enhancing the etch rate perpendicular to the sample surface compared to the lateral direction. Positive ions accelerate toward the sample, because the sample surface tends to charge negatively with respect to the plasma. The DC bias is the plasma parameter that is a measure of the potential difference between the sample and the plasma.

In the jargon of dry etching, *plasma etching* has come to refer to etching dominated by the chemical component alone, whereas *reactive ion etching (RIE)* has come to refer to directional, ion-enhanced chemical etching. (Note that "reactive ions" are not involved in RIE; reactive neutrals are, and the small ion component adds directionality.) Either mechanism may be preferentially selected through the choice of RF electrode configuration, power, and pressure. For example, ion enhanced etching (RIE) is favored by high-power and low-pressure conditions with the sample placed on the powered electrode. These conditions generate a high DC bias (in the range of 100 to 600 V).

Sputter etching is a third category of dry etching in which physical sputtering is the dominant etch mechanism. The conditions that favor sputter etching are an inert plasma gas, typically argon, and high DC bias. This mechanism uses energetic argon ions to "sandblast" the material from the sample surface. This type of dry etching is not selective to different materials and has a much slower etch rate. However, it may be useful for surface cleaning where a thin layer of contaminant needs to be removed. The three main dry etch categories are summarized in Table 6.1.

Plasma etch reactors all share some basic elements, and the process or FA engineer can adjust the etch characteristics based on a short list of parameters. The main parameters of interest are (a) the type, relative proportion, and total flow of gases, (b) the pressure, and (c) the power and DC bias. The power and DC bias may not be independently controllable, depending on reactor design.

The gas and pressure are basic parameters common to all reactor types. As noted previously, oxygen (O_2), Freon 14 (CF_4), and argon (Ar) are probably the most useful gases for package applications. A simple O_2 plasma at higher pressure (200 to 500 mtorr) will generate reactive O atoms that are very useful for chemically removing organic contamination from package pins. Alternatively, an argon plasma at lower pressure (50 to 100 mtorr) can be used for physical sputtering to remove inorganic contaminants. O_2 and CF_4 mixtures are preferred for

Table 6.1 Dry Etch Categories

	Plasma etching	Reactive ion etching (RIE)	Sputter etching
Dominant etch mechanism	Chemical	Ion enhanced chemical	Physical sputtering
Etch directionality	Isotropic	Anisotropic	Anisotropic with feature rounding
Etch selectivity	High selectivity depending on materials	Medium selectivity depending on materials	Low selectivity
Suitable reactor type	Parallel plate, downstream, decoupled plasma source	Parallel plate, decoupled plasma source	Parallel plate, decoupled plasma source
Pressure	>100 mtorr	<100 mtorr	<100 mtorr
Power and DC bias	Low	High	High
Gases	O_2, CF_4, SF_6, Cl_2, etc.	O_2, CF_4, SF_6, Cl_2, etc.	Ar

etching polyimides and other polymers used as package materials, and they are useful in decapsulation applications.

The method in which the RF power is coupled to the plasma gas is the main factor that distinguishes different reactor types. A wide variety of configurations have been developed,[62] but there are three main commercially available reactor types useful for the FA laboratory: the downstream reactor, the parallel plate reactor, and the decoupled plasma source reactor.

The downstream reactor is designed to keep the plasma away from the sample. It is useful for purely "chemical" etching, with no ions impinging on the sample at all, thus eliminating plasma damage. High gas flow and pressure are used, and the power (RF or microwave) is applied near the gas entry, keeping the plasma region away from the sample. Reactive neutral species formed in the plasma are carried toward the sample by the high gas flow. The sample is placed on a grounded chuck near the pump port and can be heated, if desired, to increase etch rate of organic materials. An interesting innovation for this reactor type is the atmospheric downstream etcher, or plasma torch, which eliminates the need for a vacuum system and uses higher gas feed pressures. Higher concentrations of reactive F and O atoms can be generated, which increases the etch rate. This type of downstream reactor may become useful for decapsulation as well as backside silicon etching.

The parallel plate reactor has been the most widely used type and can operate in all three dry etching modes, depending on parameter choice. The sample is usually placed on the powered electrode and is in contact with the plasma. The sample surface develops a negative charge with respect to the plasma, which attracts a positive ion flux to the surface. The magnitude of the sample surface potential (as measured by the DC bias) determines the degree of ion enhanced etching (RIE) versus chemical etching. The DC bias can be altered by adjusting power and pressure. The disadvantage of this reactor design is that if the power is increased to generate more reactive neutral species, and hence a faster etch rate, the DC bias also increases. If the DC bias is too high, the sample is subjected to increased ion damage. This can also increase sputtering of materials around the sample being etched, which can redeposit on the sample and micromask the desired etch region. This problem is commonly referred to as *RIE grass*.[64] Hence, the interdependence of DC bias and power in the parallel plate reactor limits the etch rate to avoid increasing the plasma damage to the sample.

The decoupled plasma source reactor allows the use of high-density plasmas to achieve higher etch rates while allowing independent control of the DC bias to minimize plasma damage. A variety of designs are commercially available, but a common approach uses an inductively coupled plasma (ICP) source to control the plasma density and a separate DC bias supply connected to the chuck that holds the sample. Hence, a high RF power on the ICP source can be used to generate more reactive species for faster etch rates, while the DC bias can be adjusted separately to control the degree of ion enhanced etching (RIE) and minimize damage to the sample. This is the most flexible design, and it allows the analyst maximum choice of etch conditions. This is likely to be the tool of choice for future application of dry etching to package FA applications.

In conclusion, dry etching is a complementary alternative to wet etching. Its applications to package FA thus far have been mainly for cleaning metal contact pins. A limited amount of decapsulation work has been done. It is expected to become more useful as package complexity increases. The increased availability of high-density plasma tools will give the analyst the flexibility to develop new recipes that are of use for controlled deprocessing of advanced package designs.

6.4.4 Wirebond Pull and Shear

6.4.4.1 Pull test. The wirebond pull test is the most common method of measuring the quality of the wirebond process. It has been used since the 1960s. The test is performed by positioning the pull test hook under the loop of the wire and applying a measured force to pull up on the wire as schematically shown in Fig. 6.56. This test is destructive.

The force is typically measured in grams force. If the hook is placed closer to either the first or second bond, a greater portion of the force is applied in the vertical or peel direction.

Typically, the hook is placed mid-span between the first and second bonds. The pull force is depends on the ratio of the loop height to the space between the two bonds. The higher the loop height, the higher the pull force. The minimum acceptable pull force is depends on wire diameter and wire length.

The nondestructive pull test is a variation of the standard pull test where the maximum force applied to the bond loop is limited to a predetermined value. This test is limited to devices that have adequate space between wires so that the pull hook does not cause damage to adjacent wires. There have also been concerns about long-term reliability of wires that have been subjected to the nondestructive pull test.

The failure mode of the wire is important, even if the force to cause failure is adequate. Common failure modes are bond lift, wire break mid-span, heel break, and neck break. Normally, pull test data with a normal distribution have failure modes of this type. Heel breaks and bond lift can often be corrected by adjusting bonder parameters.

Other failure modes include bond pad metallization failure, delamination of layers under the pad, and cratering (partial or total fracture of the Si underneath the bond pad). These failure modes can indicate more deeply rooted problems than bonder machine setup. Pull test data with out-of-control distributions typically have failure modes of this type. Pull test failures occur in the weakest link of the system, which ideally should be in the wire.

Pulling the wires with tweezers (without measuring force) can reveal gross problems, such as poorly formed bonds and die delamina-

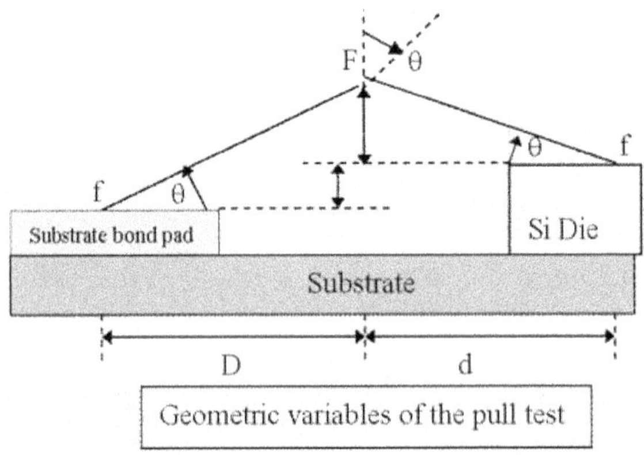

Figure 6.56 Schematic of pull test.

tion issues. This test can be performed quickly on a large number of wires.

The pull test is a good test for wedge bonds and to determine if the neck of the ball bond (recrystallization zone directly above the ball) is weak, but the shear test is more valuable for the ball bond. Since the welded area of a ball bond is 5 to 10 times the cross-sectional area of the wire, the wire would normally be expected to break before even a poorly formed ball bond will lift.

6.4.4.2 Shear test. As shown in Figure 6.57, the shear test consists of placing a chisel shaped tool beside a bond, applying enough force to push the bond off the bond pad, and recording the shear force. Considerations for the test method include shear tool height above the bond pad, shear tool width, and speed of the tool during testing.

For adequate assessment of ball bond strength, the shear strength must be correlated to the ball bond diameter. A typical value is 5.5 grams per square mil.

As in pull testing, the failure mode at shear test is important. Some of the common failure modes are as follows:

- Ball lift
- Failure within the metal of the bond
- Failure in the intermetallics
- Bond pad peel
- Cratering

At shear test, the preferred failure mode is failure within the metal of the bond. If there are several occurrences of the other failure modes listed, the data often does not have a normal distribution.

For more complete information on wirebond testing refer to George Harmon's book, *Wire Bonding in Microelectronics*.[65]

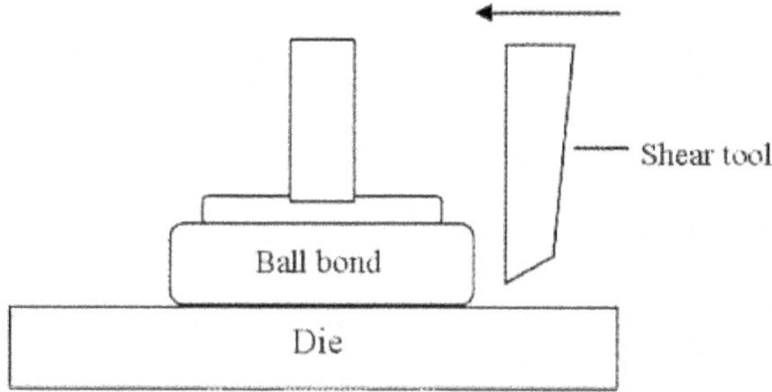

Figure 6.57 Schematic of shear test.

6.5 References

1. Semmens, J. E., L. W. Kessler, "Application of Acoustic Fourier Domain Imaging for the Evaluation of Advanced Micro Electronics," 2002 ISTFA, pp. 55–60.
2. Dias, Rajen, "Investigation of Interfaces with Analytical Tools," IEEE Transactions on Device and Materials Reliability, 3(4), 2003, pp. 179–183.
3. IPC/JEDEC Joint Industry Standard, J-STD-020B, "Moisture/Reflow Sensitivity Classification for Plastic Integrated Circuit Surface Mount Devices," July 2002.
4. Yip, L., T. Massingill, et al., "Moisture Sensitivity Evaluation of Ball Grid Array Packages," *Proc. 1996 Electronic Component and Technology Conf.,* pp. 829–835.
5. Pecht, M. G., "Moisture Sensitivity Characterization of Build-Up Ball Grid Array Substrate," *IEEE Transactions on Advanced Packaging,* Vol. 22, No. 3, August 1999, pp. 515–523.
6. Shook, R. L., V. S. Sastry, "Influence of Preheat and Maximum Temperature of the Solder-Reflow Profile on Moisture Sensitive ICs," *Proc. Electronic Component and Technology Conference,* 1997, pp. 1041–1048.
7. Shook, R. L., "Moisture Sensitivity Characterization of Plastic Surface Mount Devices Using Scanning Acoustic Microscopy," *Proc. International Reliability Physics Symposium,* 1992, pp. 157–168.
8. Bin, R., E. Balckshear, et al., "Moisture Induced Package Cracking in Plastic Encapsulated Surface Mount Components During Solder Reflow Process," *Proc. 1988 IRPS,* pp. 83–89.
9. Shook, R. L., D. L. Gerlach and B. T. Vaccaro, "Moisture Blocking Planes and Their Effect on Reflow Performance in Achieving Reliable Pb-Free Assembly Capability for PBGAs," *Proc. Electronic Component and Technology Conference,* 2001, pp. 74–79.
10. Qian, Z., M. Lu, et al., "Fatigue Life Predication of Flip-Chips in Terms of Non-linear Behaviors of Solder and Underfill," IEEE/ECTC, 1999.
11. Drexler, E. S., "Reliability of a Flip-Chip Package Thermally Loaded Between –55 and 125," *Journal of Electronic Materials,* Vol. 28, No. 11, 1999.
12. Zribi, A, R. R. Chromik, et al., "Solder Metalization Interdiffusion in Microelectronic Interconnects," IEEE/ECTC, 1999.
13. Wolf, Robert K., Tejpal K. Hooghan, et al., "Assessment of Flip Chip Interconnect Integrity Using Scanning Acoustic Microscopy," *Proc. 2000 Electronic Components and Technology Conference,* pp. 633–639.
14. Canumalla, S., "Critical Issues in Thin Layer Acoustic Image Interpretation and Metrology for Microelectronics," *Proc. 2002 Electronic Component and Technology Conference,* pp. 205–218.
15. Philippe, Richard, et al., "Thin Film Adhesion Investigation with the Acoustic Microscopy," *Proc. 1994 Ultrasonic Symposium,* pp. 1425–1428.
16. Naderman, J., F. W. Ragey, et al., "Thermal Resistance Degradation of Alloy Die Attached Power Devices During Thermal Cycling," *Proc. 1998 IRPS,* pp. 248–253.
17. Goings, J., S. Haque, et al., "Characterization of Interfacial Thermal Resistance by Acoustic Microscopy Imaging," *2000 Microelectronic Reliability,* pp. 465–476.
18. Walcott, J., Compix, Inc., "Thermal Analysis for Today's Complex Designs" at www.compix.com/articles.eeast93b.html.
19. Department of Material Science and Engineering, University of Maryland, A. James Clarker School of Engineering, "Scanning Infrared Microscope."
20. Devaney, J. R., *Scanning Electron Microscopy and Energy Dispersive X-Ray Analysis" in Microelectronic Failure Analysis,* 4th ed., R. J.Ross, C. Boit, D. Staab, Eds., ASM International, 1999.
21. Goldstein, J., D. Newbury, D. Joy, C. Lyman, P. Echlin, E. Lifshin, L. Sawyer, and J. Michael, *Scanning Electron Microscopy and X-Ray Microanalysis,* 3rd ed., Kluwer Academic/Plenum Publishers, 2003.
22. Loretto, M. H., *Electron Beam Analysis of Materials,* Chapman and Hall, 1984.
23. Edington, J. W., N. V. Philips' Gloeilampenfabrieken, Eindhoven, 1976, *Practical Electron Microscopy in Materials Science.*

24. Williams, David B., and C. Barry Carter, *Transmission Electron Microscopy: A Textbook for Materials Science,* (4 Volumes), Plenum Press, 1996.
25. Private discussions with Dr. Jon Orloff, University of Maryland at College Park, Dept. of Electrical Engineering.
26. Carleson, Peter, "Integrated Circuit Modification," course notes, The Essentials of Focused Ion Beam Technology, University of Maryland, October 5, 6, 1997.
27. Hooghan, K. N., K. S. Wills, P. A. Rodriguez, and S. J. O'Connell, "Integrated Circuit Device Repair Using FIB System: Tips, Tricks and Strategies," *Proc. ISTFA,* 1999.
28. Hooghan, K. N., S. Nakahara, R. Privette, and S. N. G. Chu, "FIB as a TEM Specimen Preparation Tool for III-V Compound Semiconductor Materials," *Proc. State-of-the-Art Program on Compound Semiconductors XXXVI and Wide Bandgap Semiconductor for Photonic and Electronic Devices and Sensors II,* The Electrochemical Society, Inc., Pennington, NJ, PV 2002-3, 2002, p. 276.
29. Hooghan, T. K., K. Hooghan, S. Nakahara, R. K. Wolf, R. W. Privette, and R. S. Moyer, "Failure Analysis of Flip Chip Bumps after Thermal Stressing," *Proc. 26th ISTFA*, November 2000, pp. 41–48.
30. Hooghan, T. K., S. Nakahara, K. Hooghan, R. W. Privette, M. A. Bachman, and R. S. Moyer, "Observation of Amorphous Chromium in Modified C4 Flip Chip Solder Joints After Thermal Stress Testing," accepted for publication in *Thin Solid Films,* 2003.
31. Post, D. B. Han, and P. Lfju, *High Sensitivity Moiré,* F. F. Ling, Ed., Mechanical Engineering Series, Springer-Verlag, 1994.
32. ESM Technology, Inc. web site http://www.expsolidmech.com.
33. Cho, S. M., S. Y. Cho, and B. Han, "Real-time Observation of Thermal Deformations in Microelectronics Devices: A Practical Approach Using Moiré Interferometry," *Experimental Techniques*, Vol. 26, No. 3, 2001, pp. 25–29.
34. Verma, K., B. Han, S. B. Park, and W. Ackerman, "On the Design Parameters of Flip-Chip PBGA Package Assembly for Optimum Solder Ball Reliability," *IEEE Trans. Components and Packaging Technologies*, Vol. 24, No. 2, 2001, pp. 300–307.
35. Cho, S. M., B. Han and J. Joo, "Temperature Dependent Deformation Analysis of Ball Grid Array Package Assembly under Accelerated Thermal Cycling Condition," *J. Electronic Packaging, Trans. ASME*, Vol. 126, pp. 1–7, 2004.
36. Han, B., "Thermal Stresses in Microelectronics Subassemblies: Quantitative Characterization using Photomechanics Methods," *J. Thermal Stresses,* Vol. 26, 2003, pp. 583–613.
37. Han, B., D. Post, and P. Ifju, "Moiré Interferometry for Engineering Mechanics: Current Practice and Future Development," *J. Strain Analysis,* Vol. 36, No. 1, 2001, pp. 101–117.
38. Han, B., "Recent Advancement of Moiré and Microscopic Moiré Interferometry for Thermal Deformation Analyses of Microelectronics Devices," *Experimental Mechanics,* Vol. 38, No. 4, 1998, pp. 278–288.
39. Han, B., and Guo, Y., "Thermal Deformation Analysis of Various Electronic Packaging Products by Moiré and Microscopic Moiré Interferometry," *J. Electronic Packaging, Trans. ASME,* Vol. 117, 1995, pp. 185–191.
40. Han, B., Guo, Y., Lim, C. K., and Caletka, D., "Verification of Numerical Models Used in Microelectronics Packaging Design by Interferometric Displacement Measurement Methods," *J. Electronic Packaging, Trans. ASME,* Vol. 118, 1996, pp. 157–163.
41. Han, B., and Guo, Y., "Determination of Effective Coefficient of Thermal Expansion of Electronic Packaging Components: A Whole-field Approach," IEEE Trans. Components, Packaging and Manufacturing Technology, Part A, Vol. 19, No. 2, 1996, pp. 240–247.
42. Han, B., "Deformation Mechanism of Two-Phase Solder Column Interconnections Under Highly Accelerated Thermal Cycling Condition: An Experimental Study," *J. Electronic Packaging, Trans. ASME,* Vol. 119, 1997, pp. 189–196.
43. Cho, S. M., Cho, S. Y., and Han, B., "Observing Real-Time Thermal Deformations in Electronic Packaging," *Experimental Techniques,* Vol. 26, No. 3, 2002, pp. 25–29.

44. Rastogi P. K., Ed., "Photo-mechanics," *Topics in Applied Physics,* Vol. 77, Springer, 1999.
45. Tummala, R. R., Ed., *Fundamentals of Microsystems Packaging,* McGraw-Hill, 2001.
46. Kirtley, J. R., and J. P. Wikswo, Jr. "Scanning SQUID Microscopy," *Annual Review of Materials Science,* 29, 1999 pp. 117–148.
47. Neocera, Inc., Beltsville, MD, 20705.
48. Dias, R., L. Skoglund, Z. Wang, and D. Smith, "Integration of SQUID Microscopy into FA Flow," ISTFA, 2001, and L. A. Knauss, et al., "Backside Fault Isolation Using a Magnetic Field Imaging System on SRAMs with Indirect Shorts," *ISTFA,* 2000, p. 503.
49. Vallett, D. P., "Scanning SQUID Microscopy for Die Level Fault Isolation," *ISTFA,* 2002, p. 391.
50. Microelectronics Failure Analysis Techniques: A Procedural Guide.
51. Goldstein, J., D. Newbury, D. Joy, C. Lyman, P. Echlin, E. Lifshin, L. Sawyer, and J. Michael, *Scanning Electron Microscopy and X-Ray Microanalysis,* 3rd ed., Kluwer Academic/Plenum Publishers, New York, 2003.
52. Lowry, R. K., and L. C. Wagner, Eds.,"Auger Electron Spectroscopy," in *Failure Analysis of Integrated Circuits,* Kluwer Academic Publishers, New York, 1999.
53. Feldman, L. C., and J. W. Mayer, *Fundamentals of Surface and Thin Film Analysis,* North Holland-Elsevier, New York, 1986.
54. Bindell, J. B., and S. M. Sze, Eds., "Analytical Techniques," in *VLSI Technology,* 2nd ed, McGraw-Hill, New York, 1988.
55. Coates, J., "A Diamond-Based Sampling Accessory for Solid and Liquid Sampling in Infrared Spectroscopy," *American Laboratory,* April 1997.
56. DeVore, J. A., "Practical Quantitative Solderability Testing," *Proc. Aerospace and Electronics Conference,* Vol. 4, May 1989, pp. 2027–2034.
57. Lea, C., "Solderability and Its Measurement," *Engineering and Science Education Journal,* Vol. 2, No. 2, April 1993, pp. 77–84.
58. Sattiraju, S. V., B. Dang, R. W. Johnson, Y. Li, J. S. Smith, and M. J. Bozack, "Wetting Characteristics of Pb-Free Solder Alloys and PWB Finishes," *IEEE Trans. Electronics Packaging Manufacturing,* Vol. 25, No. 3, July 2002, pp. 168–184.
59. Kwoka, M. A., and P. D. Mullenix, "The Association of Component Solderability Testing with Board Level Soldering Performance," *IEEE Trans. Components, Hybrids, and Manufacturing,* Vol. 13, No. 4, Dec. 1990, pp. 704–712.
60. IPC/EIA/JEDEC J-STD-002B, "Solderability Tests for Component Leads, Terminations, Lugs, Terminals and Wires," February 2003.
61. Thomasi, Dean, "Plasma Decapsulation Techniques," in *Microelectronics Failure Analysis,* 3rd ed, T. W. Lee and S. V. Pabisetty, Eds., 1993, p. 75.
62. Manos, D. M., and D. L. Flamm, Eds., *Plasma Etching, An Introduction,* Academic Press, New York, 1989.
63. Coburn, J. W., and H. F. Winters, "Plasma Etching—A Discussion of Mechanisms," *J. Vac. Sci. Tech.,* 16(2), 1979, p. 391–403.
64. Shutz, R. J., "Reactive Plasma Etching," S. M. Sze, Ed., *VLSI Technology,* 2nd ed., McGraw-Hill, New York, 1988, p. 211.
65. Harmon, George, *Wire Bonding in Microelectronics: Materials, Processes, Reliability, and Yield,* McGraw-Hill, New York, 1998.

Chapter

7

Examples of Failure Modes Common in Organic IC Packages

Case Histories and Root Cause Analyses

Charles Cohn
Ronald J. Weachock
Weidong Xie
Barry J. Dutt

7.1 Introduction

As IC complexity increases and feature sizes continue to shrink, package geometries are getting smaller, with denser circuits and higher pin counts. The resulting IC devices are prone to more frequent failures. When the failures are found and identified, and after corrective actions are implemented, the lessons learned from these failures will help improve the long-term reliability of the product. Finding and identifying such failures is greatly dependent on the analyst's understanding of the construction and operation of the given device as well as the individual's expertise in performing the failure analysis. Selecting the best-suited failure analysis techniques for a given application is critical in determining the correct cause of failure. The most common techniques used in failure analyses are described in detail in Chapter 6.

This chapter provides real-life examples of different failure modes found in various organic IC package technologies. Most of the failures shown are the result of subjecting the IC devices to reliability or qualification testing (Table 3.2). Forcing failures through rigorous qualification testing and then correcting any package deficiencies has helped promote product quality. The failures can be attributed mostly to IC package materials behavior (such as thermal mismatches) or manufacturing processes.

For each failure mode described in this chapter, the case history and physics behind the failure mechanism is presented along with the root cause analysis and corrective actions where applicable. The failures are categorized according to the IC package type in which the failures occurred.

In many cases, the corrective actions were implemented with excellent results, revealing that the type of failures previously encountered in the given IC packages did not recur in subsequent qualification tests. For other IC packages, with other material sets, die sizes, or substrate designs, the corrective actions described/implemented may have different results.

7.2 Failures in Dual In-Line Packages

7.2.1 Open Contacts—Lifted Ball Bonds

7.2.1.1 Case history and description of failure. A 28-pin dual in-line package (DIP) failed electrical testing for opens after 500 temperature cycles of Condition C (–65 to 150°C). A C-mode scanning acoustic microscope (C-SAM), in pulse-echo mode, revealed delamination at the corner of the die between the die surface and the mold compound (see Fig. 7.1*a*). The DIP was then scanned using a C-SAM in through-mode analysis, which also detected delamination at the die attach region concentrated around the perimeter of the die (Fig. 7.2*a*). Both delaminations were confirmed by cross sectioning and scanning electron microscope (SEM) analysis, as shown in Figures 7.1*b* and 7.2*b*, respectively. A circular area in the center of the die remained in good contact. The DIP was decapped, and the die surface was visually inspected. It showed three lifted (sheared) ball bonds in the corner of the die (Fig. 7.3) corresponding to the delaminated area identified by C-SAM. The delamination propagated along die surface, shearing the ball bonds. Both stress damage and voiding were seen at the ball bond interface. Several balls exhibited mushrooms at the gold/intermetallic interface. Ball bonds that have been completely detached during temperature cycling were dislodged during polishing. Passivation cracks were also seen in the corner region of the die (Fig. 7.4).

7.2.1.2 Corrective actions. The failures encountered in the above DIP can be attributed to the material properties of the molding compound used. As a result, it was recommended that the mold compound be

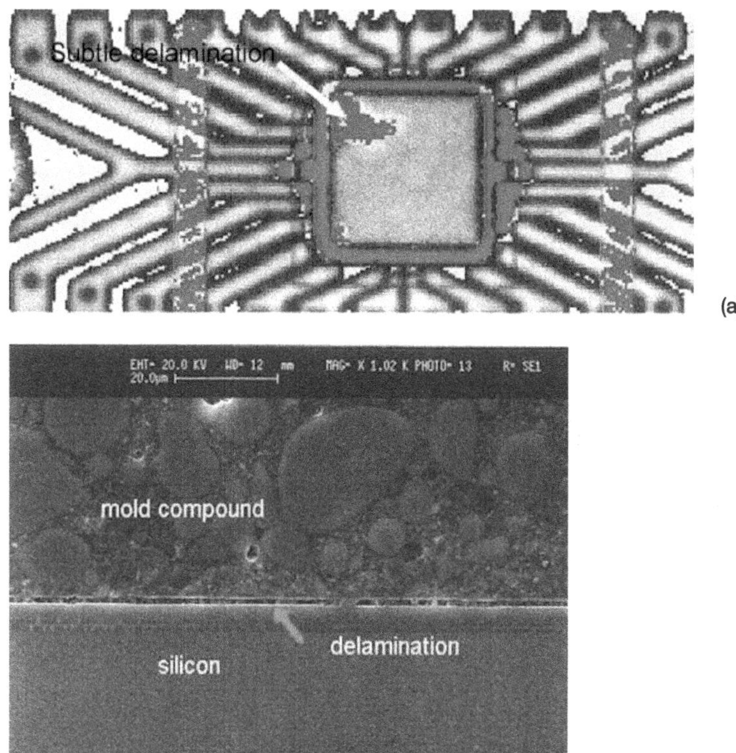

Figure 7.1 C-SAM and SEM results showing die corner delamination. (*a*) Pulse-echo delamination at the corner of the die surface and (*b*) corresponding delamination seen with SEM.

changed to one having material properties that will improve its adhesion to the die surface, absorb less moisture, and have a lower coefficient of thermal expansion (CTE). Once the change in the molding compound was implemented, no further failures of this type occurred.

7.3 Failures in Quad Flat Packs

7.3.1 Open Contacts—Sheared Ball Bonds

7.3.1.1 Case history and description of failure. A 216-lead thin quad flat pack (TQFP) failed for open contacts after being subjected to a qualification temperature cycling test of Condition C (–65 to 150°C).

A C-SAM, in pulse-echo mode, analysis showed no significant delamination on the die surface (Fig. 7.5). Decapsulation and inspection re-

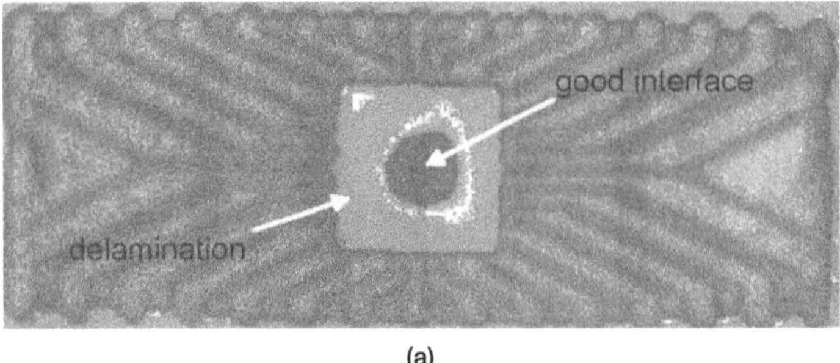

(a)

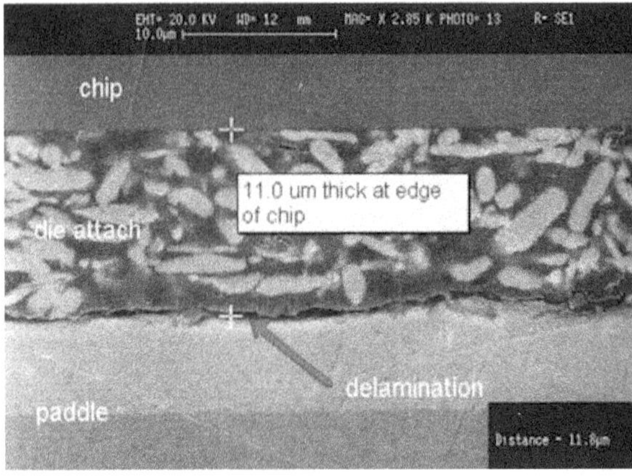

(b)

Figure 7.2 C-SAM and SEM results showing die attach delamination. (*a*) Through mode revealed delamination on die attach and chip interface and (*b*) corresponding delamination seen with SEM.

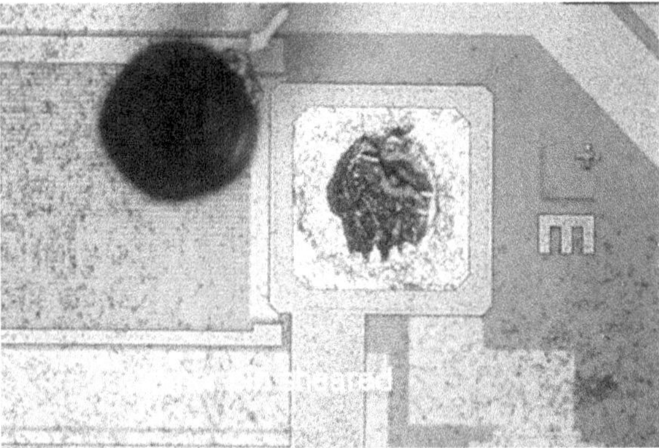

Figure 7.3 Sheared ball bonds on die surface.

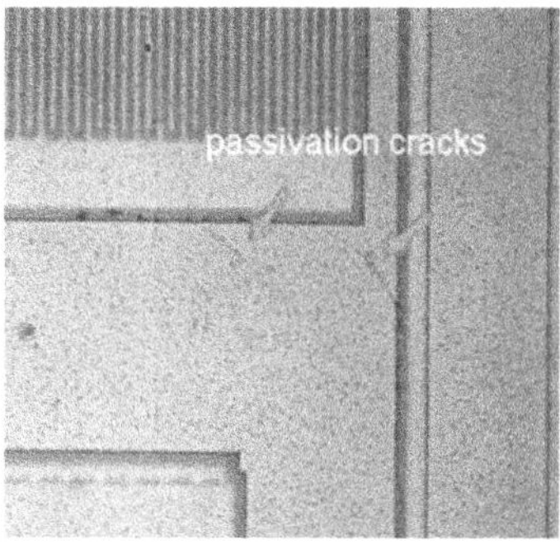

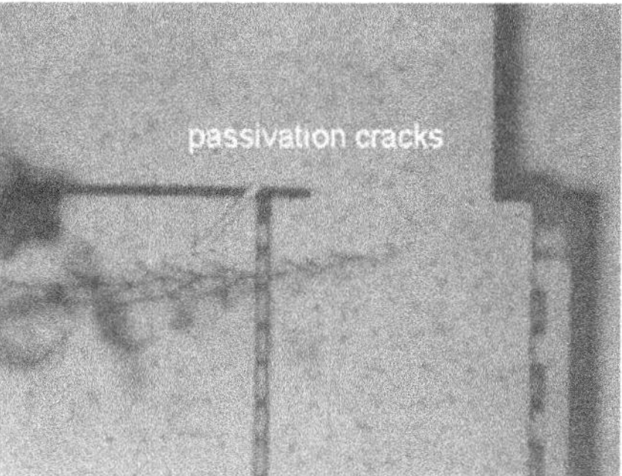

Figure 7.4 Passivation cracks on die surface.

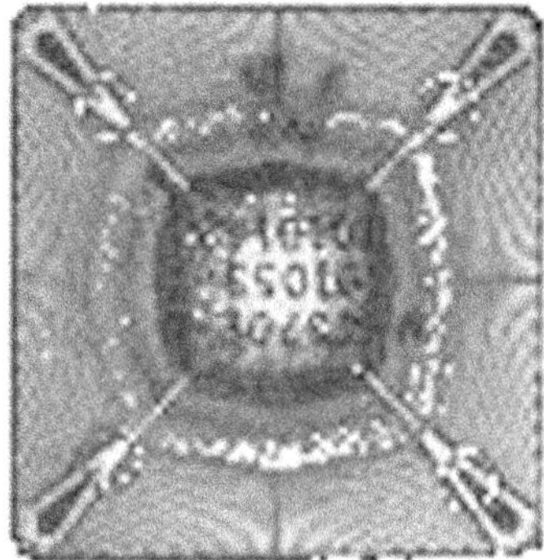

Figure 7.5 C-SAM results showing minimal delamination.

vealed extensive damage to the die surface and sheared ball bonds (Fig. 7.6). Good intermetallic formation was observed on the broken ball bonds. Cracking and displaced aluminum were also observed around the die edges. All cracks indicated radial stress applied to the die and ball bonds toward the die center. This suggests that the molding compound suffered severe shrinkage, causing damage to the ball bond connections and the circuitry under the bond pads. The absence of delamination correlates with the damage to the die surface. The failure of the device was the result of high stress on the die surface and bond wires, resulting in intermittent opens and functional failure.

7.3.1.2 Corrective actions. The failures encountered in the above QFP can be attributed to the material properties of the molding compound used. As a result, it was recommended that the mold compound be changed to one having material properties that will improve its adhesion to the die surface, absorb less moisture, and have a lower coefficient of thermal expansion (CTE). Once the change

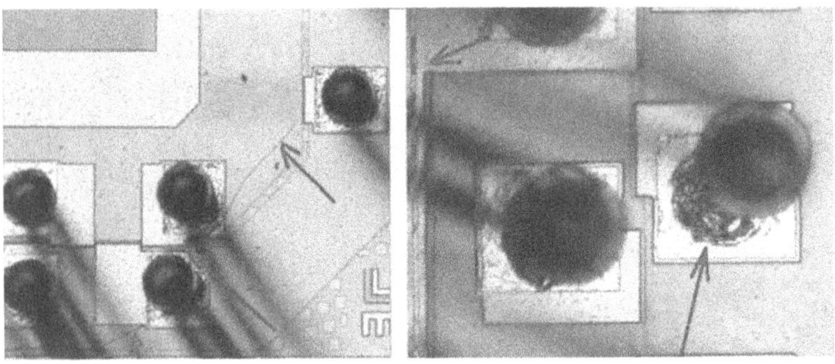

Figure 7.6 Extensive damage to die surface—sheared ball bonds.

in the molding compound was implemented, no further failures of this type occurred.

7.3.2 Open Contacts—Sheared Ball Bonds

7.3.2.1 Case history and description of failure. A 144-lead TQFP failed electrical test for opens after subjecting the device to 1000 temperature cycles of Condition C (–65 to 150°C). C-SAM, in pulse-echo mode, data showed delamination or stress pattern in the top right corner of the die. Visual inspection of the die surface, after decapping, revealed metal crazing, passivation cracking, mold compound cracking, and ball bond shear consistent within the data provided by C-SAM (Fig. 7.7). Removing the top metal busses on the die revealed surface cracks that propagated downward into metal busses and adjacent circuitry. Cross-sectioning the device showed mold cracking at the corner of the die (Fig. 7.8). The shear stress damage observed is consistent with typical temperature-cycling failures.

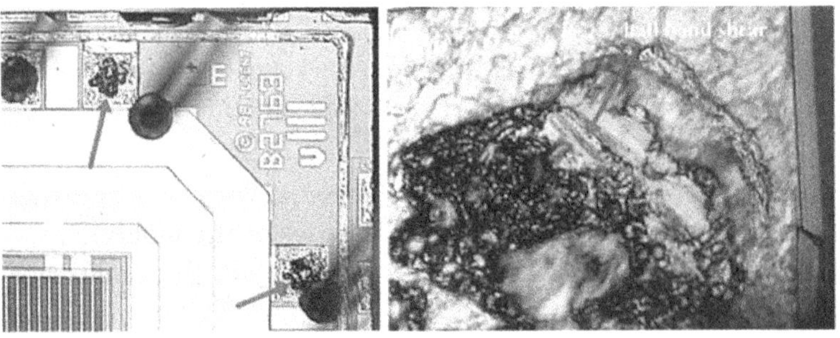

Figure 7.7 Ball bond sheared on die surface.

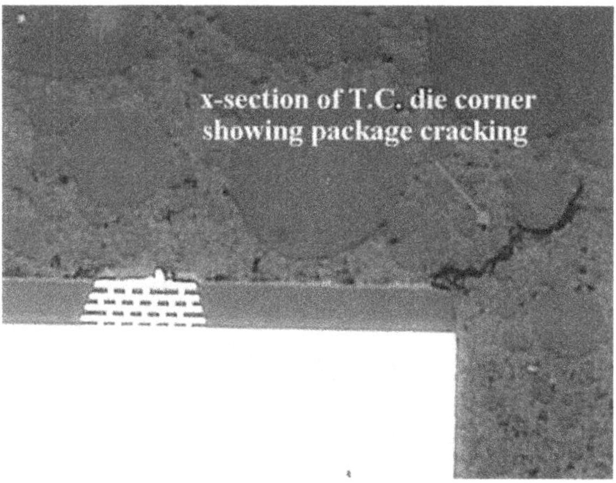

Figure 7.8 Mold compound cracking near die corner.

7.3.2.2 Corrective actions. The failures encountered in the above TQFP can be attributed to the material properties of the molding compound used. As a result, it was recommended that the mold compound be changed to one having material properties that will improve its adhesion to the die surface, absorb less moisture, and have a lower coefficient of thermal expansion (CTE). Once the change in the molding compound was implemented, no further failures of this type occurred.

7.3.3 Opens and Shorts—Intermetallic Growth

7.3.3.1 Case history and description of failure. A 64-lead TQFP failed the contact tests for opens and shorts after high-temperature operating bias (HTOB) qualification testing. The C-SAM, in pulse-echo mode, analysis showed a die surface delamination (Fig. 7.9). A small discoloration on the package exterior indicated overheating originating from the die. The device was decapsulated with significant difficulty, because the molding compound was hardened over the die area and required extensive re-etching to remove. Decapsulation and inspection revealed that the device had been exposed to high temperature (>180°C) for an extended period of time. Significant intermetallic growth was observed to extend under sincap levels, which caused cracking in both the sincap and dielectric levels. Intermetallic appearance indicated that the ball bonds were compromised and failed due to voiding caused by high heat (Fig. 7.10). The above observations suggest that the device suffered from thermal runaway, most likely due to improper socketing or bias in the HTOB environment.

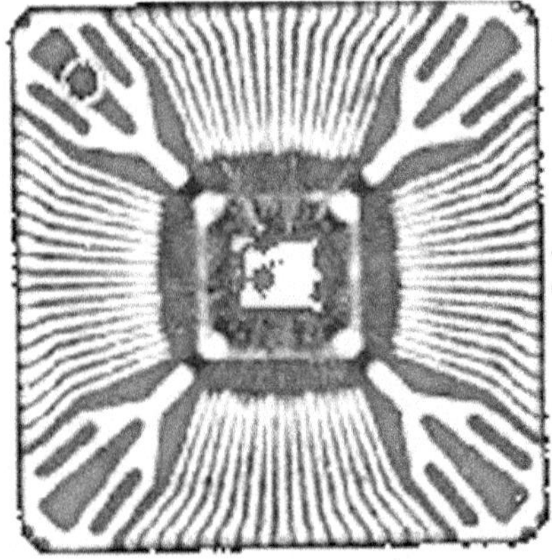

Figure 7.9 C-SAM results showing die surface delamination.

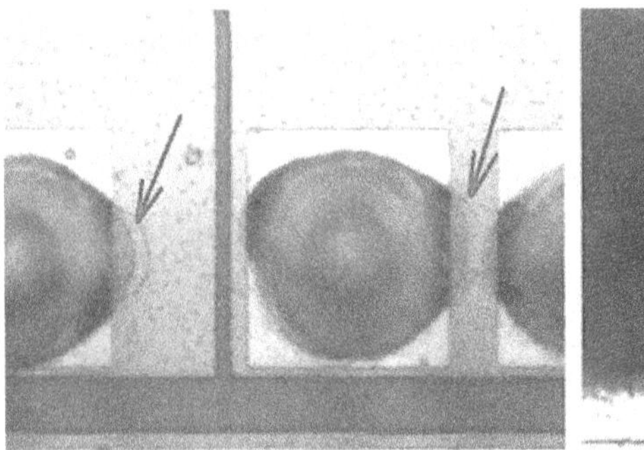

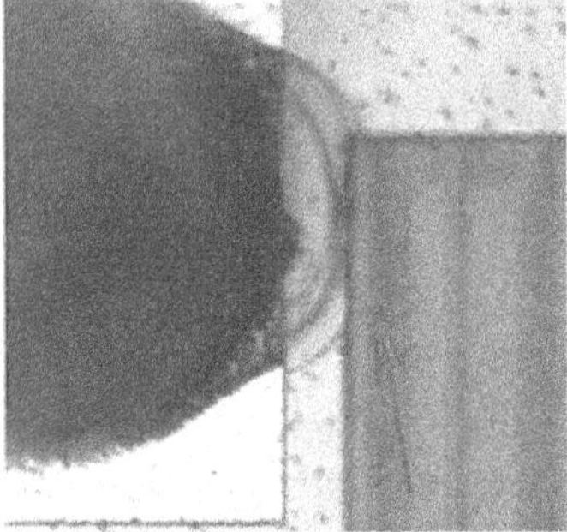

Figure 7.10 Intermetallic growth extends under sincap level.

7.3.4 Shorted Leads—Foreign Material between Leads

7.3.4.1 Case history and description of failure. A 208-lead shrink quad flat pack (SQFP) (Fig. 7.11) was field returned because of a power supply failure. A low-power microscopic inspection (prior to decapsulation) showed no external defects. A C-SAM, in through-mode, analysis determined that there was package but no die top delamination. Real-time X-ray image analysis indicated nearly shorted bond wires (leads 138 and 139). A curve trace confirmed a high leakage/short between the power plane (V_{DD}) and the ground plane (V_{SS}). High-power microscopic inspection (post-decapsulation) detected an anomaly around the bond pad perimeter of lead 9. The appearance suggested that the passivation was cracked at the edge.

"Freeze spray" was used to determine if a package-related defect was causing the high current. A package-related thermal event was

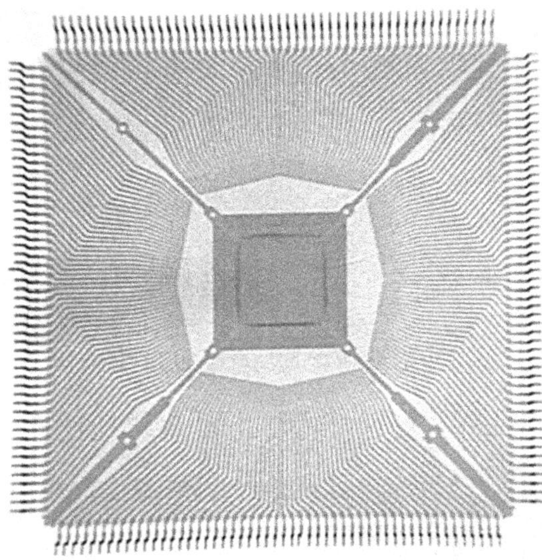

Figure 7.11 X-ray of a 208-lead SQFP.

revealed by the freeze spray, which indicated the possibility of a package-related defect as the cause of device failure. Bond wires of both leads 6 and 7 were removed from the die to isolate the die from the package leadframe. A curve trace of package lead 7 to package lead 6 revealed a 6-Ω short, confirming a package related defect.

Analysis continued by polishing the package (planar). Before the suspected area of the leadframe was fully exposed, a liquid crystal analysis was repeated to obtain the exact location of the short; the resistance was measured again as 14 Ω. The short was not in or near the Kapton® tape region. Further polishing, followed by optical and SEM analysis, identified a shiny anomaly (Fig. 7.12). A scanning electron microscope/energy dispersive spectrometer (SEM/EDS) analysis indicated the "particle" composition to be carbon. The package was polished further to expose more of the particle and to determine if there was any other material in it. During polishing, the copper finger (lead 6) dislodged (lifted) in the shorted region only, apparently weakened and thinned from the high current flowing through it, either during operation or thermal analysis.

This explains why the finger itself seemed to heat up in a small region rather than the particle. At this point, it was decided to remount the sample and perform a cross section so as to see the full extent of the anomaly. The cross section, with the leadframe removed, confirmed the carbon content and shows the full depth of the particle (Fig. 7.13).

Figure 7.12 SEM revealed a foreign particle between leads.

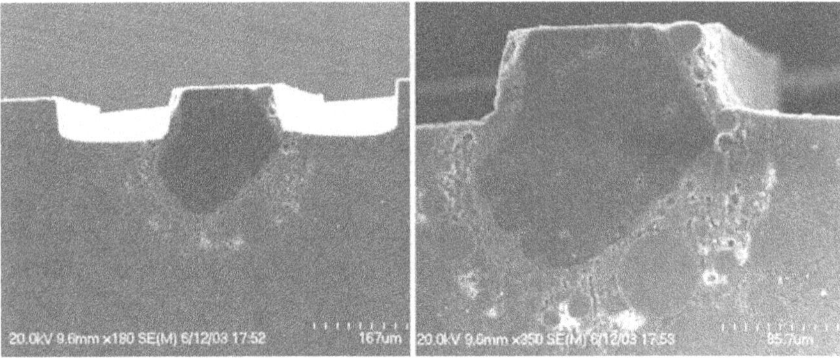

Figure 7.13 Cross section, with leadframe removed, showing full depth of foreign particle.

7.3.4.2 Corrective actions. Apparently, the particle was mixed into the molding compound and filler material before molding of the package and eventually shorted the adjacent leads. The mold compound supplier has instituted tighter control on carbon black particle sizes sifted for use as filler for pigmentation. For more detail on carbon black in mold compound, see Chapter 8.

7.3.5 Shorted Leads—Foreign Material between Leads

7.3.5.1 Case history and description of failure. A 100-lead TQFP (Fig. 7.14) was field returned for failing a contact test that showed two shorted leads. Many different failure analysis (FA) techniques were used to isolate the location of the shorts, as described below.

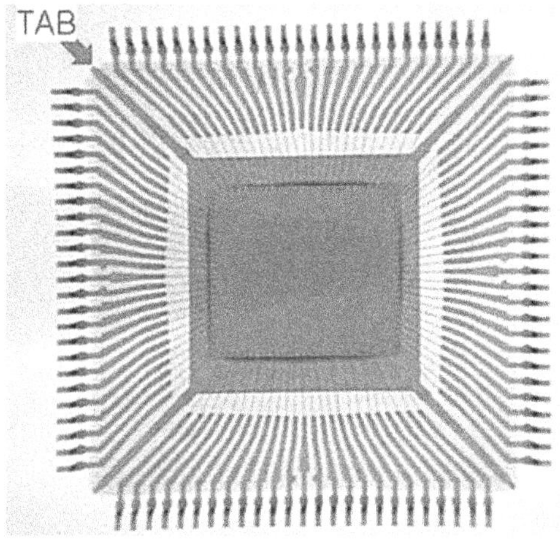

Figure 7.14 X-ray of a 100-lead TQFP.

Low-power microscopic inspection (prior to decapsulation) revealed that there were no external defects. Real-time X-ray image analysis indicated that there were no opens or shorts. C-SAM, in through-mode, analysis determined that there was package delamination, and the curve trace confirmed a lead-to-lead short between leads 19 and 20.

After decapsulation, diagnostic equipment was unable to identify a defective site on the die. No abnormal light or thermal emission was detected. A package short was strongly suspected. Wires of leads 19 and 20 were removed from bond pads. The short was still present. This indicated that a conductive path existed between leads 19 and 20 in the external package and not on the die. Although significant delamination was noticed in the area of the short (Fig. 7.15), X-ray analysis revealed no obvious defect. Liquid freeze spray was applied to the backside of the package, and a thermal site in the package was identified in the leadframe/wedge bond are of the leads 19 and 20. Fluorescent microthermal imaging (FMI) analysis was also performed on the package, and a thermal emission site was found.

A liquid crystal thermal analysis was performed after attaching leads to the device and parallel polishing to expose the Kapton tape band. An anomaly was observed under the Kapton tape, and the liquid crystal analysis (on the thinned package) confirmed a short in the anomaly area (Fig. 7.16). After pulling the Kapton tape off, the short was viewed more easily (although the adhesive was still intact). SEM/EDS analyses confirmed the short to be mostly Al composition (which explains why it could not be observed on the X-ray). Further polishing fully exposed the particle that was wider than the copper finger spacing (Fig. 7.17).

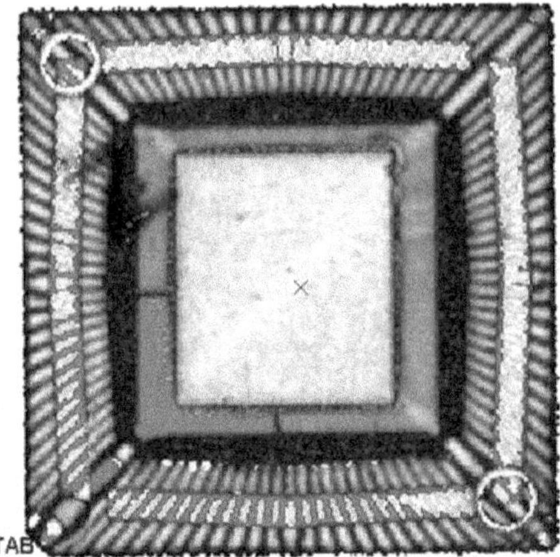

Figure 7.15 C-SAM results showing extensive delamination in the area of the short.

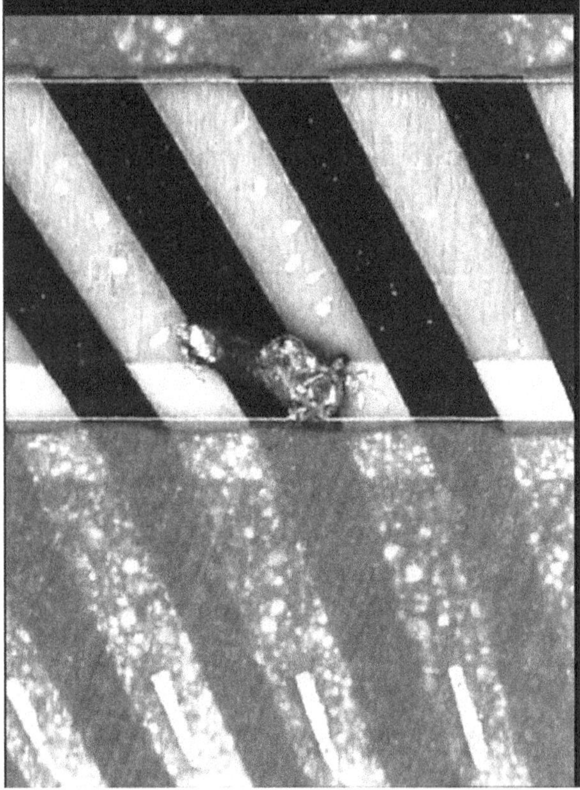

Figure 7.16 Short caused by foreign material.

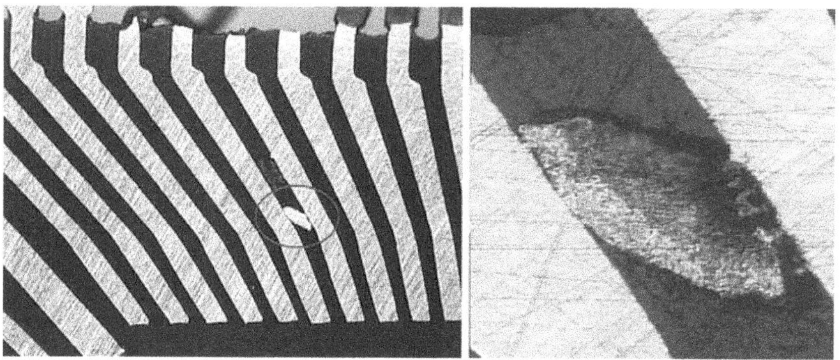

Figure 7.17 Foreign particle was wider than lead spacing.

7.4 Failures in Plastic Ball Grid Array Packages

7.4.1 Open Contact—Via Barrel Cracking

7.4.1.1 Case history and description of failure. During qualification testing of an 851 PBGA device, an electrical open was found after exposing the device to 1000 temperature cycles of Condition B (–55 to 125°C). The open was verified by using a curve tracer, followed by a time domain reflectometry (TDR) analysis, which identified the net in the BGA substrate where the open occurred. Cross-sectioning confirmed the failure to be caused by a crack in the Cu barrel of a plated-through via (Figs. 7.18 and 7.19), one of 904 150-μm diameter vias contained in the substrate. The crack was found at the interface between the outer prepreg layer and the core, extending all around the barrel, thus separating the upper portion of the Cu barrel from the lower. Cross-sectioning also revealed that the same solder mask material used in covering the external substrate surfaces also filled the vias. The voids found in the via solder mask are a common occurrence for this type of process, but it is believed that their presence have a negligible effect on barrel cracking. Inspection of the Cu plating in the subject via showed the minimum thickness (15 μm), grain structure, and overall quality of the via to be within specifications.

The 35-mm square substrate is made of a bismaleimide triazine (BT)/glass-reinforced material and consists of four metallization layers (prepreg + foil laminated to each side of a double-sided core). The inner Cu layers are 35 μm thick and are primarily used as power and ground planes and for spreading the heat dissipated by the chip. The vias have no inner layer pads.

Figure 7.18 Cross section showing barrel crack.

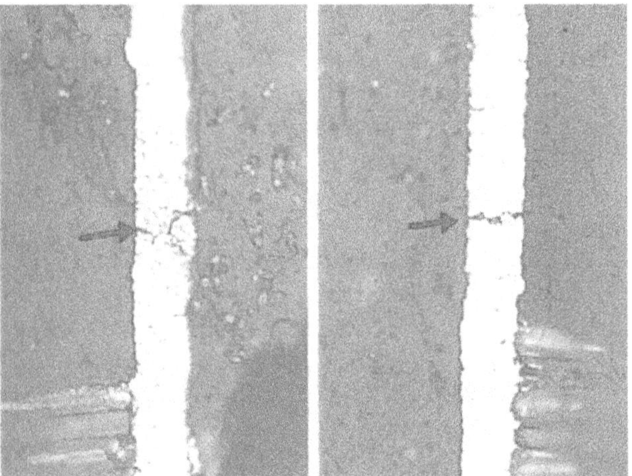

Figure 7.19 Enlarged view of crack in the via wall.

7.4.1.2 Corrective actions. Cross-sectioning the via helped identify the cause for the electrical open and revealed the probable reasons why the crack developed in the via barrel. A number of factors may have contributed to the barrel cracking, such as the following:

a. Thin Cu plating in the via (<15 μm)

b. Absence of inner layer via pads when Cu thickness of the inner metallization layers > 18 μm

c. Material property of the Cu (low ductility)

d. Type of material and fillers used in filling the via

e. Filler content in the glass-reinforced prepreg

f. Severity of the temperature cycling test (temperature range and number of temperature cycles)

g. Large aspect ratio of the drilled via, greater than 3:1 (substrate thickness versus via diameter)

Each of the above factors was evaluated as to the possibility of having caused the crack in the subject via. The thickness and ductility of the plated Cu was examined and found to be within specifications (*a*) and (*c*). Some of the other factors, such as the filler content in the prepreg (*e*), the temp cycling parameters (*f*), and the aspect ratio of the drilled vias (150 μm via diameter in a 0.56-mm thick substrate) (*g*), are constrained by the design and cannot be altered. Thus, only factors such as design (*b*) or via filling material (*d*) can be considered for change to prevent future barrel cracking.

The solder mask within the via, when exposed to temperature cycling, expands and contracts more rapidly than the surrounding Cu barrel or the BT dielectric. This results in Cu fatigue and forms a crack at any location along the via barrel where a stress concentration exists. Thus, it was recommended that the substrate supplier implement a separate filling of the vias with an epoxy containing a filler that will reduce the CTE to closely match the BT substrate. The conventional solder mask should be used only to cover the external surfaces.

As a result of implementing this process modification, the devices passed the temperature cycling requalification test of 1000 temperature cycles at Condition B (–55 to 125°C) without any barrel cracking failures.

Another factor affecting via barrel cracking is the absence of pads around vias on inner metallization layers, when the Cu thickness exceeds 18 μm (*b*). This has been evident by the barrel cracks that were observed at the interface between the prepreg and core, where a resin reach area forms (Fig. 7.20). During lamination, only the resin from the prepreg flows into the crevices formed in the pattern of the inner Cu layers, whereas the glass fibers remain on the Cu surface. Here again, the resin behaves similarly to the solder mask in vias, where it expands and contracts much more than the Cu or glass-reinforced areas, thus contributing to the fatiguing of the Cu in the via barrel. The thicker the inner Cu layer, the thicker the resin reach area around the via barrel and the greater the probability for via cracking. Thus, the other corrective action recommended for this design was to add via pads on the inner Cu layers. Most substrate suppliers discourage the

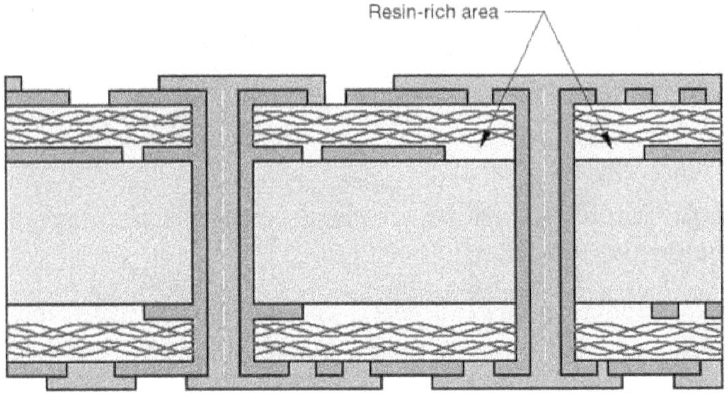

Figure 7.20 Typical cross section through four-layer PBGA substrate.

use of inner via pads, because the drill bits wear out faster when drilling through two additional Cu layers.

7.4.2 Open Contacts—Cracked Signal Traces in Substrate

7.4.2.1 Case history and description of failure. A 680 PBGA device with an embedded heat sink failed for open contacts after 800 temperature cycles of Condition B (–55 to 125°C). All failures were concentrated in one corner of the device. A C-SAM, in pulse-echo mode analysis, showed good die surface to mold compound adhesion (Fig. 7.21). The TDR narrowed the location of the failures to the substrate. Visual inspection of the substrate revealed topside solder mask cracks at the edge of the overmold, which severed several traces (Fig. 7.22). The severity, concentration, and location of the damage indicated a handling- or assembly-related defect.

7.4.3 Open Contacts—Via Knee Cracking

7.4.3.1 Case history and description of failure. During qualification testing of a 680 PBGA device, electrical opens were found after exposing the device to 1000 temperature cycles of Condition B (–55 to 125°C). The 680 PBGA has a 35 × 35 mm, four-layer BT substrate with a heat spreader embedded in the overmold compound. The mechanically drilled through holes in the substrate are 150 μm in diameter and filled with a special plug-in material to reduce the CTE mismatch between the Cu via barrel/plug-in material and the surrounding BT substrate. TDR and C-SAM tests indicated the opens to be substrate

Figure 7.21 C-SAM results showing no delamination on die surface.

related. Cross-sectioning confirmed the failures to be caused by a fatigue crack in the knee area where the Cu via pad and barrel meet. These fatigue cracks were observed on both the top and bottom of the vias (Fig. 7.23). All failed vias were found to be located in the vicinity of the die edges, mold body edges, or the heat spreader footprint edges. The via knee fatigue crack initiated at the corner of the outer wall of the via barrel and the bottom of the pad surface, adjacent to the BT prepreg. The crack then propagated vertically along the outside wall of the via barrel during temperature cycling. The crack eventually separated the via barrel from the via pad to cause an electrical open. Lap-down cross-sectioning revealed that, in some cases, a cylindrical crack developed in the solder mask over the failed vias.

7.4.3.2 Corrective actions. In addition to cross-sectioning the failed samples, a finite element analysis (FEA) was conducted to help understand the failure mechanism and identify the root causes. A 1/8 3-D model was constructed that included substrate stack-up, mold body, copper heat spreader, and a through-hole via structure (Fig. 7.24). The simulations were guided by a design of experiment (DoE) plan that was tailored to identify the significant factors affecting the stresses in the via structure. The factors considered in the DoE included material properties of the substrate (including the via plug-in material), the

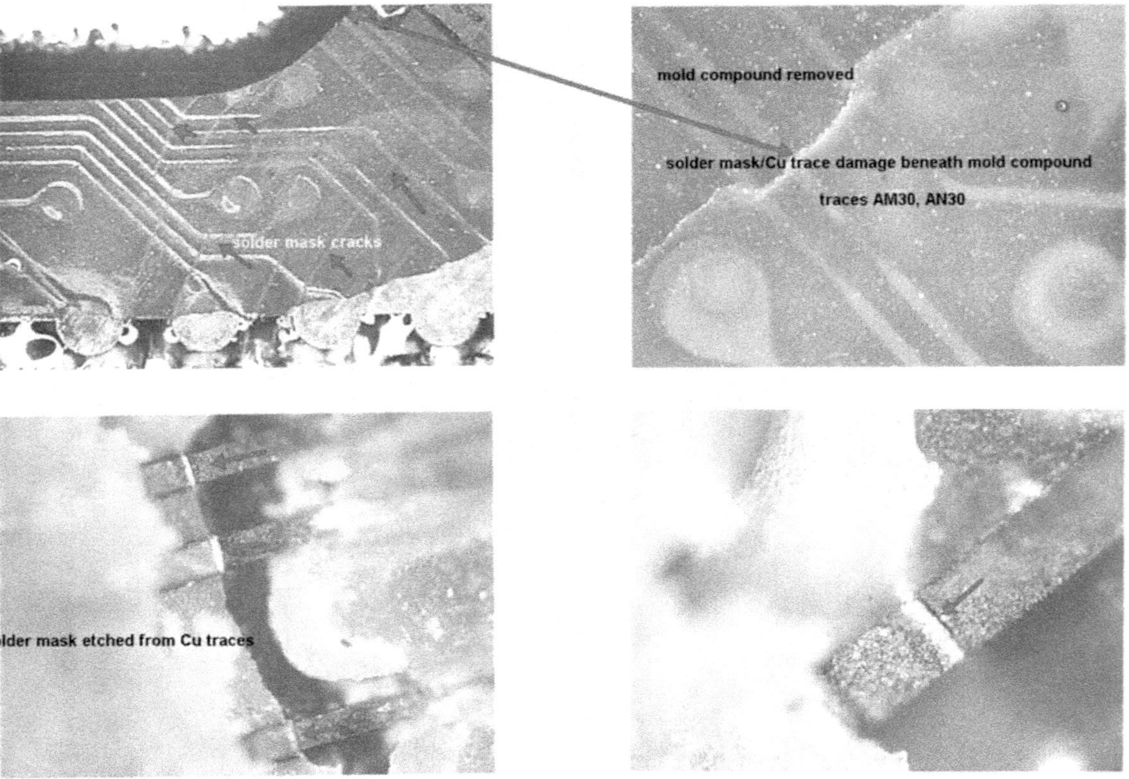

Figure 7.22 Solder mask cracks at edge of overmold, which severed several traces.

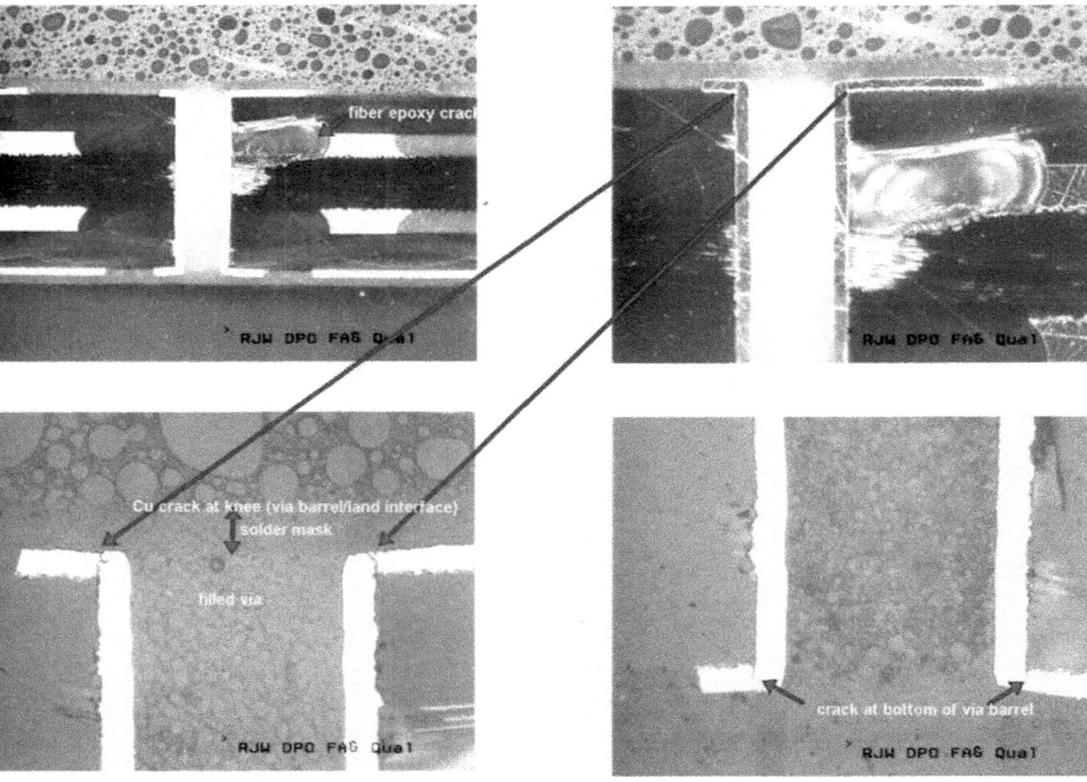

Figure 7.23 Fatigue crack on top and bottom of via barrel.

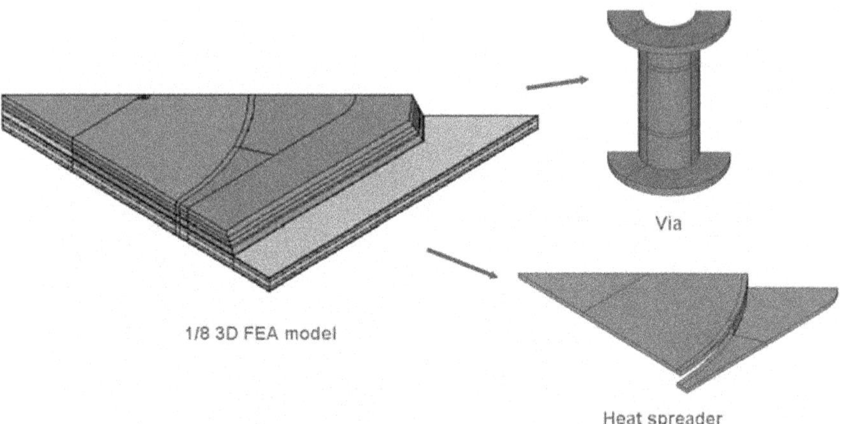

Figure 7.24 Finite element analysis (FEA) model.

overmold and die material properties, and the substrate/overmold/die geometries (including the via drill size). Other factors, such as the bonding condition between via barrel and via plug-in material, were also considered. The DoE results revealed that the drill size has a significant impact on the stress level in the Cu via barrel and pad. FEA results also showed higher stresses at the corner of the via knee during the cooling portion of temperature cycling; thus, it is prone to crack initiation at these locations. The heat spreader footprint edges, along with the die edges and mold body edges, contribute incrementally to the via stresses (Fig. 7.25).

The via plug-in material used to fill the vias has a CTE of 24 ppm/°C (in the x, y, and z directions), in contrast to the surrounding BT dielectric, which has a CTE of 55 ppm/°C in the z direction, through the substrate thickness. The FEA simulation results indicated that the CTE mismatch between these two materials could induce very high tensile stresses at the corner of the via knee during temperature cycling. Furthermore, cross-sectioning showed that the plug-in material adhered well to the via barrel wall, resulting in a stiffer via structure that accelerated the crack propagation in the via knee area as a result of the CTE mismatch. Cross-sectioning also found cracks in the mold body near the heat spreader footprint edges where some of the failed vias were located. Therefore, mold resin shrinkage may also have contributed to the failure of vias in those areas.

The following corrective actions were taken:

- Review the current substrate design and move the affected vias, if possible, as far away as possible from the high-stress regions,

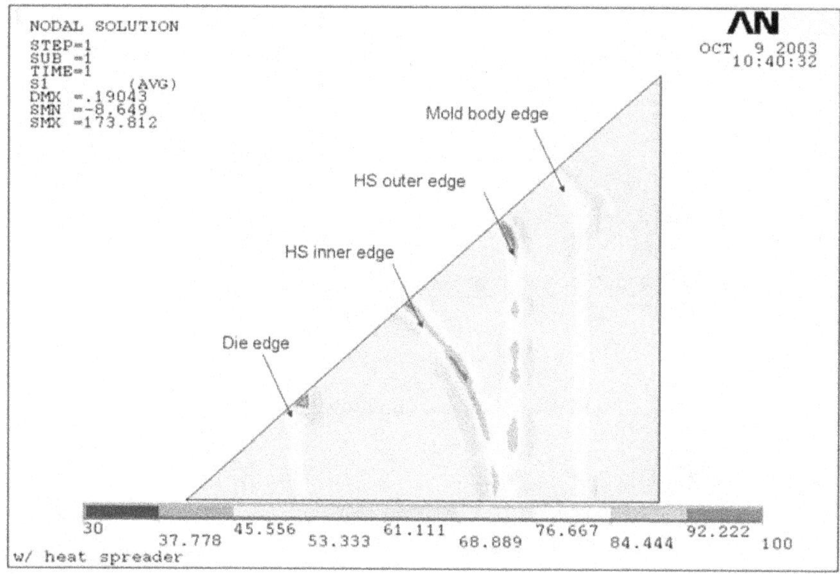

Figure 7.25 First principal stress on the top side of the substrate, showing edge effect.

such as die edges, heat spreader footprint edges, and mold body edges.

- Recommend that the substrate supplier use a new via plug-in material that has a higher CTE to reduce the CTE mismatch between the via structure and the surrounding BT material.

As a result of implementing these corrective actions, the devices passed requalification of 1000 temperature cycles of Condition B (–55 to 125°C) without failures.

7.4.4 Open Contacts—Lifted Wedge Bonds

7.4.4.1 Case history and description of failure. A field return of a 316 PBGA device that failed electrical test was analyzed to determine the location and cause of failure. An electrical test verified opens in two separate signal paths. Cross-sectioning the device through the mold compound and substrate (Fig. 7.26) revealed a slight kink upward in the body of the substrate. At the top of the kink, adjacent to the affected wedge bond pad, a crack initiated in the solder mask and propagated into the overmold material (Fig. 7.27).

The mold compound crack propagated over the wedge bond pad, causing the gold wire wedge bond to separate from the bond pad (U8).

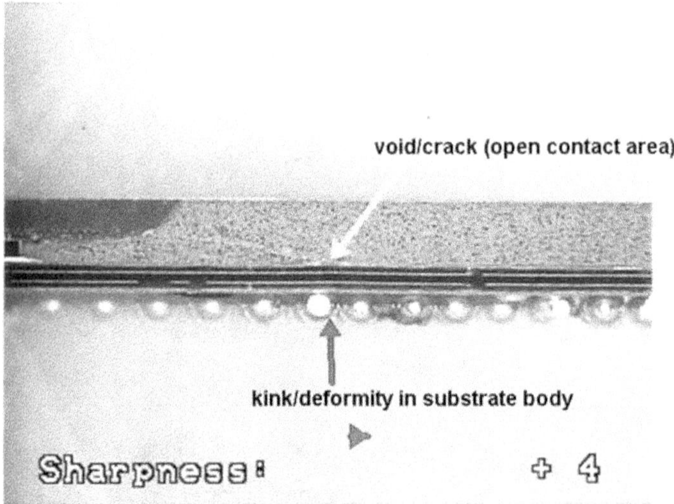

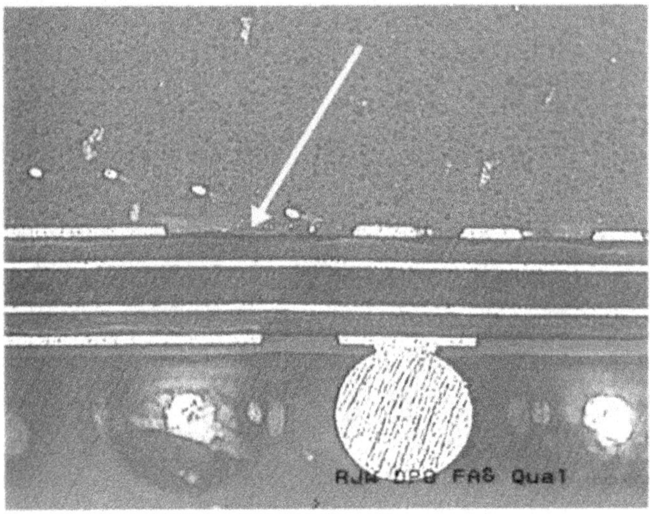

Figure 7.26 Cross section showing slight kink upward in the substrate.

A similar type of failure also occurred at wedge bond pad T10 (Fig. 7.28).

Because the crack occurred right above the kink in the substrate, it is believed that the force exerted by the kink on the solder mask/overmold material caused the crack. The kink in the substrate is thus the culprit, which may have occurred during overmold assembly or as a result of improper test socketing.

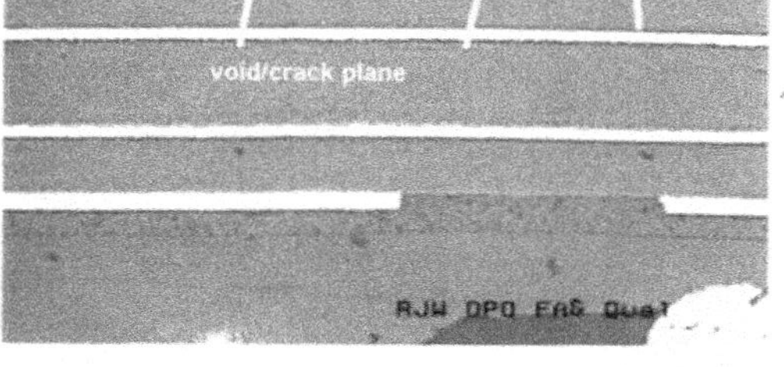

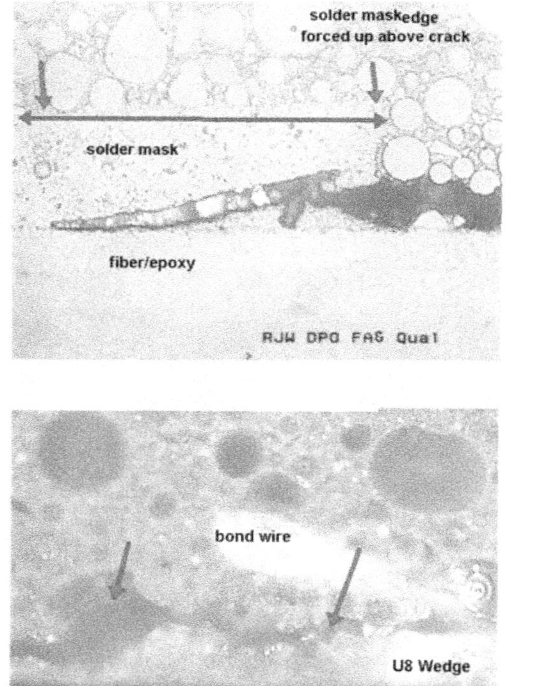

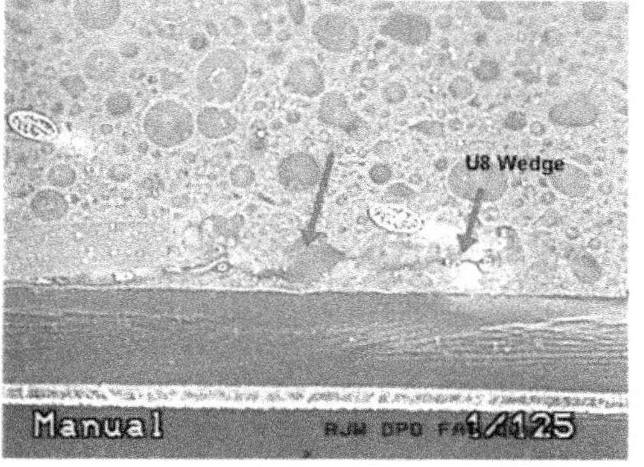

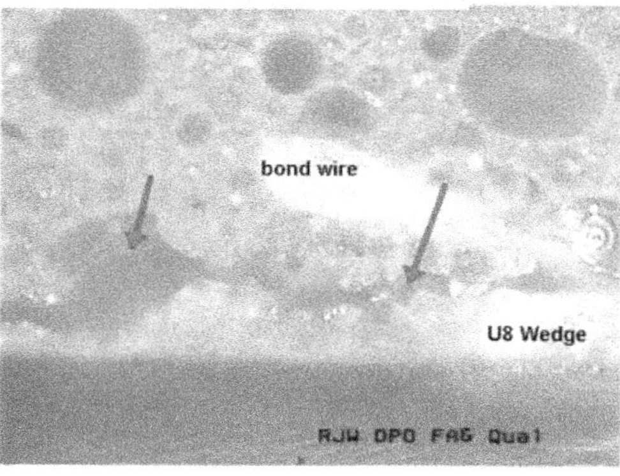

Figure 7.27 A crack initiated in the solder mask propagated in the overmold material.

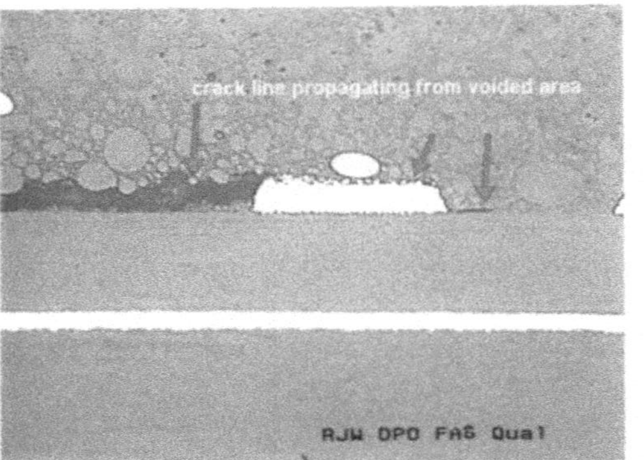

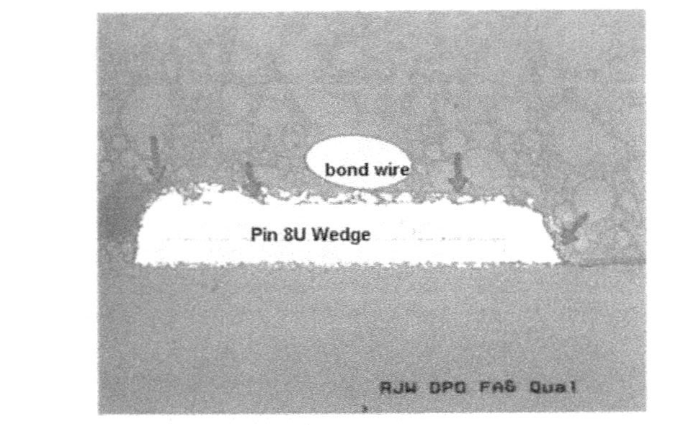

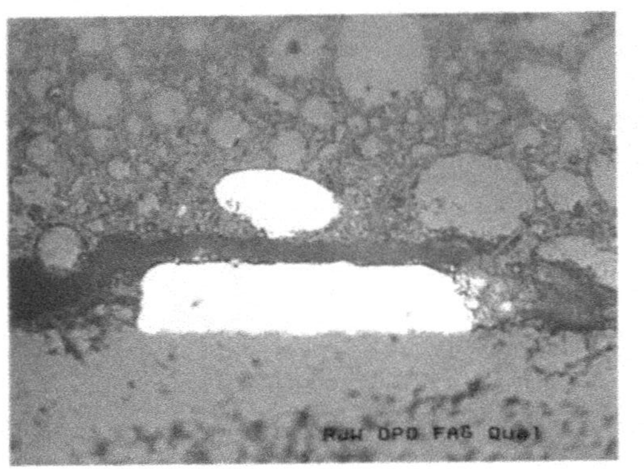

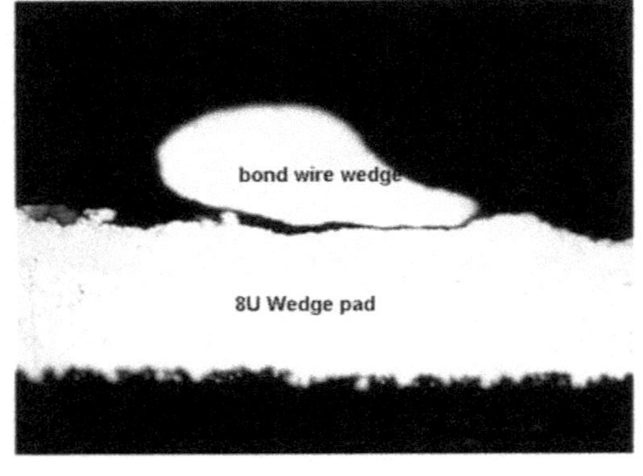

Figure 7.28 Mold compound propagated over the wedge bond pad, separating the gold wire wedge bond.

7.4.5 Open Contact—Cracked Via Barrel

7.4.5.1 Case history and description of failure. An 851 PBGA device failed for a single open contact after 500 temperature cycles of Condition B (–55 to 125°C). The open was confirmed with a curve tracer, isolating it to solder ball R33. A cross-sectional analysis (Fig. 7.29) showed that the via associated with ball R33 failed due to temperature cycling fatigue. A fatigue crack was observed adjacent to the fiber pull-out regions in the substrate (Fig. 7.30). The via barrel had a minimum Cu plating thickness of 12 μm.

7.4.5.2 Corrective actions. The vias in this particular device were filled with a solder mask material that has a CTE much greater than that of the surrounding substrate material (BT). These CTE differ-

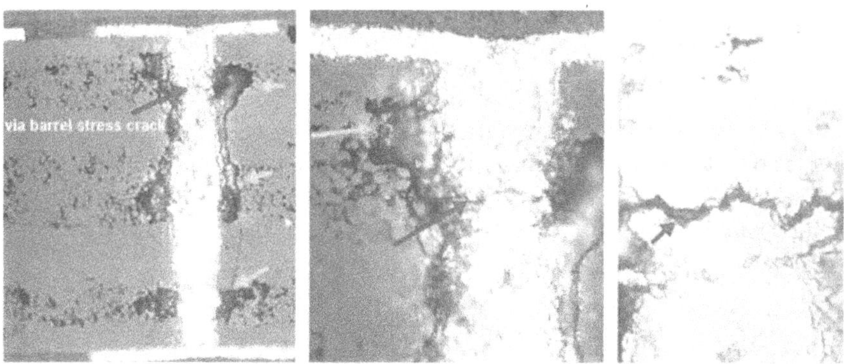

Figure 7.29 Cross section showing fatigue crack in via barrel and drill pull-out fiber.

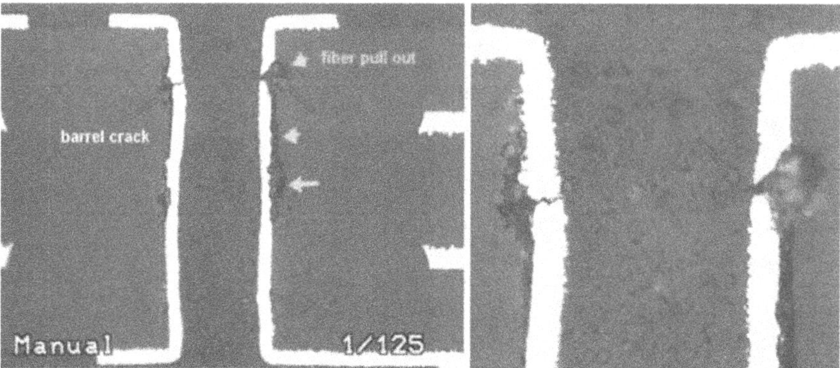

Figure 7.30 Barrel crack adjacent to fiber pull-out region.

ences caused the Cu barrel to expand and contract with the plugged via material, cracking the barrel. The Cu barrel separated from the substrate walls during contraction, which resulted in fracturing some of the fibers. The voids seen around the Cu barrel (Fig. 7.30) are the result of fractured glass fibers falling out at cross-sectioning. To prevent this type of failure in the future, it was recommended that the solder mask via plugging material be replaced with an epoxy/filler material having a CTE close to that of the BT substrate.

7.4.6 Open Contact—Cracked Signal Trace in Substrate

7.4.6.1 Case history and description of failure. A 217 PBGA device failed electrical test for an open contact following exposure to 600 temperature cycles of Condition B (–55 to 125°C). The open was verified on the curve tracer, which isolated the failure to the C3 corner region trace. A C-SAM, in through-mode analysis, was used to evaluate the integrity of the package. Peak amplitude scans showed no obvious interlevel delaminations; however, the overmold/solder mask interface did show degradation in signal intensity (Fig. 7.31). Inspection of this interface revealed cracks in the top-side solder mask. Excessive vertical cracking was also seen at the edge of the overmold.

Cross-sectional results of the failed net showed that the crack in the solder mask continued through the copper signal trace into the sub-

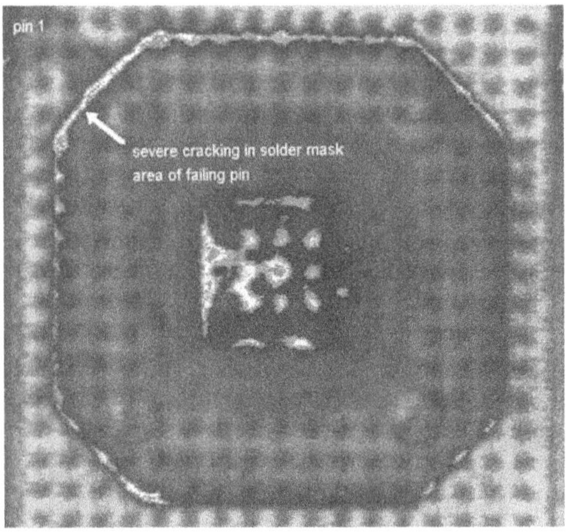

Figure 7.31 C-SAM results showing delamination at the overmold/solder mask interface.

strate core (Fig. 7.32). The crack had initiated below the overmold/solder mask interface.

The substrate exhibited the same degree of solder mask cracking typically seen following exposure to temperature cycling. However, the deformation of the corner BGA balls and vertical cracking at the edge of the overmold/solder mask interface suggested that the damage originated from an external mechanical force such as automated test equipment (ATE) socketing or clamping during the assembly process, and it was not due to temperature cycling.

7.4.6.2 Corrective actions. It was recommended that the mold clamping force be reduced to prevent cracking. The magnitude of the force is a compromise between a low force that may cause an excessive mold flash and a larger force that may initiate substrate cracking.

7.4.7 Open Contact—Lifted Ball Bonds

7.4.7.1 Case history and description of failure. Three 680 PBGA devices failed for a single open contact in each following exposure to 500 temperature cycles of Condition B (–55 to 125°C). TDR analyses on the failed devices indicated that the open paths in each device were in the vicinity of the wirebond ball/chip interface. C-SAM, in pulse-echo mode, analyses on the three devices showed that the failures exhibited a common delamination pattern not typical of temperature cycle stressing (Fig. 7.33). A C-SAM in through-mode, analysis was also performed on the devices to examine the substrate interface integrity. No obvious underlying package defects or anomalies were observed. One of the devices was decapped to expose the die surface. Stress cracks in

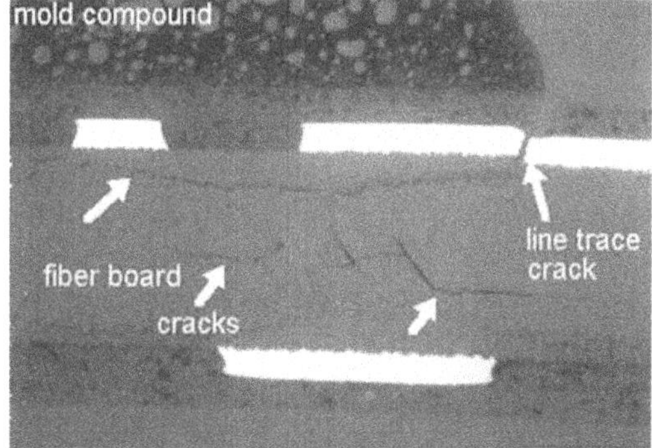

Figure 7.32 Cross section showing the crack propagating through the signal traces.

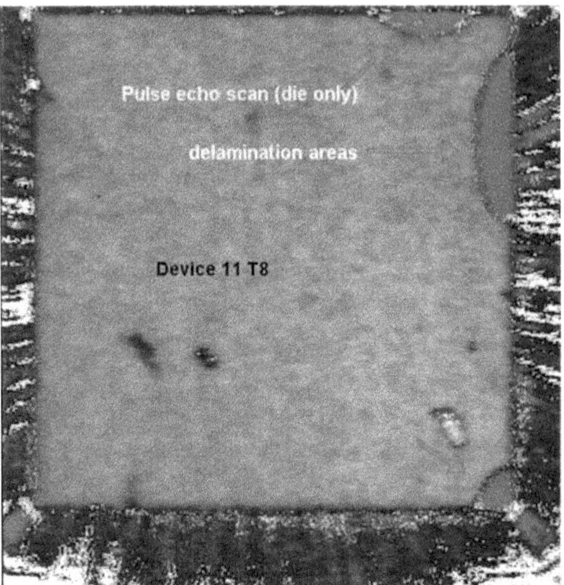

Figure 7.33 C-SAM results showing delamination on die surface.

the passivation correlating to the delamination pattern were observed. The multiple lifted ball bonds were consistent with the delamination location accounting for the open contact on the ATE.

The second device was delayered horizontally to examine the top surface of the die for evidence of foreign material. After removing all but a few micrometers of mold compound from the surface, the delamination areas were viewed optically. The edge of the delamination site showed a rainbow color pattern beneath the thin layer of mold compound (at the die surface), suggesting that a contaminant was present at the chip surface/mold compound interface. No other discoloration sites were observed across the surface of the die.

The third device was vertically cross-sectioned to examine the die-attach profile in the region of the delamination sites. Examination of the first delamination site, which looks into the face of the die attach fillet, showed excessive fillet height (Fig. 7.34). Die-attach epoxy and silver filler were observed between the bulk silicon and chip surface (test pattern) in the saw street of the die edge. The presence of the epoxy and silver filler in this area indicated that the height of the die attach fillet was at or near the surface of the die. A cross-sectional view of the die attach fillet profile showed the epoxy separated from the silver filler and crested over the edge of the die surface (Fig. 7.34). Both regions possessing excessive die attach fillet height were consistent

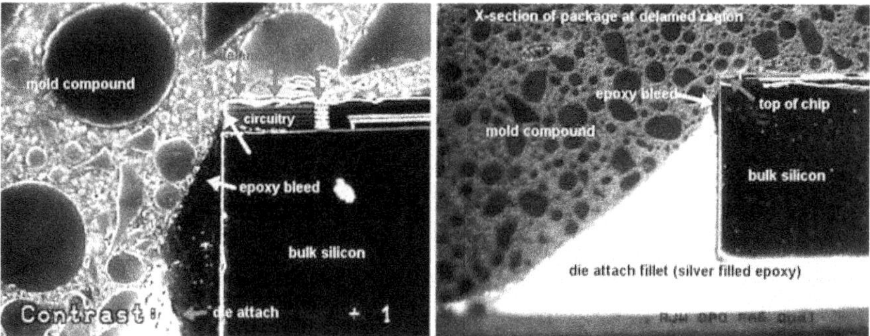

Figure 7.34 Cross section showing excessive die attach fillet height.

with die surface delamination regions observed on the C-SAM data scans.

7.4.7.2 Corrective actions. The devices examined failed as a result of die-attach epoxy contamination on the die surface. Die surface delamination patterns were consistent among the failed devices and were not typical of a temperature cycle failure. Package analysis showed that the delamination patterns, passivation cracking, and lifted ball bonds observed along the edge of the die were in line with excessive die attach fillet heights. A cross section of the fillet profile showed residual epoxy crested over the die surface. Areas of the chip surface with normal die attach profiles at the edges did not show die surface delamination, passivation cracking, or ball bond lifting. The current fillet height specification is 90 percent maximum of the die thickness. Based on the analyses, it seems that it was the resin in the paste that had crept up to the die surface, not the silver. The current fillet height inspection covers only the top and bottom sides of a unit in a strip. This is because the left and right sides of the unit are blocked by adjacent units and are unable to be seen clearly under the 'scope. Even for the end units, either the left or the right side would not be seen, depending on at which end the device is located (Fig. 7.35).

A cross section through the die showed the fillet heights for both the top and bottom sides of the device to be below the die surface, at a safe region estimated to be about 60 percent to 70 percent of the die thickness. On the other hand, the fillet height on the right side of the device where delamination occurred was high, almost touching the die surface. The fillet height on the left side was low. It is believed that, during die attach, the relative placement of the die on the dispensed paste pattern could have been slightly offset, resulting in the difference in the paste fillet heights along the four sides.

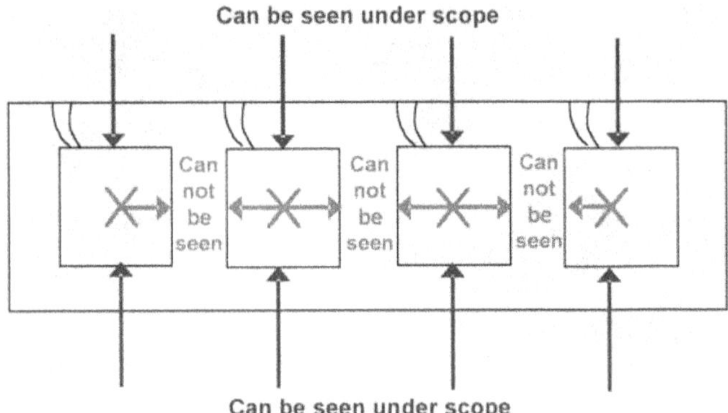

Figure 7.35 Die attach fillet inspection of PBGA in a strip.

To prevent future die surface contamination due to fillet height, the current inspection process was changed so that all four sides of the die are inspected for epoxy fillet height. In addition, the maximum allowable epoxy fillet height was changed to 75 percent from the current 90 percent.

7.4.8 Open Contact—Broken Bond Wire

7.4.8.1 Case history and description of failure. A 208 PBGA failed an electrical test for open contacts after exposing the device to 500 temperature cycles of Condition B (–55 to 125°C). TDR analysis pointed toward the ball bond/die surface interface or a broken wire as the most likely location where the failure occurred. A crack was observed on the surface of the overmold (Fig. 7.36) which, after horizontal delayering, showed that the same crack had penetrated the mold compound all the way to the die surface (Fig. 7.37). Further cross-sectional analyses revealed that the crack occurred near slivers of "resin material" (Fig. 7.38). The resin material appears to be mold compound (due to presence of filler in some) that is suspected to have been cured prior to the mold shot. These pieces of precured compound may have then mixed with the "regular" compound prior to or during the mold cycle, and they appear as inclusions in the molded body. Because the mold compound will not adhere well to the surface of compound that has already cured, the interface between these inclusions and the bulk compound will be structurally weak and could lead to cracking under temperature-cycling stresses. The inclusions did not only occur in the region of the molded body above the die, where the package cracks occurred; they were also present in other regions of the molded body. Due to the uniform thickness of the inclusions (~9 μm), they may be

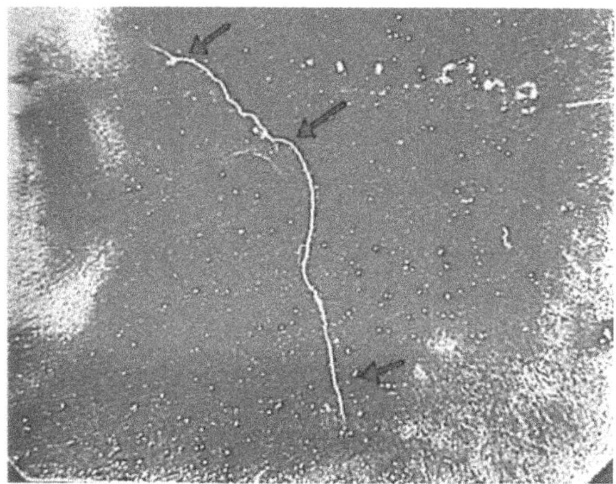

Figure 7.36 Crack on the surface of the overmold.

vent flash (cavity or runner) or plunger flash, which was inadequately cleaned from previous mold shots.

7.4.8.2 Corrective actions. Because the mold plates were checked many times after the cleaner arm finished its cleaning cycle and found to be clean of any flash, it is suspected that a small amount of flash may have accumulated instead at the plunger tips. An improved plunger cleaning process, after every mold shot, has resolved this issue.

7.4.9 Open Contact—Cracked Die

7.4.9.1 Case history and description of failure. A 700 PBGA device failed electrical test for opens during board test assembly. The device has a 35 × 35 mm, four-layer BT substrate with solder balls on 1.0-mm pitch. The TDR found the failure to be in the die, and C-SAM indicated that it was the result of a crack (Fig. 7.39). The substrate and the die-attach material were removed, exposing the backside of the die. Visual inspection revealed a crack that initiated from the needle plunger mark, formed during the removal of the die from the wafer tape (Fig. 7.40). Propagation of the crack occurred during solder reflow, when the thermal stress is at its peak, resulting in a latent electrical failure.

7.4.9.2 Corrective actions. A larger-size needle plunger is now being used to remove the die from the wafer tape.

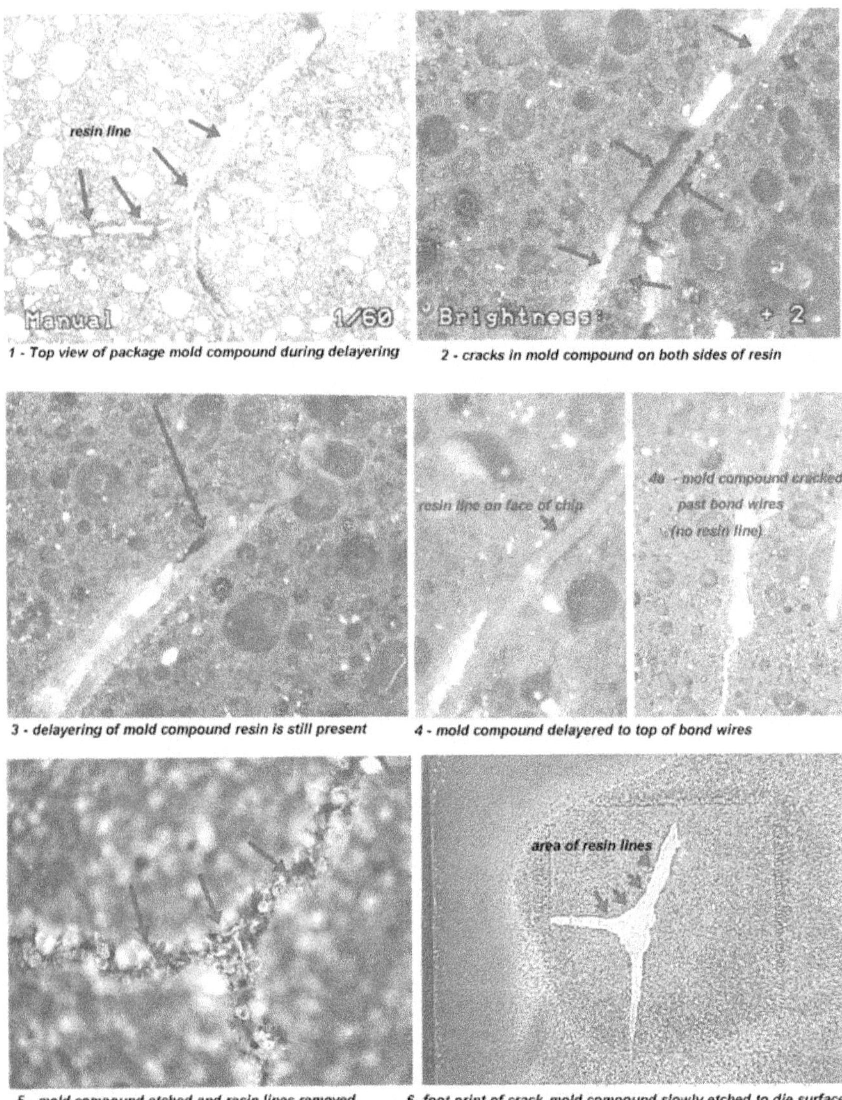

1 - Top view of package mold compound during delayering

2 - cracks in mold compound on both sides of resin

3 - delayering of mold compound resin is still present

4 - mold compound delayered to top of bond wires

5 - mold compound etched and resin lines removed

6- foot print of crack mold compound slowly etched to die surface

Figure 7.37 Crack penetrated through the overmold, reaching the die surface.

7.4.10 Shorted Contacts—Die Attach Material on Exposed Cu

7.4.10.1 Case history and description of failure. A 310 PBGA device failed electrical test for shorted contacts after device assembly. It was found that some I/O signals and power (V_{DD}) were shorted to ground (Fig. 7.41). The lap-down cross section showed that the shorts were

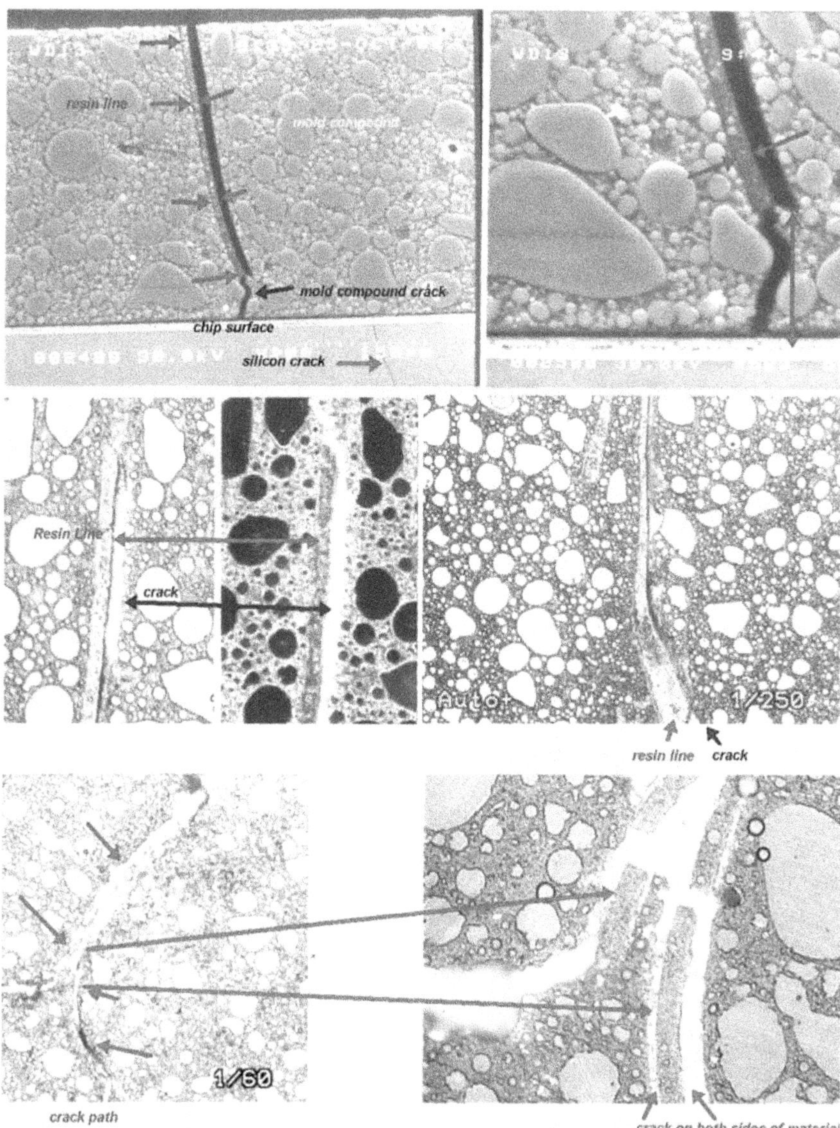

Figure 7.38 Crack penetrated through the overmold, reaching the die.

caused by the die-attach conductive silver filled epoxy filling the uncovered etchback windows (Fig. 7.42).

The BT substrate consists of four layers, with electroplated Ni/Au finish on the wedge bond pads on the top side and BGA pads on the bottom side. To electroplate Ni/Au over exposed Cu surfaces requires the use of plating traces, which provide electrical connection to the iso-

Figure 7.39 Crack location reference.

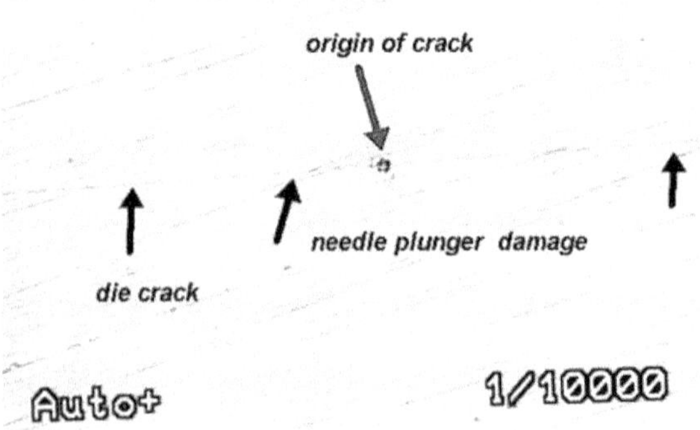

Figure 7.40 Needle plunger damage on the backside of the die.

lated bond pads and BGA pads. To improve the electrical performance during high-frequency operation, the plating traces are etched back after the Ni/Au plating process. Windows are formed in the solder mask, on top and bottom of the substrate, which uncover the plating traces to be etched. After etchback, the windows remain uncovered. In this substrate design, some of the uncovered etchback windows were placed under the die where, during die attachment, the conductive die-attach material filled the uncovered windows, shorting the tips of the remaining plating traces.

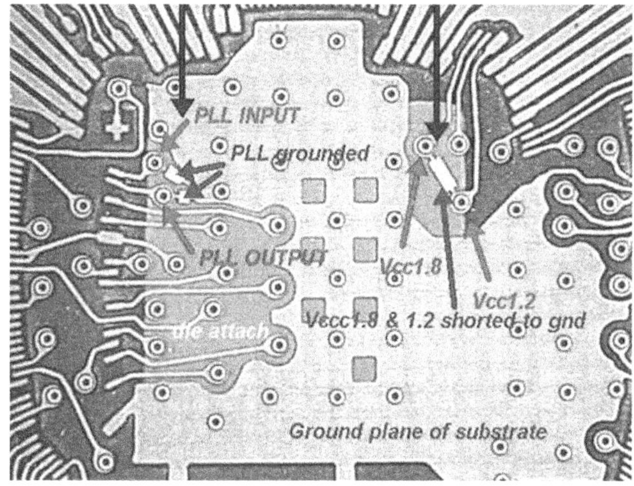

Figure 7.41 Etchback windows under the die filled with conductive die-attach material-created shorts.

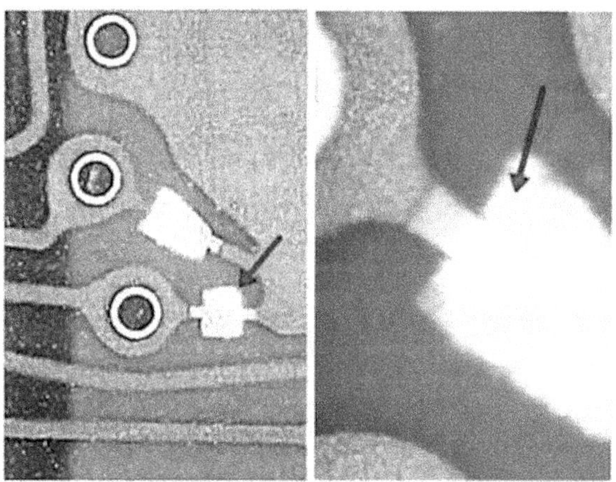

Figure 7.42 Close-up view of shorted traces through etchback windows.

7.4.10.2 Corrective actions. Corrective actions taken to fix the problem are as follows:

1. Request the substrate supplier cover the etchback windows with solder mask. Because this requires additional process steps, the substrate cost will most likely increase.

2. Use nonconductive die attach epoxy, if replacing the conductive silver filled epoxy will not have a detrimental effect on thermal performance.

7.5 Failures in Matrix Ball Grid Array Packages

7.5.1 Open Contact—Via Knee Cracking

7.5.1.1 Case history and description of failure. A 100 MBGA failed electrical test for open contact after exposure to 300 temperature cycles of Condition B (–55 to 125°C). The failure was verified using a curve tracer, which isolated it to a particular net containing an open via.

The device was cross-sectioned vertically through the via barrel to observe any interconnect damage (Fig. 7.43). The device was ground through a sequence of progressively finer abrasive papers to expose the target area. The suspected area, where the open occurred, was polished to 0.5 µm, and the Cu structures were micro-etched prior to evaluation. The analysis found that the open was the result of stress cracks developed at the knee of the via barrel (Fig. 7.44). In addition, the via showed evidence of Cu nodule formation, nonuniform plating, localized Cu thinning, and discontinuity in the Cu plating (Fig. 7.45).

7.5.1.2 Corrective actions. The most likely cause for the anomalies described above was poor drilling, i.e., drill bit wear and improper drill speed or feed.

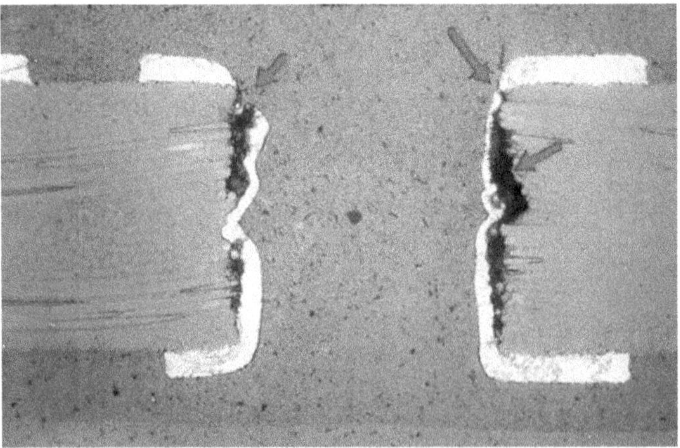

Figure 7.43 Cross section through the via barrel showing cracks and nonuniform Cu plating.

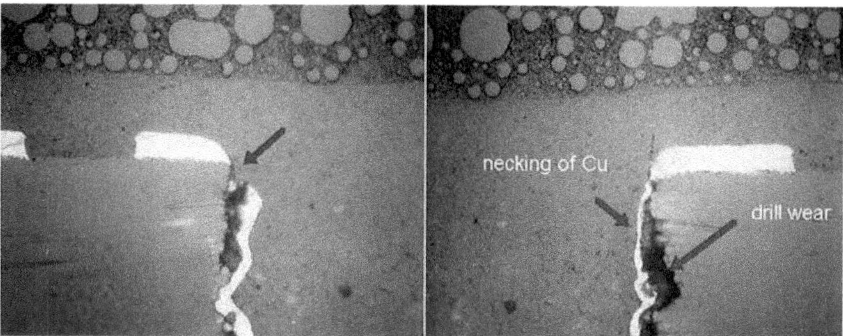

Figure 7.44 Stress cracks at the knee of the via barrel.

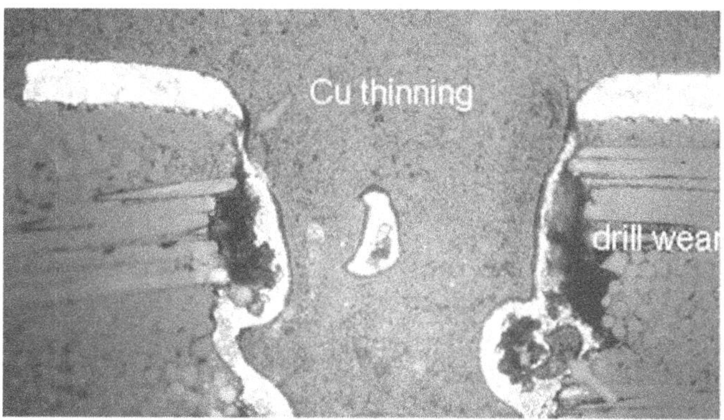

Figure 7.45 Cu nodule formation and nonuniform Cu plating.

7.5.2 Substrate Opens—Cracked Signal Trace in Substrate

7.5.2.1 Case history and description of failure. A 100 MBGA device failed electrical test for an open contact after it was subjected to 500 temperature cycles of Condition B (–55 to 125°C). The open contact failure was verified using a curve tracer. The MBGA has a 10 × 10 mm, two-layer BT substrate with vias in pads. C-SAM, in pulse-echo mode, scans did not show die surface delamination; however, through-mode analysis revealed interlevel delamination (Fig. 7.46) occurring at both the chip/die-attach interface and die attach/solder mask interface. C-SAM (through-mode) at the area of interest showed that the open contact was the result of a cracked signal trace in the substrate (Fig. 7.47). Cross-sectioning revealed that the open contact (crack) was physically located at the structural interface beneath the edge of the die (Fig. 7.48).

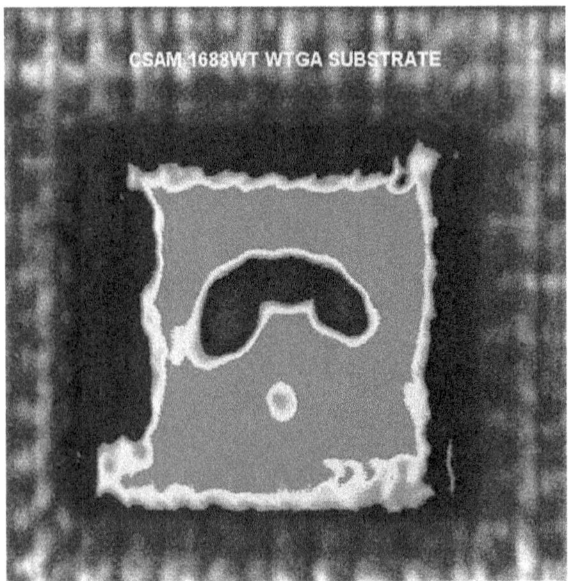

Figure 7.46 C-SAM results showing delamination at the die attach/solder mask interface.

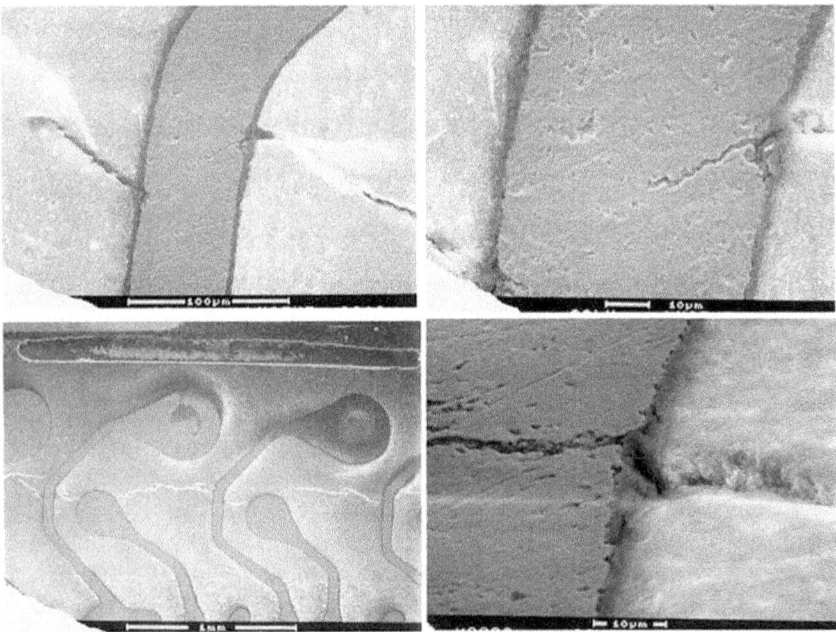

Figure 7.47 SEM results showing cracked signal traces.

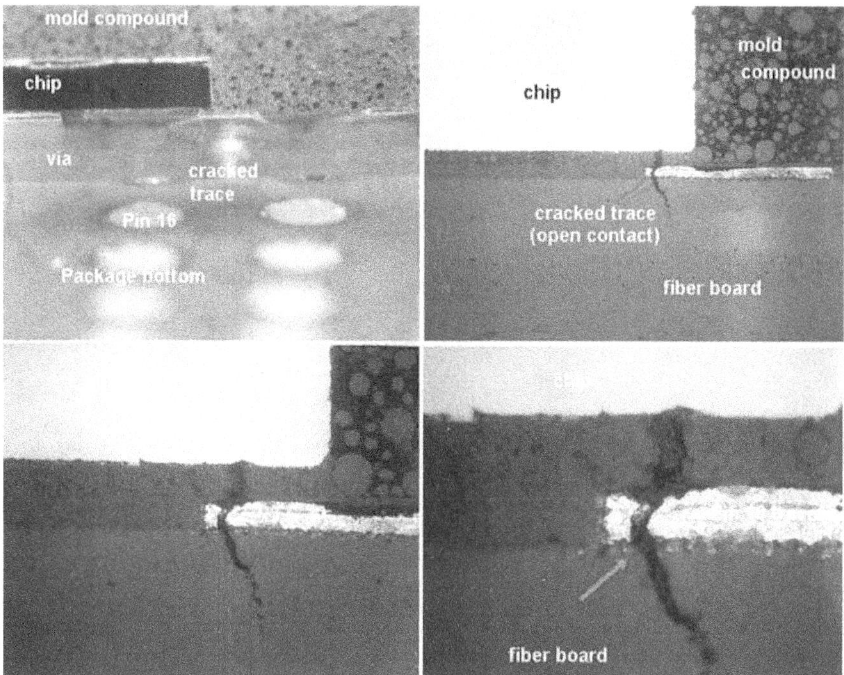

Figure 7.48 Cross section showing cracked trace under edge of die.

7.6 Failures in Plastic BGA Multichip Modules

7.6.1 Intermittent Open Contacts—Cracked Signal Traces

7.6.1.1 Case history and description of failure. A 448 PBGA multichip module, containing three chips, failed a test for open contact after 500 temperature cycles of Condition B (–55 to 125°C). A curve tracer and TDR analysis showed intermittent opens on a signal trace in the substrate. The six-layer BT substrate is 31 mm square, with traces running under the three chips. By delayering the module, the origin of the open was traced to a cut in a Cu signal trace on layer 1, at the edge of the die-attach fillet. The cut in the trace was caused by a crack in the solder mask that developed at the same edge of the fillet during temperature cycling (Fig. 7.49). Cross-sectioning the module showed that the crack initiated at the surface of the solder mask and propagated downward into the substrate, cutting traces in its way on layers 1 and 2 (Fig. 7.50). In addition, the cross section also revealed die corner damage, most likely caused during wafer dicing. What appears as a

Figure 7.49 Cut in the trace was caused by a crack in the solder mask.

die-attach fillet/solder mask delamination (Fig. 7.51) is really an epoxy reach area. It was also observed that some of the vias were not centered in their pads (Fig. 7.52). This is acceptable in the industry as long as the vias do not break out of the pads.

7.6.2 Open Contacts—Cracked Signal Traces

7.6.2.1 Case history and description of failure. A 1152 PBGA multichip module failed the open contact test following moisture preconditioning exposure during qualification testing. The preconditioning test included exposure to 60°C/60 percent RH for 52 hr (Level 3) followed by three solder reflows at 220°C. The 45-mm square BT substrate has six metallization layers and contains four chips. The opens were verified using both the curve tracer and a TDR analysis, which showed that the open signals were substrate related. C-SAM, in pulse-echo mode, scans on the chips found the chip/mold compound interface to be solid. No signs of chip delaminations or package anomalies were observed. A cross-sectional analysis of the substrate revealed that the open signal paths were caused by a cracked Cu signal trace on layer 1 (Fig. 7.53). The open signal path for both contacts

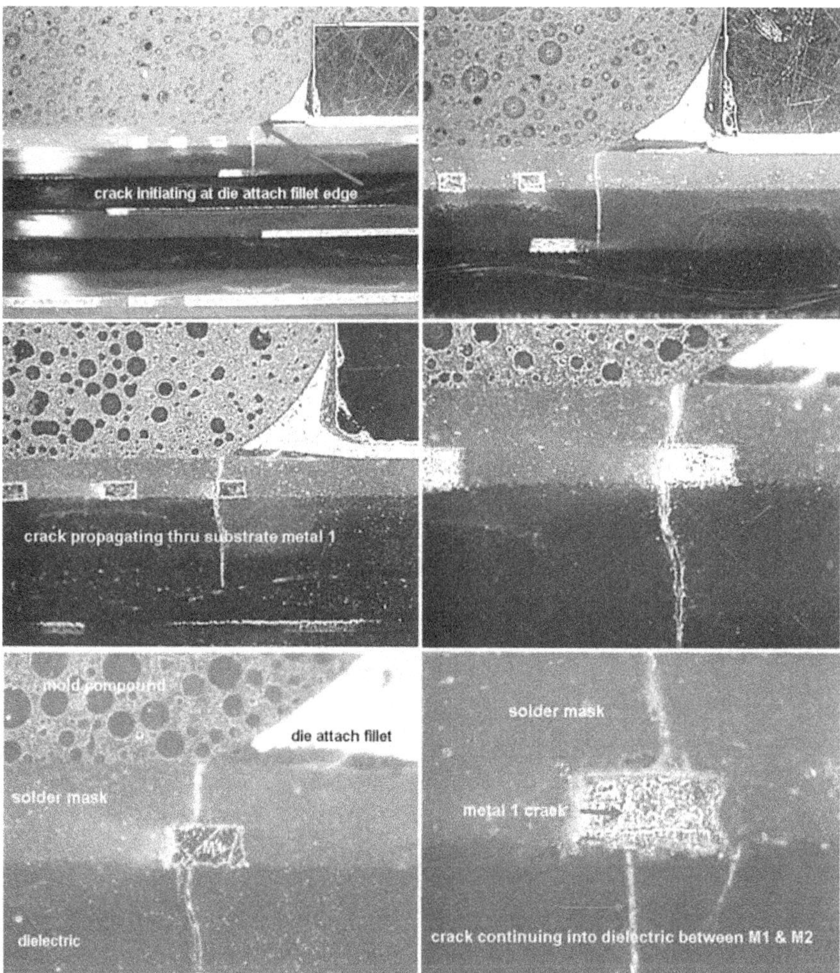

Figure 7.50 Crack initiated at the surface of the solder mask and propagated downward into the substrate.

were severed by a crack that started as a mold compound to solder mask separation at the chamfered corner of the overmold compound and propagated downward into the substrate layers (Fig. 7.54). It appears that both open contacts were the result of an external stress applied to the corners of the package/substrate. Microcracks were also present in the solder mask on the bottom side (BGA side), in corners of the substrate (Fig. 7.55).

The findings are consistent with stresses exerted on the substrate by breaking (singulating) the individual modules from the strip after assembly. Further investigation found that the singulation (breaking

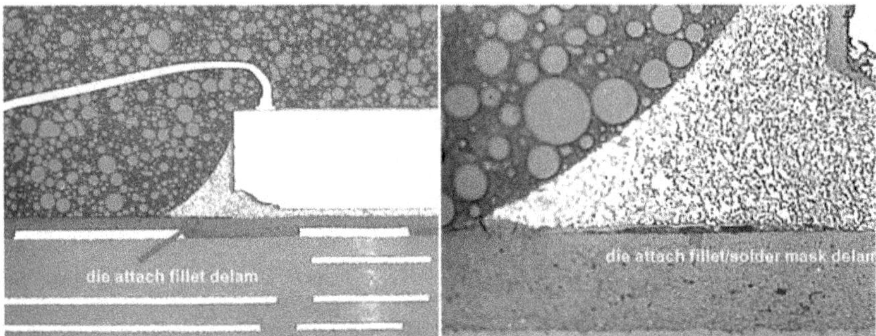

Figure 7.51 What appears to be a die-attach fillet/solder mask delamination is an epoxy reach area.

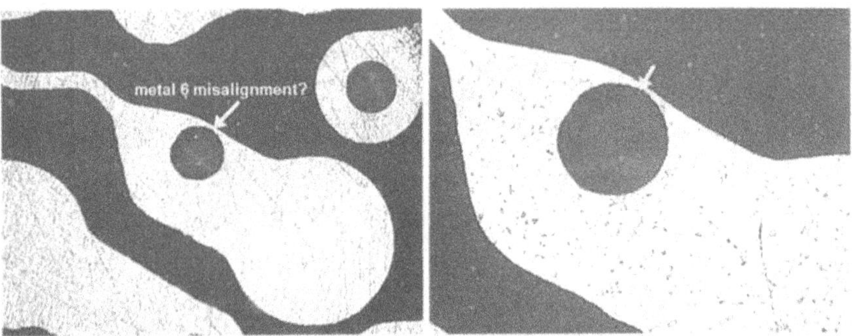

Figure 7.52 Misaligned vias in pads.

the corners) was done manually, and it was the cause of the cracks and eventual opens in the traces.

7.6.2.2 Corrective actions. Instead of manually breaking the corners to singulate the devices from the strip, an automatic singulation tool was installed that shears the four corners. The change in the singulation process solved the cracking issue, and no further failures of this type occurred.

7.6.3 Functional—Dendritic Growth

7.6.3.1 Case history and description of failure. A 318 BGA multichip module (Fig. 7.56) failed qualification testing after 114 hr of temperature-humidity bias (THB). The module failed for leakage between a signal and a ground pad. The 25-mm square, four-layer BT substrate

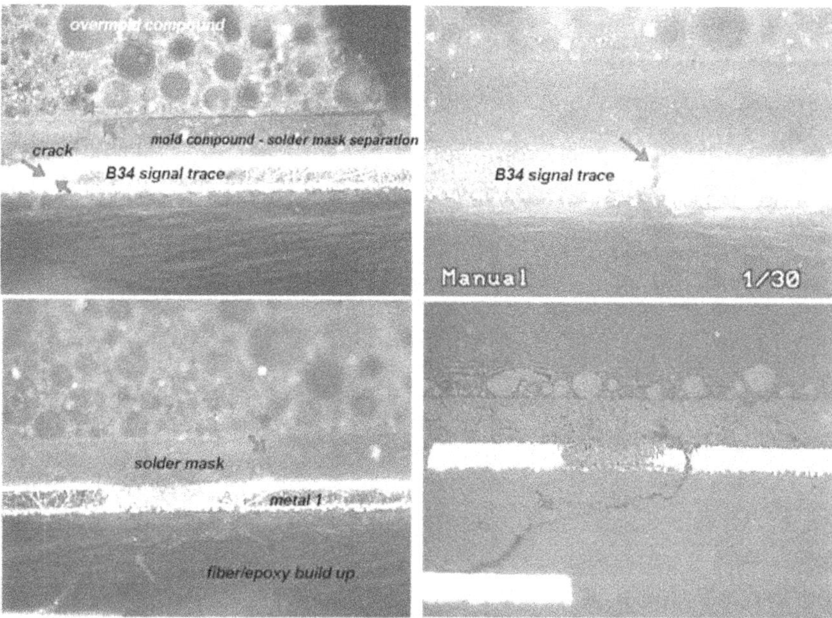

Figure 7.53 Cracked signal trace on layer 1.

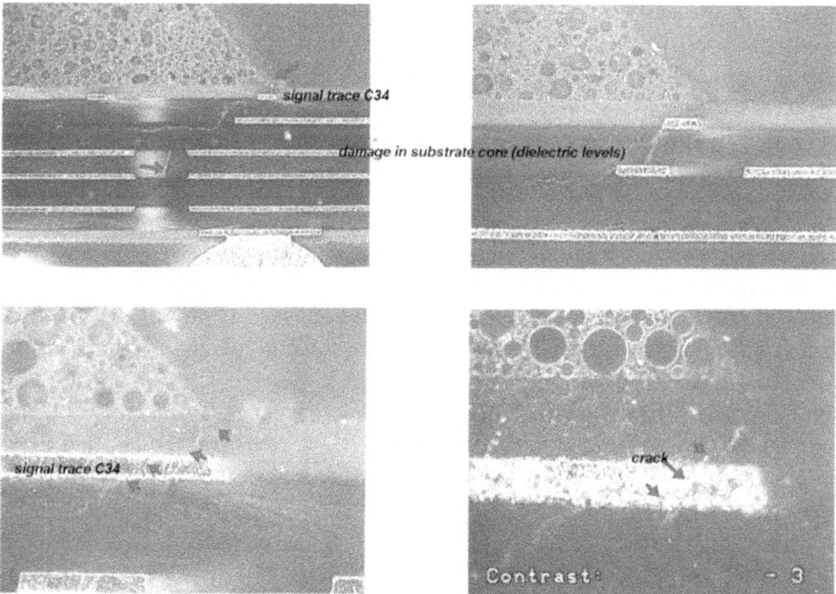

Figure 7.54 Crack started at the chamfered corner of the overmold and propagated downward into the substrate.

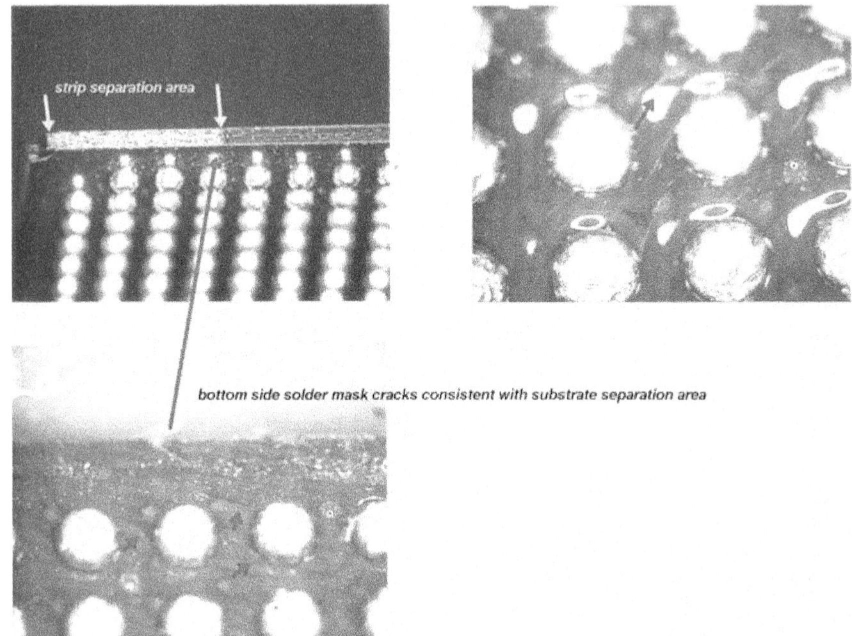

Figure 7.55 Microcracks in the solder mask on BGA side.

contains 4 wire-bonded chips and 126 passive/active surface-mounted components. A close examination of the substrate revealed dendritic growth at the filter component between a signal and a ground pad (Fig. 7.57). In addition, some solder beads were also found close to the filter, which may also have contributed to the leakage (Fig. 7.58).

7.6.3.2 Corrective actions. The combination of moisture, bias, and ionics in the flux material under the filter have contributed to the dendrite growth and the leakage path between signal pads and ground. Because of the geometry of the filter, removing the flux residue completely under the component was difficult due to the assembly and cleaning techniques used. A change to no-clean surface mount technology (SMT) solder paste was implemented, which greatly reduced the ionics and thereby removed one of the causes of dendrite growth.

In addition, the placement and reflow operation process of the SMT components was changed such that the filter is placed and reflowed last, after the other components are in place. This allows for very effective cleaning and removal of solder beads around the filter before it is attached to the substrate. There was concern that converting to a no-clean flux would increase the occurrence of solder beads and that

Figure 7.56 318 BGA multichip module.

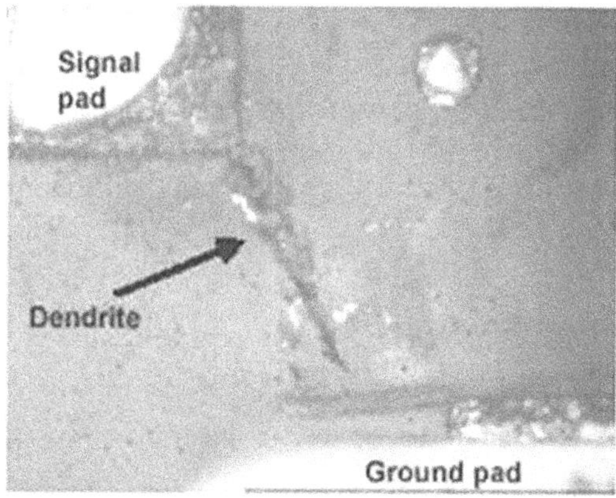

Figure 7.57 Dendritic growth between a signal pad and a ground plane.

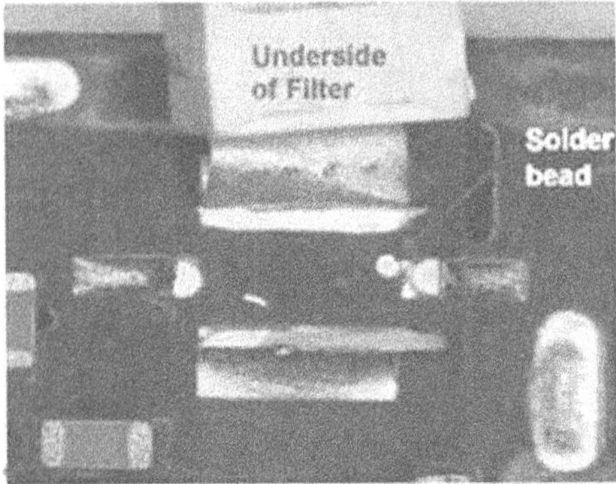

Figure 7.58 Solder bead under filter.

the new assembly process might affect the placement accuracy of the filter on pads that have a reflowed solder finish.

7.6.4 Shorted Contacts—Die Attach Material on Exposed Cu

7.6.4.1 Case history and description of failure. A 448 PBGA multichip module failed the functional test on ATE, indicating a possible short. Examination of the substrate module assembly showed that the solder mask was severely voided, both under the die and in other open areas (Fig. 7.59). The areas under the dies are the most critical, since many of the voids bridge two adjacent signals. When the voids are filled with a conductive die-attach material, the signals within the void short to each other. Vertical cross-sectioning confirmed that the die-attach material extended from the back of the die through the void, contacting the signal traces.

7.6.4.2 Corrective actions. Incoming inspection of the substrates failed to detect the voids, which are not allowed under the die. A change in the incoming inspection procedure will prevent this from happening again.

7.7 Failures in Plastic Enhanced Ball Grid Array, Cavity-Down Packages

7.7.1 Open Contacts—Broken Bond Wires

7.7.1.1 Case history and description of failure. A 352 EBGA (Fig. 7.60) was field returned for a contact test that showed opens. Low-power microscope inspection (predecapsulation) found a crack or blister in

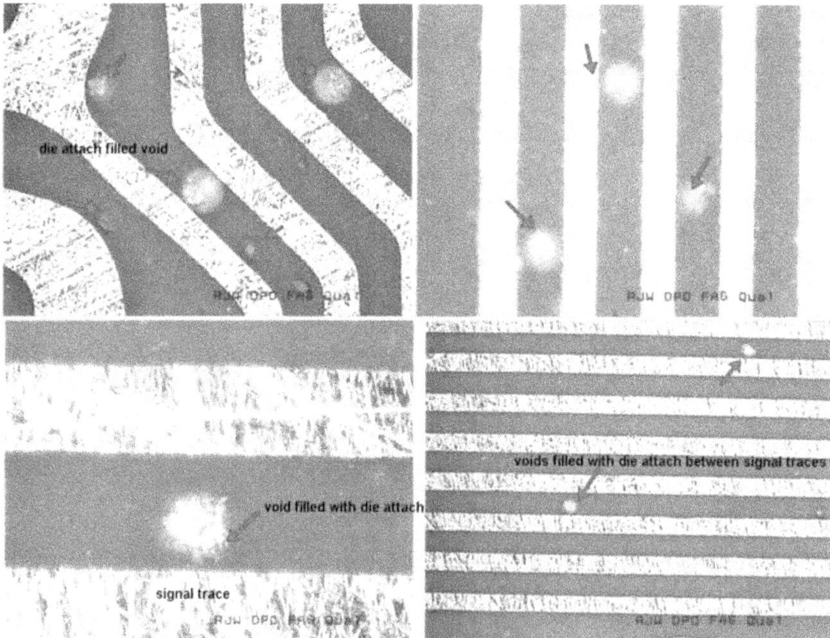

Figure 7.59 Solder mask is severely voided, both under the die and in other open areas.

Figure 7.60 352 EBGA.

the molding compound that covers the die and wirebonds. A C-SAM, in pulse-echo mode, analysis detected evidence of delamination over the die surface. X-ray analysis revealed damaged bond wires (Fig. 7.61). The damage can be attributed to the crack or blister in the mold compound (Fig. 7.62). Curve trace analysis confirmed open circuits in the package, and the TDR analysis of the package circuit appeared to indicate that the defect was on or near the die surface. High-power optical inspection (post-decapsulation) found lifted ball bonds and wires that were broken just above the ball bonds (Fig. 7.63). Mechanical im-

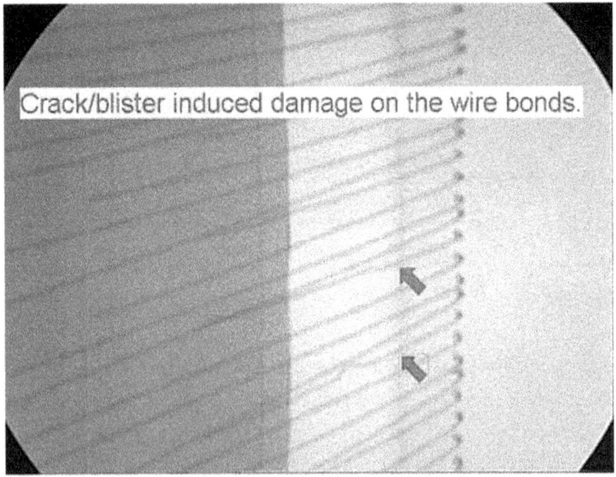

Figure 7.61 X-ray showing damaged bond wires.

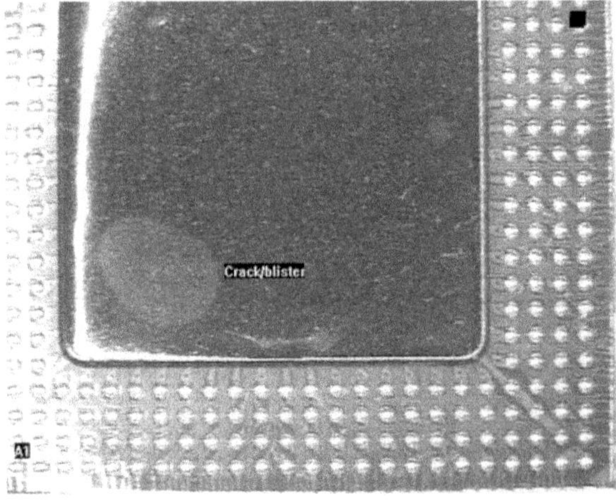

Figure 7.62 Crack in the mold compound.

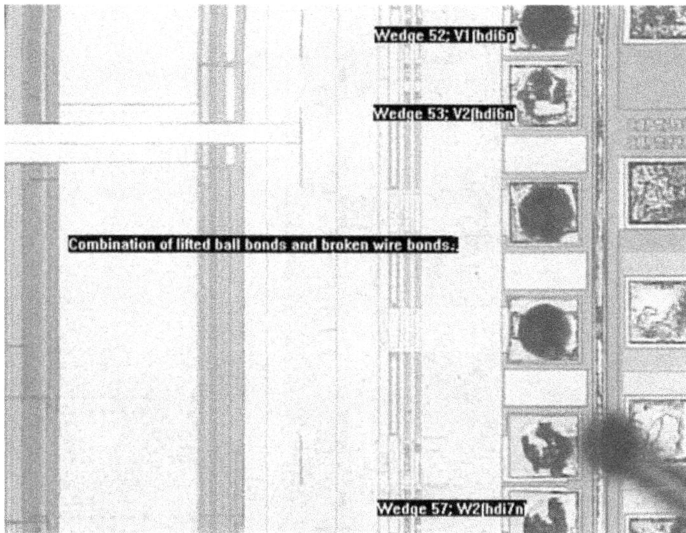

Figure 7.63 Lifted ball bonds and wires.

pact damage caused small cracks in the passivation in various places. No obvious electrical overstress (EOS) damage was noted.

7.7.2 Open Contacts—Sheared Ball Bonds/Damaged Signal Traces

7.7.2.1 Case history and description of failure. A 520 EBGA device failed the electrical continuity test for open contacts after it was subjected to 500 temperature cycles of Condition B (–55 to 125°C). The C-SAM (pulse-echo) test showed severe die surface/mold compound corner delamination (Fig. 7.64). Visual inspection after decapping revealed multiple ball bond shears, passivation cracking, and cracked/damaged metal busses and signal trace routing (Fig. 7.65). Electrically overstressed flashed metal, resulting from displaced metal features, was also observed. The damage observed was consistent with the stress pattern/delamination shown by the C-SAM analysis.

7.8 Failures in Plastic Flip Chip Ball Grid Array Packages

7.8.1 Open Contacts—Cracks around FC Bump Pads

7.8.1.1 Case history and description of failure. An electrical open failure was found in a 1012 FCBGA, 40 × 40 mm device after subjecting the device to 1000 temperature cycles of Condition B (–55 to125°C).

Figure 7.64 Die surface/mold compound corner delamination.

Figure 7.65 Multiple ball bond shears, passivation cracking, and cracked signal trace.

The 3-2-3 flip chip build-up substrate consists of three build-up layers on each side of a double-sided BT core. A TDR analysis confirmed the open to be within the substrate, very near the bump pad. C-SAM analysis revealed some underfill anomalies. Since the detected anomalies were not in the vicinity of the failed bump, it was believed and then confirmed by a follow-up cross-section that the anomalies were not related to the substrate cracking that caused the open failure. X-ray inspection did not show any obvious defect in the failing bump.

Cross-sectioning revealed that the crack was initiated at the corner formed by the Cu bump pad and the solder mask (Fig. 7.66). During temperature cycling, the initiated crack (1) propagated into the solder mask, forming a surface crack (Fig. 7.67) and (2) penetrated into the dielectric build-up layers beneath the pad, down toward the substrate core (Fig. 7.68). The crack, as it propagated through the build-up layers, cut through the Cu traces, resulting in the electrical open (Fig. 7.69).

All the failed bump pads had solder mask openings that were misaligned from the center of the Cu pad as a result of a shift in positional tolerance during substrate fabrication. An FEA model, schematically shown in Fig. 7.70, confirmed that such a misalignment contributed to the initiation of substrate cracking.

The FEA results (Fig. 7.71) show that the stress level at the corner of a Cu bump pad is linearly proportional to the amount of misalignment between the solder mask opening and the Cu bump pad. The maximum stress occurred when the solder mask opening was tangential to the bump pad, which is allowed by the substrate supplier's

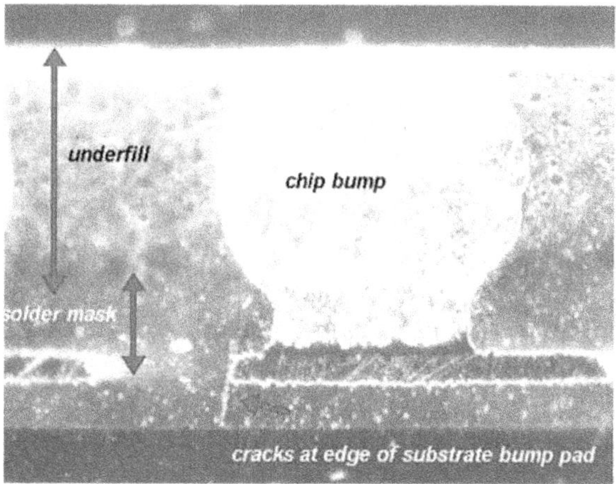

Figure 7.66 Crack initiated from the pad/solder mask interface corner.

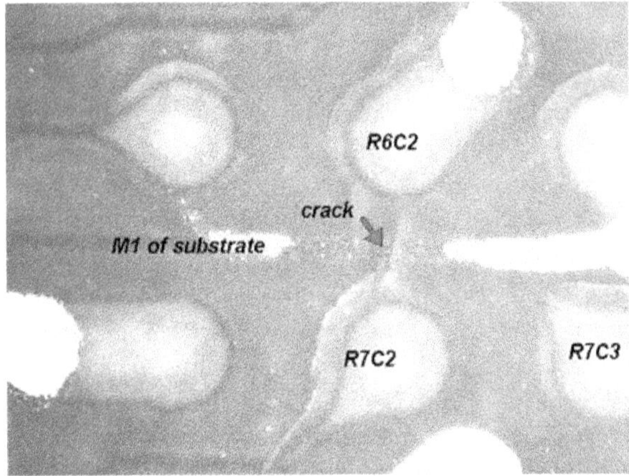

Figure 7.67 Crack seen from solder mask top surface.

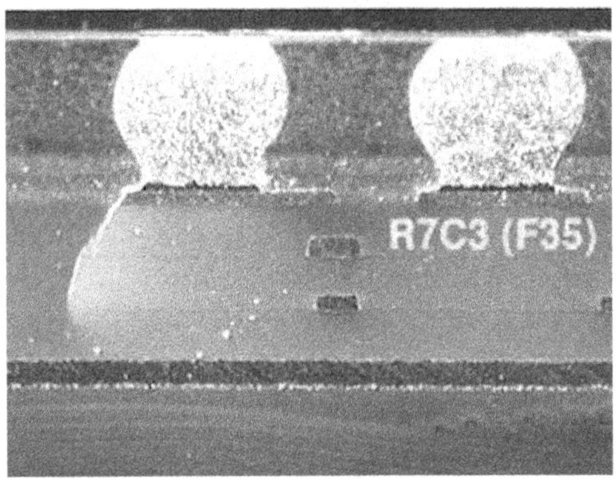

Figure 7.68 Crack propagating toward he core of the substrate.

specifications. The FEA results also suggest that the underfill properties have a significant impact on the stress level at the same location. The properties of underfill 1 and underfill 2 at room temperature (25°C) are listed in Table 7.1. Underfill 2 has a glass transition temperature, T_g, of 75°C, compared to 145°C for underfill 1. A lower T_g generally means that less stress would be locked inside the package during temperature cycling, as the underfill material becomes softer above the T_g, thus helping prevent or reduce substrate cracking.

7.8.1.2 Corrective actions. The following corrective actions were taken, based on the failure analysis discussed above:

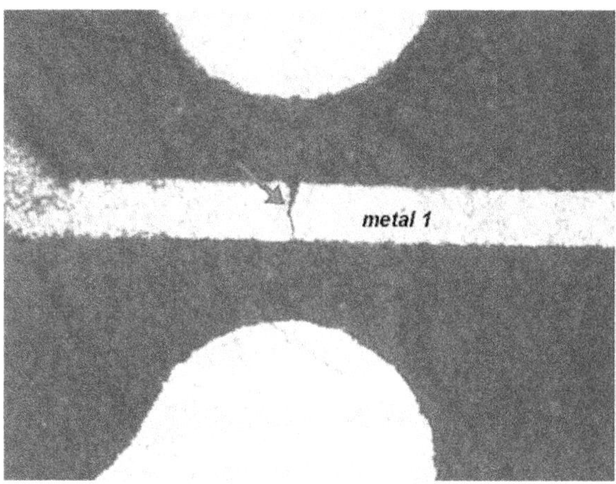

Figure 7.69 Trace broken by a crack propagating into the substrate.

Table 7.1 Underfill Properties Used in FEA

	Property		
Material	Young's modulus	CTE, ppm/°C	T_g, °C
Underfill 1	5.6	32	145
Underfill 2	6.6	32	75

1. It was recommended that the substrate supplier implement tighter control of the misalignment between the solder mask openings and the Cu bump pads during substrate fabrication.
2. All substrates are to be visually inspected and checked for a 5-μm minimum overlap of the solder mask over the top surface of the Cu bump pads.
3. A new underfill that has a lower T_g and higher modulus will be used on future assembly lots to reduce the stress level on bumps of assembled devices.

As a result of the above corrective actions, the devices passed 1000 temperature cycles of Condition B (–55 to 125°C) without failure.

7.8.2 Open Contacts—Cracks around BGA Pads

7.8.2.1 Case history and description of failure. An electrical open was detected when qualifying a 1444 FCBGA package after 300 tempera-

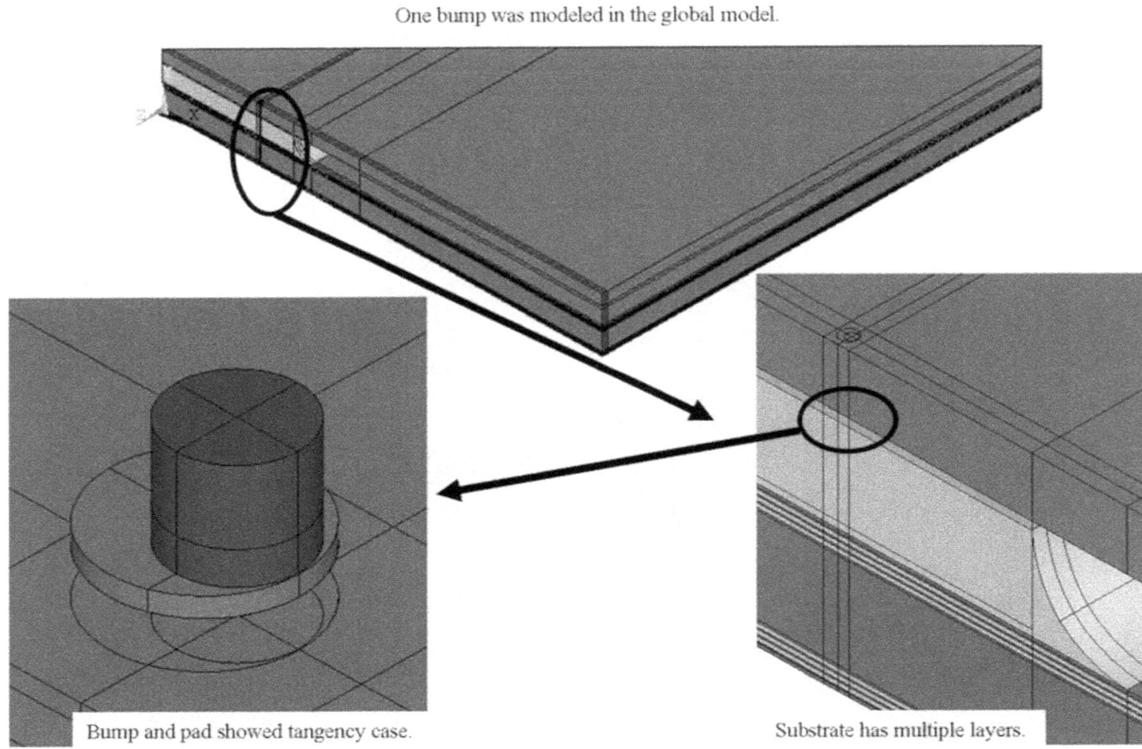

Figure 7.70 FEA model to study bump pad and solder mask opening misalignment effects.

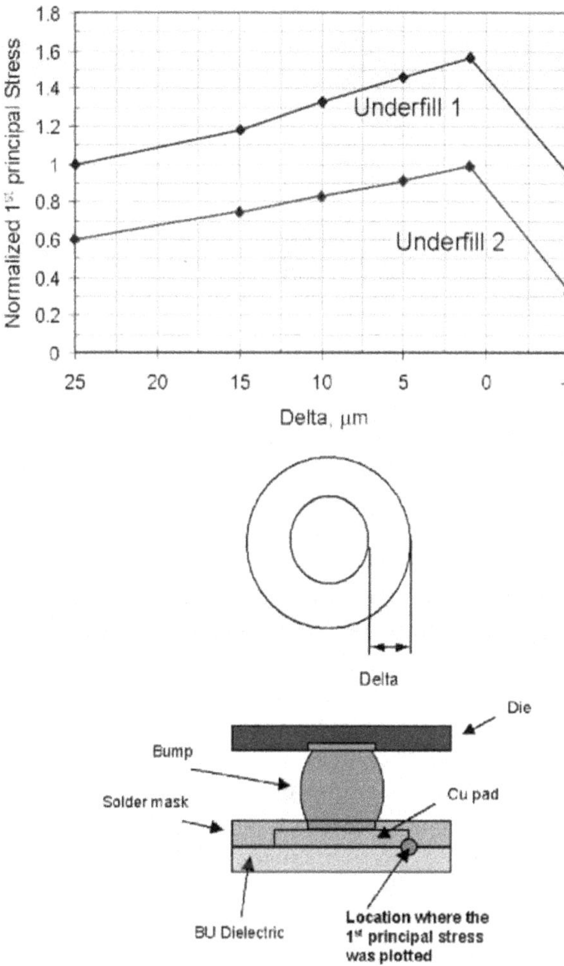

Figure 7.71 FEA results, impact from misalignment and underfill material.

ture cycles of Condition B (–55 to125°C). The FCBGA multilayer substrate was 40 × 40 mm and consisted of build-up layers made of a low dielectric constant material laminated to each side of a metal core. The die was 12 × 12 mm.

A TDR analysis confirmed that the failure was within the substrate, close to a BGA pad. Cross-sectioning revealed that a signal trace on layer 2, beneath the BGA pad, had been cut through by a propagating crack in the substrate. The failure mechanism analysis (FMA) suggested that the crack initiated in the solder mask at the corner of the BGA pad beneath the die corner. The initiated crack (Fig. 7.72,

Figure 7.72 Crack initiated from the pad/solder mask interface (*A*) and then penetrated into the substrate (*B*).

Region A) propagated into the substrate during temperature cycling (TC) and penetrated through the layers (Fig. 7.72, Region B). In some cases, the propagating crack had been observed cutting through the copper traces on the adjacent metal layer, causing electrical opens.

Cross-sectioning of incoming bare substrates before assembly showed no microcracks. The sockets used for electrical testing were suspected at first of causing these cracks. The mechanical clamp force of the sockets could have induced microcracks at the interface between the BGA pads and the solder mask. But, after the same failure mechanism was found in samples that were temperature cycled but not electrically tested, the sockets were absolved of any blame.

A FEA was performed to identify the root causes of the failures. The analysis utilized the "submodeling" technique that consists of a coarse "global" model (Fig. 7.73) and a detailed "local" model (Fig. 7.74) incor-

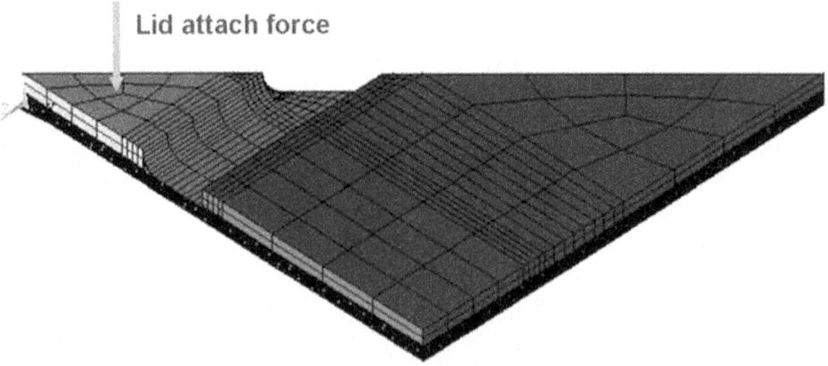

Figure 7.73 Global model for studying lid attach process effects.

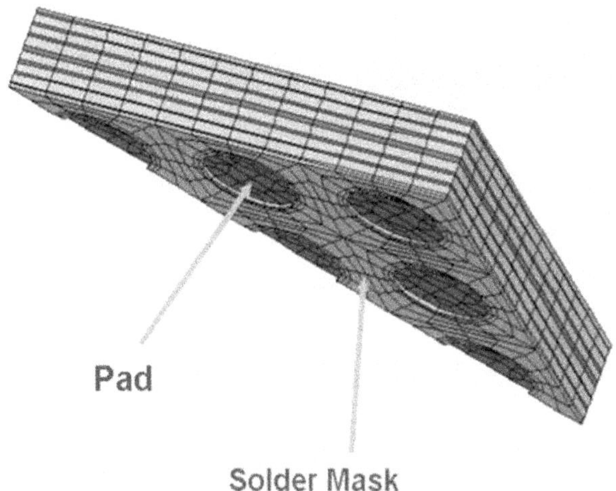

Figure 7.74 Submodel of BGA pads under the die corner area.

porating the fine features in the package such as underfill fillet, BGA pads, and metal layers in the substrate. FEA modeling showed that the lid attach process, with insufficient supports, may introduce high tensile stresses along the interface of BGA pads and solder mask (Fig. 7.75). These tensile stresses could generate microcracks if they exceed a certain level or interact with existing defects. Simulations also showed that, during temperature cycling, compressive stresses develop along the very same interfaces, combined with high tensile stresses in the dielectric layer adjacent to BGA pads (Fig. 7.76). This implies that temperature cycling could be a fatigue crack accelerator and not an initiator.

7.8.2.2 Corrective actions. Based on above analysis, the following corrective actions were taken to mitigate/eliminate the type of substrate cracking described above:

- A new tray with a solid bottom was introduced for lid-attach assembly to eliminate/reduce the generation of microcracks at the BGA pad and solder mask opening interface.
- The substrate was redesigned to have a wider gap between the silicon die and the copper ring attached on the top surface of the substrate.
- It was recommended that the substrate supplier seek a new build-up dielectric material with a higher fracture toughness to help resist fatigue cracking during temperature cycling.

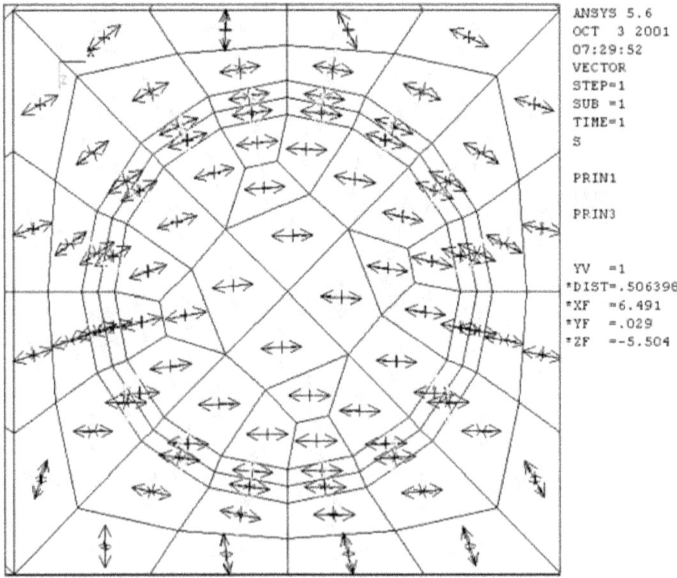

Figure 7.75 Tensile principle stresses induced by lid attach pressure indicates the potential for crack initiation.

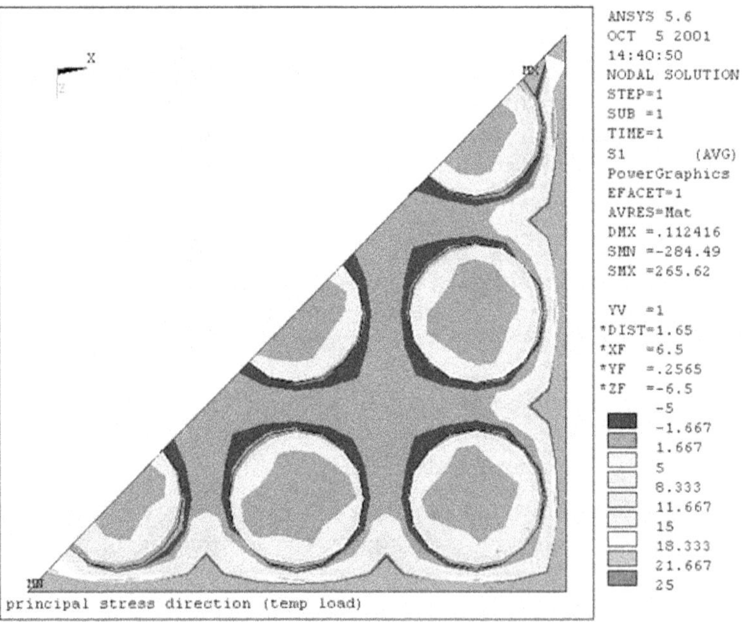

Figure 7.76 Compressive stress along BGA pad and SM interface in temperature cycling.

4. A design guideline was issued specifying that no signal bumps be placed under the die corner area. It is preferred that, whenever possible, solid copper layers be incorporated in the design under the die corner area.

No substrate brittle cracking was found on the BGA side during requalification of the device. Instead, a new device with a similar package design and technology, but with a larger die, failed for FC bump cracking. A detailed failure analysis of FC bump cracking can be found in Section 7.8.3.

7.8.3 Open Contacts—Flip Chip Bump Cracking

7.8.3.1 Case history and description of failure. Electrical opens were picked up in the early stages of temperature cycling of a 1377 FCBGA device. The device had an 18.5-mm square die with eight rows of peripheral FC bumps and fully populated stitched core bumps at a 200-μm minimum pitch. The 45 × 45 mm multilayer substrate consisted of build-up layers made of a low dielectric constant material, laminated to each side of a metal core.

Cross-sectioning revealed overstressed cracks on FC bumps in the die corner region. Such cracks could eventually separate the bump, resulting in an electrically open contact (Fig. 7.77). In addition, underfill material was found inside the cracks, implying that those cracks were generated before underfilling the device. Cracks were also observed on some electrically good bumps (Fig. 7.78), which explains why the device passed the open/short test before temperature cycling.

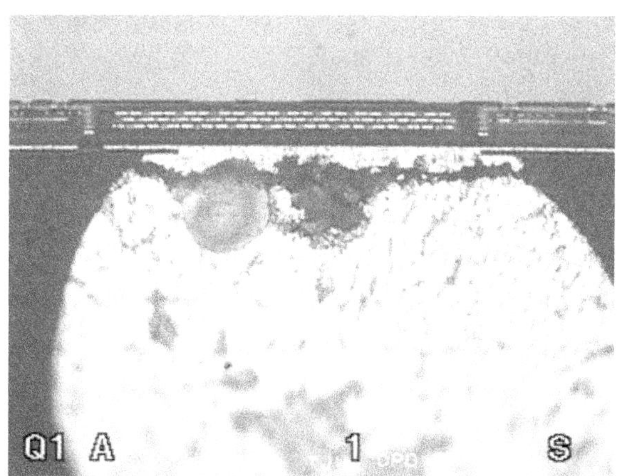

Figure 7.77 Fully cracked bump.

Figure 7.78 Partially cracked bump.

An extensive FEA was conducted to help identify the root causes of the cracked FC bumps. Figure 7.79 shows the FEA model created to simulate the die attach process. To reduce the problem scale, only eight rows of peripheral bumps were modeled. As the previous simulation showed, the presence of the core bumps has little affect on stress and strain distribution of bumps located at the die corner. Figure 7.80 shows the normalized first principal strain on the critical bump, usually the corner bump far away from the die center along the diagonal. In Figure 7.80, the baseline case represents the actual design at which failure occurred. The simulation results show that thinning the die or changing the substrate material may significantly reduce the bump strain. Compared with an eutectic solder bump (63Pb/37Sn), the high-lead (95Pb/5Sn) metallurgy has a lower yield strength but a larger elongation to rupture, which may provide some margin to resist bump cracking.

An assembly DoE was also performed at the assembly site to verify the potential solutions. Based on the simulation analysis, the fore-

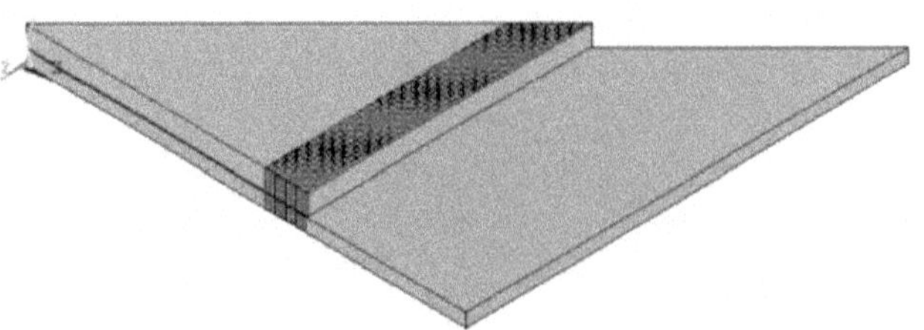

Figure 7.79 FEA model for die attach simulation.

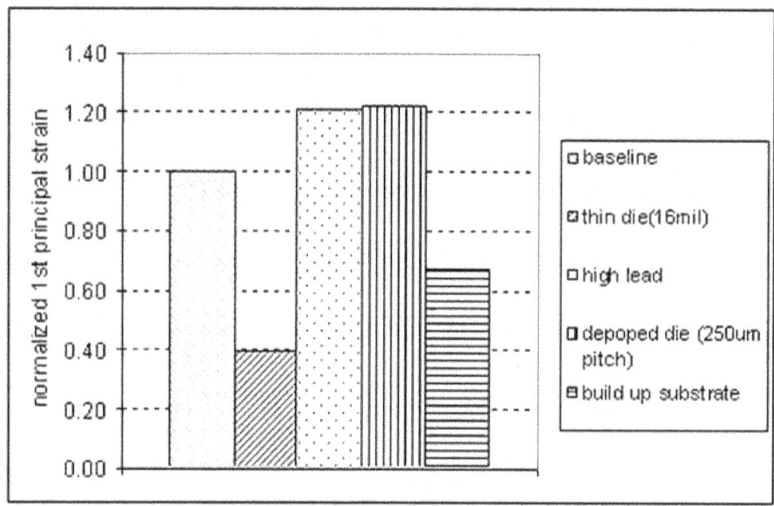

Figure 7.80 Simulation results showing major factor of bump strain.

most assembly-related factors affecting bump cracking are solder reflow profile, handling method after die attachment, and so on. Other parameters selected by the simulation results, and included in the DoE, were thinner die and use of high-lead solder bumps instead of eutectic.

7.8.3.2 Corrective actions. Based on the above analysis and recommendations, the package was redesigned in a different substrate technology, which solved the cracking issue, and no further failures of this type occurred.

7.8.4 Open Contact—Incomplete Via Shape

7.8.4.1 Case history and description of failure. A1280 FCBGA device, having a 40 × 40 mm, 3-2-3 build-up substrate, failed electrical test for open contacts after moisture preconditioning at JEDEC Level 4 and a solder reflow temperature of 240°C. The TDR analysis indicated an open-contact-related failure in the substrate at the T32 signal path. A cross-sectional evaluation of the T32 signal path identified the open as being related to the via in layer 5 (Fig. 7.81). Further examination of the open via showed what appeared to be debris between the root of the via and the land of the core via (Fig. 7.82).

7.8.4.2 Corrective actions. The supplier concluded that the open in the via was the result of an incomplete laser-drilled via shape, caused

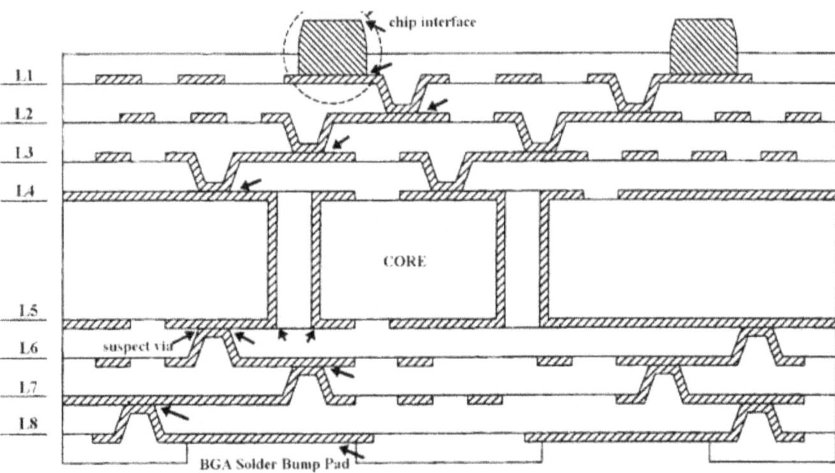

Figure 7.81 1280 FCBGA layer structure (arrows indicate the interconnects examined during FA).

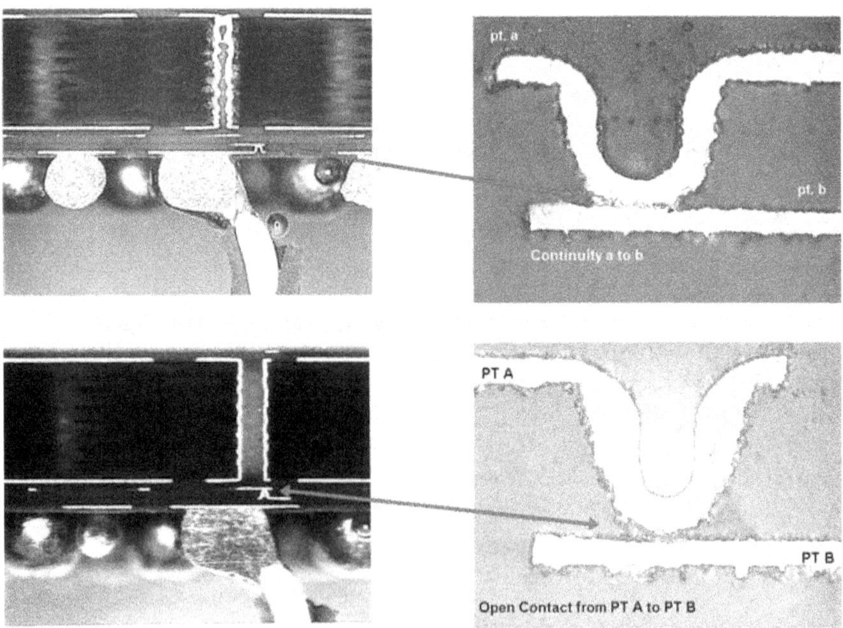

Figure 7.82 Laser drilled via that does not contact the bottom pad.

by a contaminated F-theta lens. The F-theta lens is used to focus a laser beam, directed by a moveable mirror, onto the working panel. If the lens is contaminated, its ability to focus the laser beam is impaired, so the depth and shape of the hole is compromised. The cause of the F-theta lens contamination was its usage for laser drilling of plated-through holes (PTHs) product (drilling through build-up layers and core) without applying adequate controls and maintenance for cleaning the lens after usage. It was found that the F-theta lens contaminates faster when PTH drilling than when drilling blind vias in build-up layers. Based on these observations, the F-theta lens cleaning maintenance procedure and frequency of cleaning were changed. Since the implementation of the above corrective actions, no via failures of this type have occurred.

7.8.5 Shorted Contacts—Vertical Deformation of Signal Trace

7.8.5.1 Case history and description of failure. A 1280 FCBGA device, having a 40×40 mm, 3-2-3 build-up substrate, failed electrical testing for a shorted contact after 24 hours of burn-in. Electrical characterization using a curve tracer verified the short. TDR testing located the short to be within the substrate. The device showed external package heating (substrate discoloration and melted BGA balls), which were attributed to internal substrate shorts. Horizontal lapping through the substrate as well as cross-sectioning revealed a combination of defects: (a) vertical Cu dendrites between Cu traces on one layer and the power plane in the adjacent layer (Fig. 7.83), (b) horizontal Cu dendrites between adjacent traces (Fig. 7.84), and (c) vertical signal deformation or Cu nodules protruding from signal traces on one layer to the power plane in the adjacent layer (Fig. 7.85). It is because of this last defect that internal substrate heating initiated. The dendrite growth, which was a result of substrate overheating, caused more shorting and developed additional heating sources. It is thus the nodule protrusion defect that is the root of the failure in this device, and it is believed that the causes had to originate in the lamination process during substrate fabrication.

In the substrate lamination process, two sets of stack-ups, each consisting of a release film on the outside over a semicured dielectric build-up layer, are placed on each side of a BT core and run through rollers where the stack-up is laminated under pressure and temperature. It is postulated that some foreign particles ended up on top of the release film and, when pressure was applied to the stack-up, the particles embedded into the dielectric. After exiting the rollers, the release film was lifted from the dielectric, along with the foreign particles, leaving holes in the dielectric layer (Fig. 7.86). We have seen these

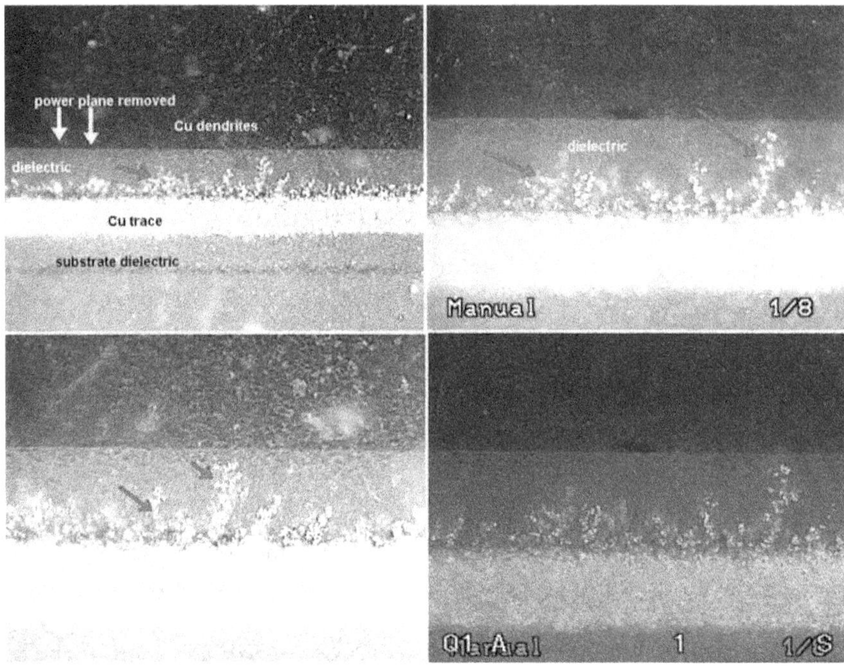

Figure 7.83 Vertical Cu dendrite growth between adjacent traces.

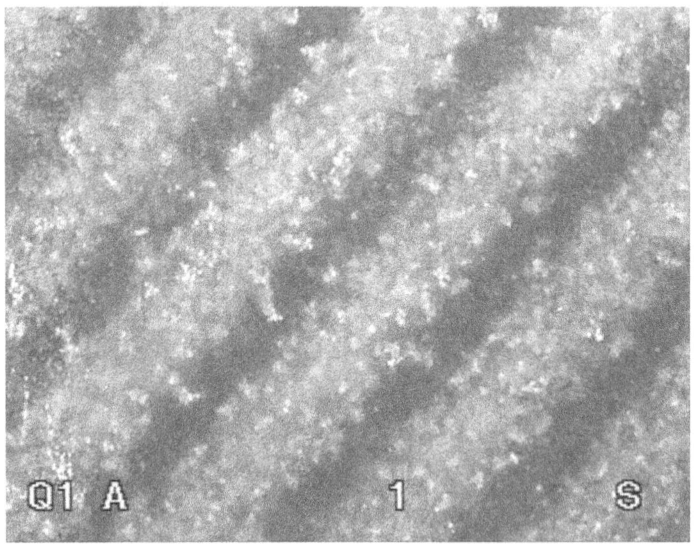

Figure 7.84 Horizontal dendrite growth between adjacent traces.

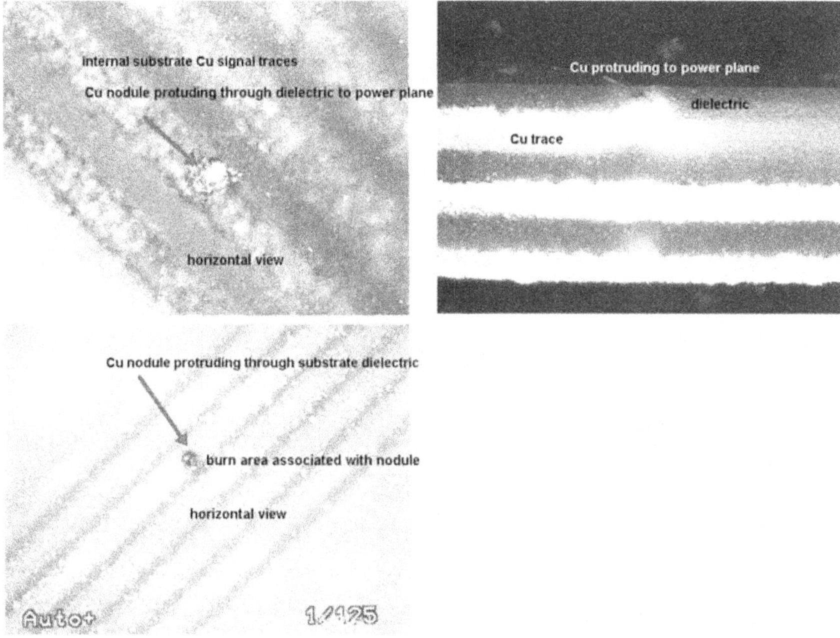

Figure 7.85 Cu nodule protruding from signal traces on one layer to the power plane on the adjacent layer.

holes only on the top layer (between metallization layers 2 and 3) where the particles settled on the release film, and not on the bottom layer where they would have fallen off, which adds further credence to the theory.

After lamination, the vias are laser drilled, and the Cu features are patterned. If one of the traces happened to be over one of the depressions created by the foreign particles, the depression is Cu plated, which may eventually short to the adjacent plane.

7.8.5.2 Corrective actions. The rollers are cleaned more frequently, and the environment around the lamination equipment is better controlled. In addition, a more sensitive electrical short test was implemented at the substrate supplier's facility that will better detect any near shorts at final substrate inspection.

7.8.6 Shorted Contact—In-Plane Cu Bridge

7.8.6.1 Case history and description of failure. A 1280 FCBGA device failed electrical test for a shorted contact after the device was subjected to qualification testing of component and lead assembly simula-

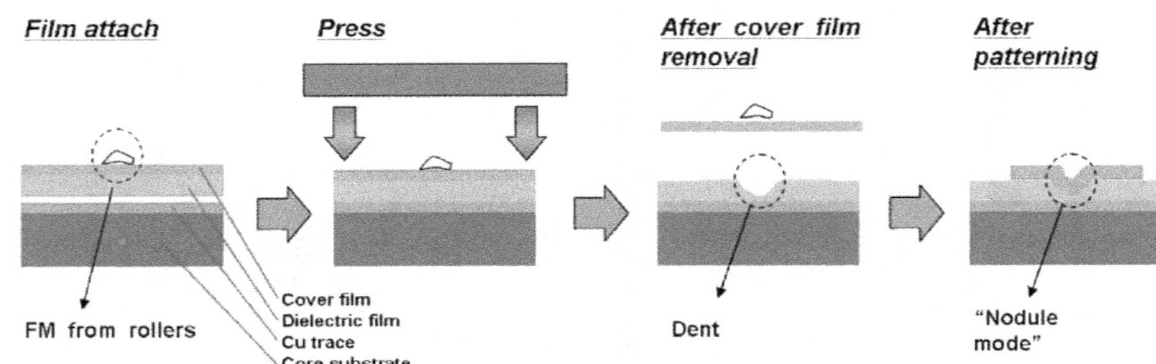

Figure 7.86 Probable cause of Cu nodule protrusion.

tion sequence (CLASS), which consists of a 30°C/60 percent RH environment for 96 hr, followed by flux, three solder reflows, and clean. The shorting condition was verified in the lab using a curve tracer. Evaluation and physical analysis of the device showed that the short was located within the substrate. After delayering, the outer metallization layer 8 (BGA side) and layer 7 revealed a short between the power plane (VDD) and a signal via land (Fig. 7.87). The short was caused by an extension of the VDD plane and not by a foreign particle. The Cu bridge was not a hard short prior to CLASS. The dielectric build-up material isolated the two connections prior to breaking down during exposure to CLASS. Vertical cross-sectioning of the Cu bridge shows the Cu thickness to be roughly the same as the plane, suggesting a possible mask-related defect (Fig. 7.88).

7.8.6.2 Corrective actions. The supplier agreed that the defect was caused by a speck of dirt on the artwork. A new cleaning process for the artwork was instituted to prevent future defects.

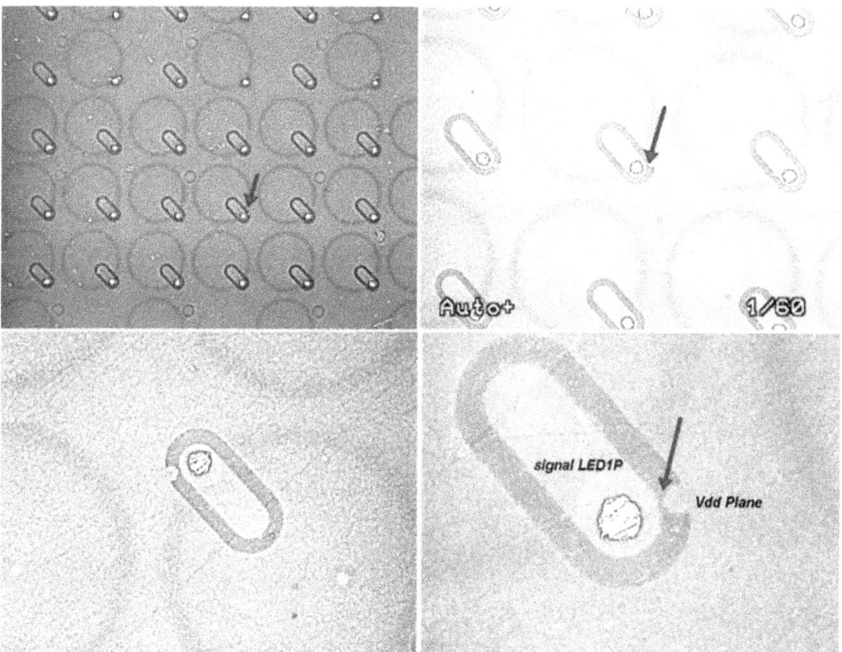

Figure 7.87 V_{DD} plane to signal pad short.

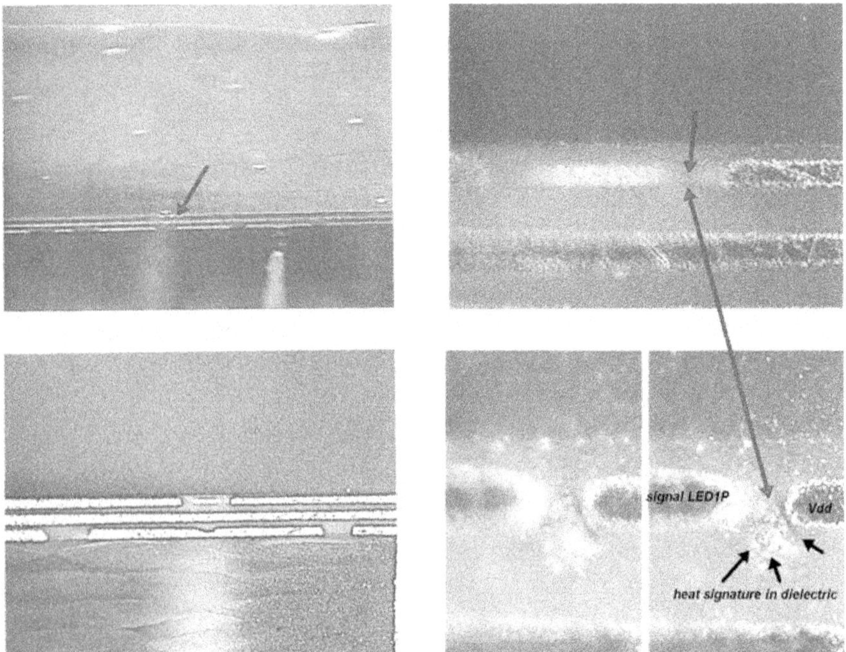

Figure 7.88 Cross section showing the thickness of the Cu bridge to be about the same as the V_{DD} plane.

Chapter

8

Emerging Assembly Materials for IC Packaging

Jason P. Goodelle
John W. Osenbach

8.1 Introduction

As the microelectronic packaging field gains maturity, new technologies are emerging that enable both increases in performance and reductions in the overall cost of the package. In many cases, the new technologies are required to simply maintain performance while reducing costs as a particular package type becomes a commodity. In any case, the application of existing materials to a new field (namely, microelectronic packaging), the introduction of new materials for specific applications, and new formulations and processes using older materials technology will be critical steps in meeting the demand for advanced package technology.

Each stage of package assembly makes use of a variety of materials that perform tasks as mundane as a structural adhesive to as complex as a critical thermal link in the dissipation of heat generated by the IC, which directly affects the performance of the circuits. Each material has specific limitations that are usually bounded by the application (such as costs, the desire to electrically insulate the circuit, stress reduction, and so on). Advances for may of these materials can be realized by performing the designated task more efficiently, at a lower cost, or by removing boundaries that are normally associated with the specific packaging element (such as increasing thermal conductivity of a thermal interface material without increasing the stress imparted on the silicon, which is normally considered a limiting boundary condition for materials used for this purpose).

Finally, it is important to consider the reliability aspect of emerging materials for the packaging industry. Increasing the performance of a particular packaging material in many cases may be realized through improvements in reliability. As the industry demands higher reflow temperatures to accommodate new lead-free interconnect alloys, longer shop floor exposure times for increased usefulness, and long failure-free lifetimes in the presence of harsh use or environmental conditions, packaging materials will have to advance to enable these demands while still keeping costs in control. To this end, the focus of this discussion has been limited to items specific to the advancement of molding compound and die attach materials in the presence of more aggressive lead-free processing conditions.

8.2 Lead-Free-Compatible Assembly Materials

8.2.1 Characteristics Specific to Lead-Free Materials

As the industrial world begins to move toward removing heavy metals and halogens from commercial goods, the packaging field is certainly not immune. The Japanese microelectronics industry began voluntary phase-out of lead in electronics manufacturing (following the 2001 electrical appliance recycling bill). The European microelectronics manufacturers are slated (via more rigid legislation) to be lead free by 2006 (as dictated by the European Union's Removal of Hazardous Substances [RoHS] initiative).[1] Although the lead-free initiative is based on reducing the environmental impact of releasing heavy metals in to the waste stream, the realized impact on the microelectronics world is to alter reflow temperatures for solder interconnect processes. The most widely used lead-free alloys are usually based on some combination of Sn, Ag, and Cu. Table 8.1 shows details the liquidus and reflow temperatures of several candidate lead-free alloys targeted to replace Sn/Pb eutectic interconnect materials. Because the melting temperature of Sn/Pb eutectic is 183°C (with reflow temperatures ranging from 195 to 235°C), the most outward difference between lead-free and conventional eutectic Sn/Pb solders is the increase in reflow temperature. This local increase in temperature translates into a temperature increase that the entire package must endure, although for brief exposure times. During board attach processes for multiple packages, some devices may be exposed to temperatures approaching 260°C to accommodate different package masses. Many polymeric materials begin to degrade at temperatures much lower than 260°C, and

Table 8.1 Liquidus and Reflow Temperatures of Candidate Lead-Free Solder Alloys for Replacing Eutectic Tin-Lead Solder*

Alloy composition	Liquidus temperature (°C)	Reflow temperature (°C)	Melting range (°C)
Sn-2Ag			221–226
Sn-3.5Ag	221	240–250	
Sn-0.7 Cu	227	245–255	
Sn-3.0Ag-0.5Cu	220	238–248	
Sn-3.2Ag-0.5Cu	218	238–248	217–218
Sn-3.5Ag-0.75Cu	218	238–248	
Sn-3.8Ag-0.7Cu	220	238–248	217–210
Sn-4.0Ag-0.5Cu			217–219
Sn-4.0Ag-1.0Cu	220	238–248	217–220
Sn-4.7Ag-1.7Cu	244	237–247	

**Source:* NIST/Colorado School of Mines online database.

certainly many polymeric materials have glass transition temperatures (the temperature at which an amorphous polymer chains relax from a "frozen" state to a more mobile state) well below this point. Polymeric materials have very temperature-dependent material properties (most markedly through the glass transition and near degradation temperature), and the performance of the polymer structures will most affected by increases in temperatures during device board mounting processes. Although metal and inorganic components of the package are usually unaffected by brief periods of elevated temperatures, increasing the exposure temperature during reflow will affect the adhesion, load bearing, moisture diffusivity, and physically stability behavior of the polymer portions of the package. Figure 8.1 shows a schematic of the temperature effects on physical properties for a typical amorphous thermosetting (cross-linked) polymer material that can be found in modern packages (plastic or otherwise).

The common failure mode observed in packages exposed to moisture followed by interconnect reflow temperatures (either real or simulated during stress testing) is referred to as *popcorning* (refer to Chap. 7) where the buildup of moisture at interfaces during exposure of plastic packages to humid conditions rapidly vaporizes during heating of the part. This causes a disbond to occur at interfaces formed by polymers and various other materials such as the silicon die, metal lead frames,

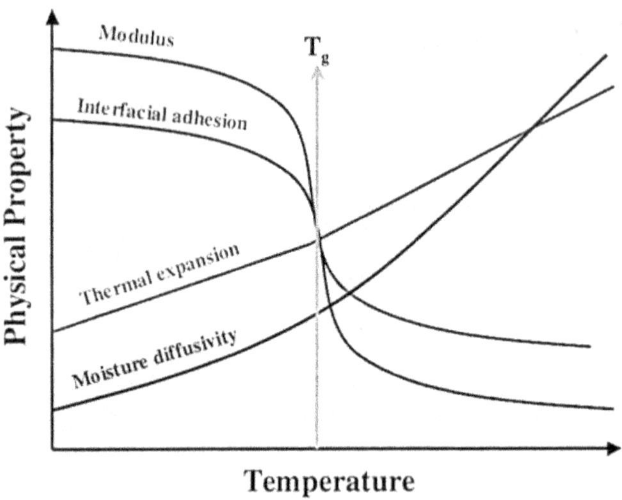

Figure 8.1 Schematic of the temperature effects on selected physical properties of a typical thermosetting (cross-linked) polymeric material found in microelectronic packaging.

and inorganic fillers. Additionally, the high processing temperatures can expose a polymer to temperatures that are much higher than their respective glass transition temperatures. A typical lead-free reflow profile is found in Fig. 8.2. Depending on the solder alloy, it is clear that the exposure temperature needed for these interconnect metallurgies is quite high, if even for a brief period of time (as mentioned above). Thermosetting polymeric materials tend to be very weak above the glass transition temperature, thus rendering them susceptible to flaw generation or outright cohesive failure. In the former case, the creation of flaws can lead to subsequent failures resulting from sub-

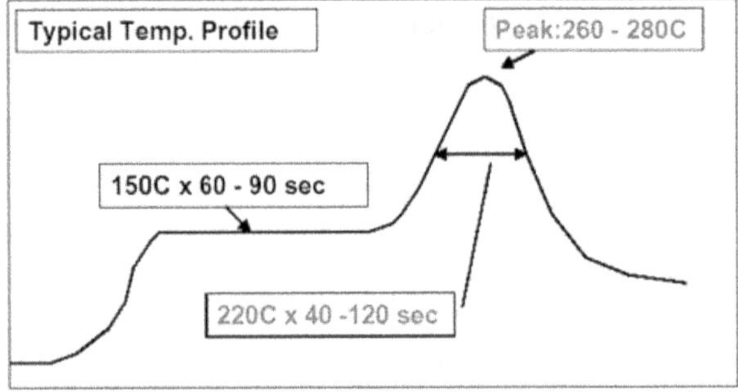

Figure 8.2 Schematic of a typical lead-free solder reflow profile.

critical crack growth during temperature (power) cycling while in use or during additional board attach excursions.

In most cases, the increase in processing temperature brought about by introducing lead-free materials to a microelectronic package must be offset by improvements in the performance of the various constituents of the assembly. Most likely, the improvements must be imparted to the various polymer components because of the temperature-dependent properties mentioned above. The following sections discuss in detail some of the advancements made in various materials used in the manufacture of microelectronic packages that are required to render them *lead-free compatible.* This term is used throughout the text to refer to materials that are generally considered capable of performing in the presence of lead-free processing temperatures. These materials should not be confused with the broader category know as "green" materials. These materials adhere to a strict guideline that specifies the removal of hazardous materials such as PCB and heavy metals such that lead-free solders are "green" in the sense that they eliminate the heavy metal lead. However, a lead-free-compatible material may not be green at all. This is an important distinction.

8.2.2 Lead-Free-Compatible Mold Compounds

Materials that are required to maintain adhesion to the silicon surface during elevated temperatures associated with interconnect reflow are the most susceptible to changes in these temperatures. The molding compound portion of a typical plastic package is a good example of a material that is susceptible to changes in the reflow temperatures.

In general, mold compounds are designed to do a few simple tasks. Specifically, they protect the chip from the damage that can be imparted by the environment, such as humidity, solder stress, and handling during assembly. In the last role, the delicate interconnects (wirebonds or other) are also protected from the same hazards. The other task that the mold compound performs is to minimize the damaging effect that the package can have on itself, such as CTE-mismatch-induced stress. While a mold compound must do all of these things in the presence of the various environmental and use stresses to which it is exposed, all of these factors constantly act to compromise the performance of the mold compound through several avenues, thus rendering the mold compound ineffective. Moving to higher reflow temperatures will only accelerate the reduction in performance or present new pathways to performance reduction. In general, the improvements that are being imparted to new mold compounds to make them "lead-free compatible" come in the areas of the following outwardly measurable properties: adhesion, fracture toughness, and moisture uptake. The improvements are usually realized through changes to the fundamental polymer chemistry, changes in the fillers used in production, or highly engineered additives, or by altering the surfaces of the materials to which they are bonded.

8.2.2.1 Overview of mold compound chemistry. To realize the improvements that are currently required or that dwell on the horizon for advanced materials for lead-free or high-reflow-temperature compatible molding compounds, it is important to understand the basic formulation of modern materials used for this role in IC packaging. Molding compounds, the most general sense, consist of mixtures of the following components:

1. A thermosetting epoxy resin
2. Accelerators
3. Curing agents
4. Inert filler
5. A flame retardant
6. Mold release
7. Colorant

The exact nature of these components and the ratios used in the mixture will depend highly on the application and largely constitute the difference between one suppler and another. Typically, they are guarded proprietary formulations. A typical formulation for a modern molding compound, with relative formulation amounts (for demonstration only), is listed in Table 8.2. However, regardless of the secrecy of the formulation, the roles of each constituent remain well understood. To achieve the thermosetting (cross-linked, not capable of melting in the classic sense) thermally stable structure, a base resin and a chemical moiety that enables the "cure" (*cure* and *cross-linking* are used interchangeably here) process are typically required ("typically" is qualified below). This enables one or more of three possible curing mechanisms as listed below:

1. Cross-linking (curing) of the matrix via various functional groups provided by the curing agent or hardener
2. Cross-linking of epoxy groups with aliphatic or aromatic hydroxyls, again provided by the curing agent or hardener
3. Direct linkage between the epoxy groups through self-polymerization

Table 8.2 Example Formulation of a Modern Molding Compound

	Raw materials	Parts by weight
Epoxy	Depends on application	100
Hardener	Depends on application	52–62*
Accelerator	Phosphonium salt	4
Filler	Fused silica	1145
Flame retardant	Modified phosphorus compound	0.7
Coupling agent	γ-glycidoxypropyl trimethoxysilane	6
Releasing agent	Carnauba wax	2.5
Colorant	Carbon black	2.5

*Depends on type of epoxy used

In the strictest sense, systems that cure via the third mechanism do not require cross-linking agents and usually only require the inclusion of accelerators. The last mechanism typically results in cure times and final mechanical properties that are not suitable for molding compounds for microelectronic applications. Typically, the epoxy portion of the formulation consists of a novolac, a biphenyl, a multifunctional, or some mixture of these types. These materials replaced the use of diglycidyl ether of bisphenol A (DGEBA) base resins in the early 1970s because of the low T_g levels associated with DGEBAs. High T_g levels are typically desirable for mold compounds to reduce the strain on the individual wires during temperature or power cycling in product use and also to retain good interfacial integrity during solder reflow cycles (refer to Fig. 8.1 for thermoset material behavior changes through a glass transition). This material characteristic becomes an important parameter to monitor when temperature increases are required in a move to lead-free interconnects. Often, high T_g materials will possess poor processing characteristics such that the "moldability" is poor. This usually is due to the relatively high viscosity that stiff, highly aromatic base resins (required for higher T_gs) will possess. Viscosity is often linked at some level to good adhesion (among other things) resulting from the linkage between adhesion of a liquid polymer and intimate contact with the surface to which it is adhering. This is true regardless of whether the adhesion mechanism is chemical or physical (i.e., mechanical interlocking) or some mixture of the two. In reality, the microscopic chemical or physical adhesion mechanism is a small contributor to overall bond strength as compared to crack tip process zone mechanisms (the "toughness" of the interface), but they are still important in that, if a weak interface is present at the microscopic level, process zone mechanisms will not have a chance to activate. In any case, viscosity during the processing of molding compounds can be an important parameter to consider when looking at materials for improved adhesion, especially at higher temperatures.

The curing agents used in the formation of the thermosetting matrix usually consist of one or more of the following basic curing agent formulations:

1. Aliphatic and cycloaliphatic amines or polyamines
2. Amides and polyamides
3. Cyclic anhydrides

The multifunctional nature of the curing agents will allow for the linkage from epoxy chain to epoxy chain (known and chain extension) though a reaction known as *ring opening* (in which the epoxy rings "open" during the carbon-hydrogen bond formation between curing agent and epoxy) as well as between chains, which stands as the basis for network formation. Figure 8.3 shows a schematic of the formation of a network structure through the reaction of an epoxy and a curing agent.

The nature of the backbone chemistry of the curing agent and the base resin will control the properties of the final cured matrix. Table 8.3 shows some examples of some of the epoxy resins and curing agents commonly used in the manufacture of molding compounds for microelectronic packaging. Many of the materials shown in Table 8.3 combine to form what would be considered "standard" molding compound. (For example, the combination of an ortho cresol novolac resin with a novolac curing agent is considered to be a mainstream molding compound technology.) However, formulations that use combinations of resins and curing agents (new and old technologies in some cases) constitute what is considered to be advanced technology in that specific properties of each of the components are being utilized for desired end results. For example, a lead-free-compatible compound

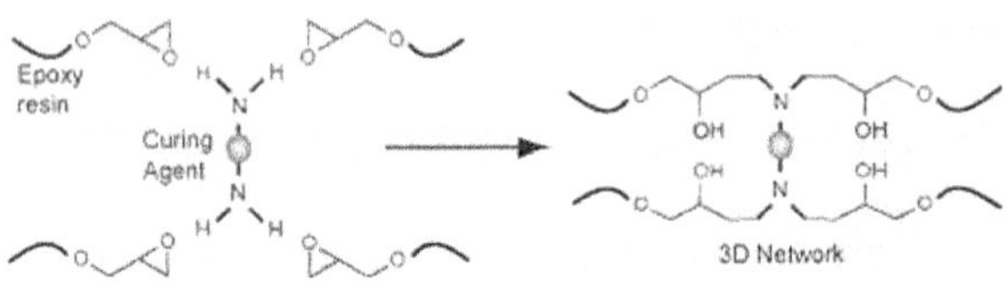

Figure 8.3 Schematic of the reaction between epoxy resin and a curing agent to form a cured network.

Table 8.3 Overview of Selected Epoxy Resins and Curing Agents Used in the Processing of Modern Molding Compounds

Epoxy type	Chemical structure	Industry status	Comments
Bisphenyl A	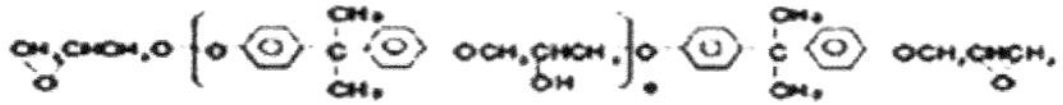	Old technology	Lower T_g, higher moisture uptake, prone to low temp. reflow problems, wire sweep
Ortho cresol novolac	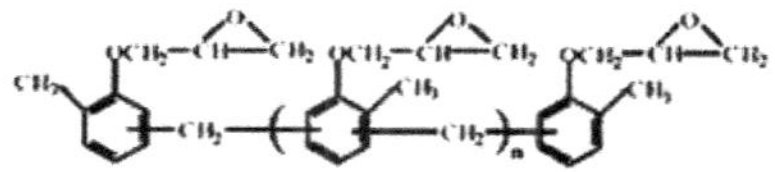	Long history in microelectronics	Good moldability, good reliability, medium water uptake
Naphthol	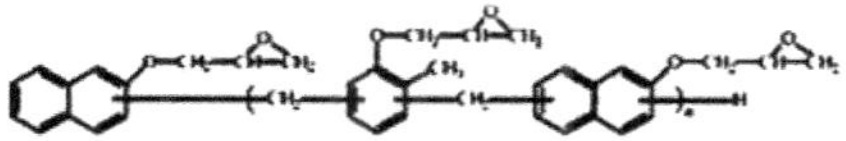	Newer technology	High heat and moisture resistance, medium T_g
Multifunctional	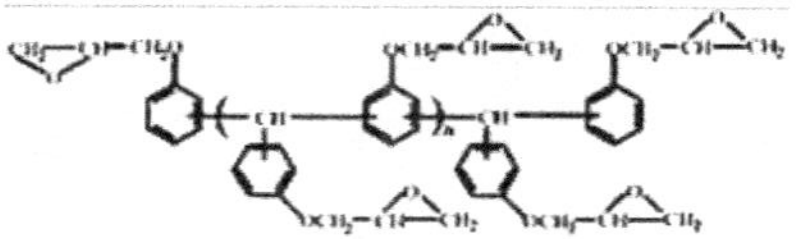	Newer technology	High strength at high temp, higher T_g, good planarity
Biphenyl	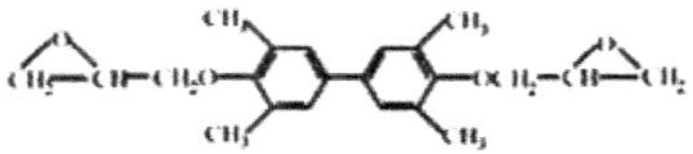	Newer technology	Higher adhesion strength, lower moisture uptake, lower viscosity
Dicylcopentadiene (DCPD)	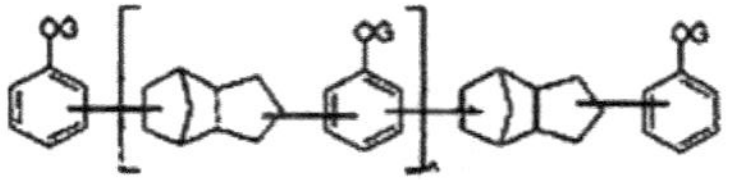	Modern/advanced	Higher adhesion strength, lower moisture uptake, lower viscosity
Curing agent type	**Chemical structure**	**Industry status**	**Comments**
Dicyandiamide	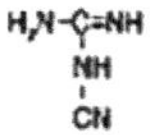	Old technology	Lower temperature cure, low transfer viscosity
Phenol novolac	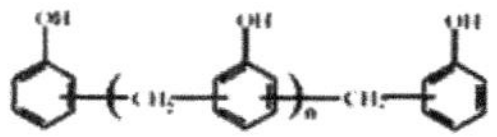	Current technology	Thermally stable

would require high T_g, high adhesion, high crack resistance, and low moisture uptake but would need to balance these needs with the desire for good moldability such that combinations of the resins (likely) and curing agents (less likely) would be used to provide the best end product. Several instances of this type of formulation work to arrive at high-reliability materials for high-reflow-temperature applications.[2] Additionally, modifications to the backbone chemistry of either the base resin or curing agent can provide the means to modify the final matrix properties. This concept will be discussed later.

8.2.2.2 The role of additives and chemistry in mold compound performance. Clearly, the role of resin properties is an important consideration in the final performance of the molding compound. However, nearly as important is the nature of the inert or active additives used in the final formulation. From Table 8.3 we can see that epoxy and curing agents are only part of the equation when it is desired to modify/control mold compound properties. By altering the content of the inert filler or second-phase stress reducers, or inclusion or modifications of the adhesion promoters, the reliability or moldability performance of a compound can be tailored. Finally, and arguably the most important effect of additives, the mechanical properties of the final cured compound can be dramatically altered by the inclusion of appropriate additives. In general, controlling the material properties of the final molded product is important for many final requirements of the package. To understand which properties become relevant to control in the presence of increased reflow temperatures, the failure modes associated with this high-temperature reflow event for plastic packages will be reviewed briefly.

In the simplest sense, most failures associated with the reflow event for plastic packages will be one of the types (many of which have been addressed elsewhere in this text) shown in Table 8.4.

Figure 8.5 shows a schematic of a typical BGA package construction with the interfaces highlighted that are subject to the failure modes listed in Table 8.4.[4] The possible failure mechanisms associated with high-temperature reflow will be aggravated only by increasing the reflow temperature in the presence of the same amount of ambient moisture exposure. More clearly, since most of the failure mechanisms listed above are intrinsically linked to the material properties (such as molding compounds), then the temperature effect on these properties will be very important. The material properties that will play the most important roles in the failure modes listed above are moisture diffusivity, adhesion (interface and material dependent and also moisture content dependent), modulus, fracture toughness (resistance to crack

Table 8.4 **Common Failure Modes in IC Packages during Solder Reflow Excursions**

Failure mode no.	Description
1	Delamination at one of the interfaces formed by the molding compound (chip, substrate wire, lead frame, and so on), resulting from the expansion of moisture at the interface, which can then lead to cracking of the compound or other package component (Delamination and expansion of moisture at the leadframe/molding compound and subsequent cracking of the mold compound is known as a classic "popcorning" failure. Figure 8.4 shows a schematic of this failure type. Popcorning in PBGA packages can involve failures at many interfaces.[3])
2	Expansion of moisture at the interface, which can then lead to cracking of the compound or other package component (also can play a role in popcorning)
3	Delamination at the inner layers of the substrate due to expansion of absorbed moisture (also plays a role in popcorning)
4	Warpage of the package at reflow temperatures that creates solder interconnect failures during board attach
5	Damage to the package in some way caused by the stress buildup resulting from a coefficient of thermal expansion (CTE) mismatch of the molding compound and other package elements

propagation once a flaw, such as a delamination, is formed), glass transition temperature, and (CTE). All of these properties are in some way dependent on the ambient temperature (or, in the case of transition temperatures, the material property changes associated with this event will be dependent on the ambient temperature) such that increasing the exposure temperature will alter or degrade the performance of the constituent material. The proper use of additives can help alleviate the mostly negative effect that increasing the exposure temperature will have on packaging materials, especially in the case of molding compounds.

For molding compound materials that are compatible with lead-free processing conditions, several opportunities present themselves for improvement of the base materials through the use of additives. For example, increasing (or altering) the filler content can reduce moisture uptake, increase the modulus, and reduce the CTE or minimize the dramatic CTE increase associated with passing through the T_g. Addition of second-phase modifiers can act to provide stress relief through reducing the modulus without significantly lowering the glass transition temperature. Also, inclusion of second-phase materials (beyond the primary filler material) can act to increase the toughness, which will be important in controlling the fracture resistance in the molding compound (an important parameter for reliable molding compounds

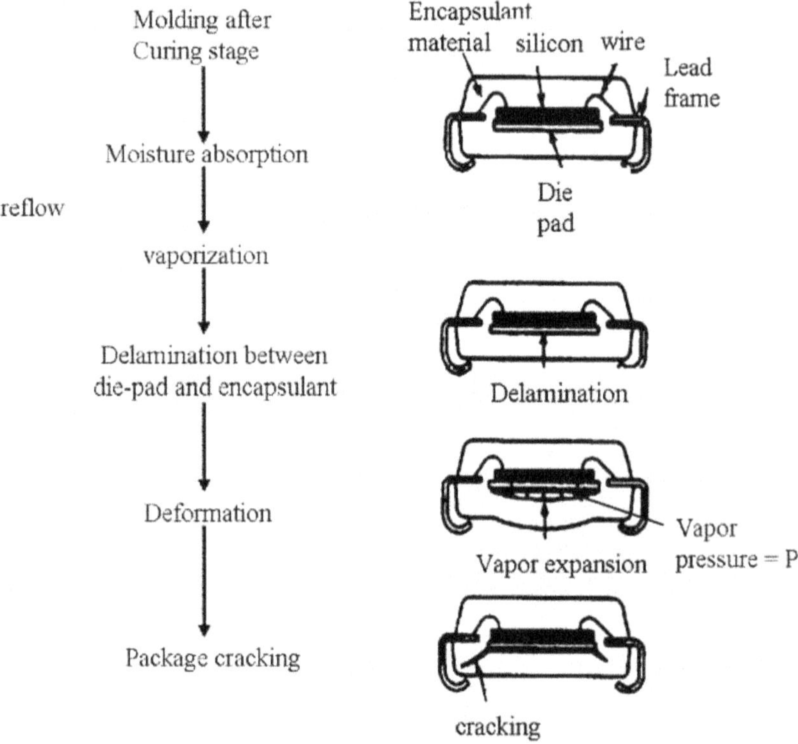

Figure 8.4 Schematic of classic popcorn failure mechanism during solder reflow. *(After Ref. 3.)*

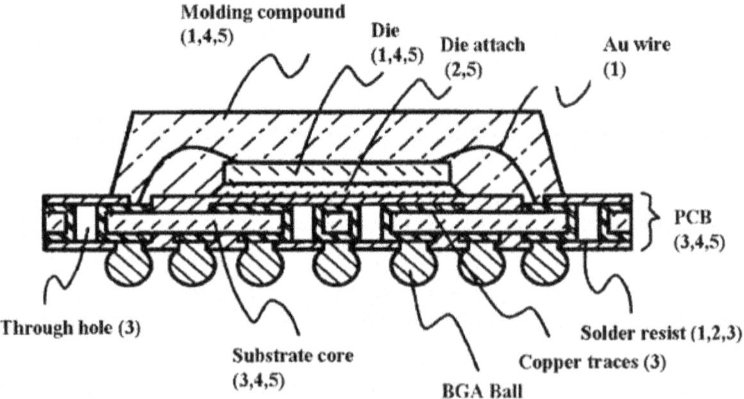

Figure 8.5 Effect of filler on the CTE and T_g of an epoxy resin (filler is spherical amorphous silica). The numbers next to each package component represent locations of the failure modes listed in Table 8.4. *(After Ref. 5.)*

subject to lead-free solder reflow temperatures in the presence of the popcorn failure mode).

Additionally, proper selection of the base resin chemistry can be even more important to the final performance of the compound. The base resin chemistry can control adhesion, and it plays a substantial role in the T_g of the system, modulus, CTE, moisture diffusivity, and many other physical-mechanical properties. As mentioned above, formulators must constantly balance the need to maintain good processability while providing the best possible reliability performance.

The following sections will briefly touch on the specific roles of common additives used in molding compounds and how the performance of molding compounds is influenced by each additive. Also, the means by which new materials can result from specific combinations of these additives (as well as new material formulations) is discussed.

8.2.2.3 Filler materials. The role that filler materials play in a molding compound is usually one or more of three distinct types. Fillers are present either to modify the physical properties of the material, to pigment it, or to reduce the overall use of resins to keep costs down. This discussion will focus on the first role. Many mechanical properties of the bulk molding compound used in electronic packaging are largely controlled by the filler content in each formulation (recalling that, according to Table 8.2, most of the material mass of a molding compound is actually filler). Most fillers used in epoxy molding compounds consist of amorphous silica with a carefully controlled particle shape and size distribution. Modifications to the material type, shape, and particle distributions can affect the end properties and processability of the materials. To understand the role of the filler particles in important physical properties of the molding compound, a review of several molding compound properties important to package reliability, especially in the presence of high reflow temperatures, is presented here.

Unfilled epoxy resins commonly used for encapsulation (referring to Table 8.3) cure into solids that typically have CTEs below the glass transition in the 50 to 100pp/°C range and have CTEs above T_g in the range of 100 to 250ppm/°C. Because the stress imposed on dissimilar materials that are adhesively bonded together will be depend on the moduli of the two materials, the differences in their CTEs, and the temperature range which to which they are subjected (provided they remain bonded), the CTEs of raw epoxies are much too high to be used unmodified in plastic packaging. Adding low-expansion fillers to the molding compound will act to reduce the volumetric expansion or contraction associated with temperature changes, provided good dispersion and adhesion to the fillers are maintained. Table 8.5 lists the

Table 8.5 Effect of Filler Content on the CTE and T_g of an Epoxy Resin (Filler is Spherical Amorphous Silica)

Filler loading (wt%)	TMA T_g^* (°C)	CTE (25–200°C) ppm/°C
60	129	34.5
65	127	28.8
70	132	24.1
75	130	20
80	Unprocessable	

*Measured using thermal-mechanical analysis.

effect of the filler content on CTE and T_g in an epoxy resin.[5] It is clear that a drastic reduction in CTE can be obtained by incremental changes in the filler content without changing the T_g of the system to any degree at all. The filler effect on the final cured properties is also portrayed graphically in Fig. 8.6. Controlling the CTE through the use of filler materials will reduce the overall stress on the package during temperature excursions and result in a more reliable assembly.

In a similar manner as the CTE, the modulus of a polymer system can be controlled through the addition of an inorganic filler. The modulus is a measurement of the stiffness and, in many senses, the load-bearing capacity of the compound. In many "low stress" or stress-sensitive systems, the modulus needs to be carefully controlled because, to a first order, the modulus will play nearly as important a role in determining the stress levels in the package as the CTE. In situations where large temperature swings are observed, as in the case of a lead-free reflow, stress levels most certainly need to be controlled to minimize failures (most of type 5, described in Table 8.4) during board attach. Modern molding compounds use a combination of carefully selected resin chemistry, filler type and content, and other modifiers (discussed below) to control the modulus and thus the stress in the package. Figures 8.6 and 8.7 show how filler content can affect the modulus of an epoxy resin as well as how this can, in turn, affect the flowability of the liquid resin prior to cure.[6] This is further evidence that a balance must be struck between mechanical properties and processability.

Because the encapsulant is designed to prevent moisture ingress into the chip, and since moisture content will limit the effectiveness of the encapsulant in its role to protect the interconnects through reflow and handling, it is desired that the solubility and moisture diffusivity

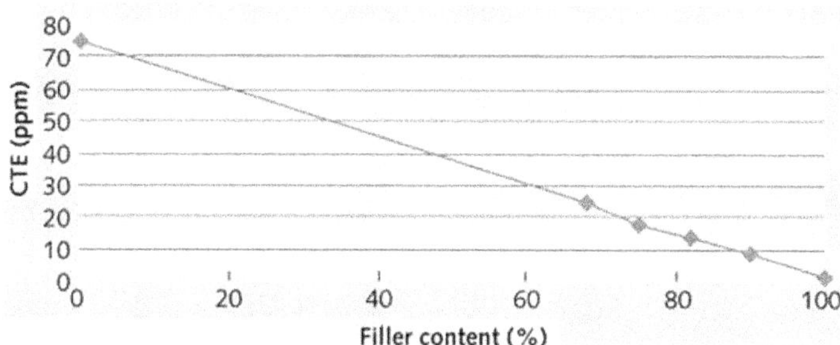

Figure 8.6 Effect of filler content on the CTE of a typical epoxy molding compound (CTE as a function of filler content, by weight). *(After Ref. 6.)*

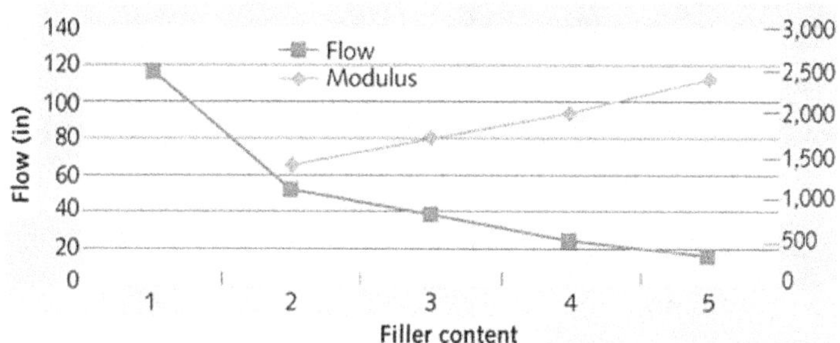

Figure 8.7 Effect of filler content on the modulus and flow of a typical epoxy molding compound. *(After Ref. 6.)*

of the molding compound be relatively low. Once again, unfilled resins would not be suitable in this role. The moisture diffusivity of a filled epoxy system can be up to 10 times lower than for the same volume of pure epoxy (depending on epoxy/curing agent combination, ambient conditions, and filler content). Likewise, the equilibrium moisture content can vary tremendously with filler content.[7] Figure 8.8 shows the effect that filler content can have on moisture diffusivity and moisture uptake percentage.

The driving force for high-temperature reflow failures in many cases will be linked to the amount of moisture content available at the interface during reflow (or how much moisture is contained in the polymer matrix). Therefore, reducing the permeability to and solubility of moisture in the epoxy encapsulant is desired for new materials destined for use in lead-free applications. Clearly, for a given volume of polymeric material, the addition of inorganic fillers, which are relatively impermeable to moisture, will act to reduce the overall weight gain of the composite polymer/filler bulk material. Additionally, the presence of the filler will increase the thermal stability and conductivity of the

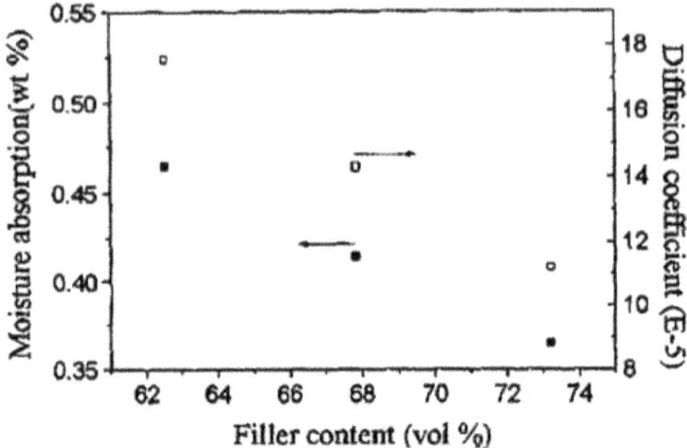

Figure 8.8 Effect of filler content on the moisture diffusion coefficient and moisture uptake in molding compounds (filler is a fused amorphous silica). *(After Ref. 7.)*

base resin, providing further areas of improvements and control to arrive at better materials for high-temperature reflow applications.

8.2.2.4 Stress reduction modifiers and tougheners. Filler content plays a large role in controlling the mechanical properties of the final molded material. However, other nonreactive/reactive second-phase additives can alter the final properties as well. For example, addition of second-phase low-modulus particles (such as silicone rubber) can act to toughen the matrix as well as reduce the overall stress on the system by flexibilizing the material. These second phases can consist of discrete inclusions in the system that simply act to alter the global material properties by expanding and contracting locally to reduce the local load transfer. Reactive second-phase materials play a greater role in modifying the mechanical properties in that they modify the mechanical properties of the overall matrix, in many cases by modifying the fundamental averaged physical properties of the continuous phase (the epoxy molding compound). Often, the increases in toughness or reductions of stress imparted by the molding compound on the system can be realized without sacrificing overall desired physical properties of the compound. It has been shown that reductions in stress and increases in toughness can be obtained without significant reductions of material strength or T_g.[8,9] This phenomenon is true mostly because of the very small amounts of second-phase materials that are required to obtain the desired effect.

Second-phase materials can act to toughen the bulk matrix by several methods. Much research over the years has been dedicated to the science of toughening epoxy systems.[8,10,11] Pure cured epoxy systems tend to be quite brittle. By simply increasing the strain-to-rupture properties of the bulk material through the inclusion of a lower-modulus, tougher material, an overall toughness increase of the system can be realized (a "strength of material" approach). This effect will depend highly on the level of interaction between the second-phase material and the continuous matrix (i.e., in this mode, a reactive second-phase material will tend to have a larger effect than an unreactive one). Toughening can also be realized by the effects that a second-phase material can have once a crack has been initiated. In this crack-resistance mode, the second-phase material acts to toughen the system via the interaction of the crack tip and the inclusion. In an energy balance approach, strain energy in the system can dissipate through crack extension. Interaction of the crack tip with the second-phase material (i.e., diverting around it, proceeding through the material, and ripping it apart or dilating it) acts as a energy-consumptive mechanism and results in a tougher system. In the wake of a crack tip passing through a tougher second-phase material (or even a stronger material used to toughen the system), the act of further hindering the opening of the crack (i.e., crack closure) can toughen the system. Many different materials can be included in epoxy matrices to impart this toughness. Because a rigid second phase is already present in most molding compounds in the filler system used, this discussion focuses on second-phase inclusions that tend to be lower modulus and can act to reduce stress and toughen the system at the same time.

Reducing the stress created by curing the molding compound, as well as the stress generated due to silicon/substrate (lead frame/molding compound), will act to increase the reliability performance (especially to thermal/power cycling excursions) of the molding compound. Increasing the toughness of the bulk material will reduce the tendency to fail as a result of crack propagation in the molding compound and subsequent cracking of metal conductors. This will be even more true in the presence of higher-temperature excursions, where the driving force for crack growth and stress evolution is higher. The presence of second-phase stress reducers can also reduce warpage of the device resulting from the modified overall physical properties of the bulk material.

An example of the effect of addition of a second phase is found in Ref. 9 (Nakamura), where a second phase of unreactive polybutylacrylate (PBA, T_g = –54°C, modulus ~ 0.001 GPa) was added at various levels to a molding compound matrix to determine the effect of the "internal stress" of the system. Internal stress is a common technique used by mold compound suppliers to rank materials. The stress is calculated in the material based on the temperature-dependent material modulus and CTE cooling from the reference temperature (in this case, the cure temperature, 175°C) to the lowest use temperature, in this case –50°C. The stress is then calculated from the following relationship:

$$\sigma = \int_{-50}^{175} E(t) \cdot \alpha(t)\,dt \tag{8.1}$$

Figure 8.9 shows how the internal stress of a molding compound can vary with inclusion of small amounts of second-phase low-modulus material. Referring to Fig. 8.9, it can be concluded that small domain sizes of the second phase are preferred and that a decent effect can be obtained with small concentrations of the additive, all of which would help maintain the processability of the material. Commercial molding compounds have incorporated second-phase low-modulus modifiers for stress reduction with much success. Formulators have been successful in reducing the total internal stress [as calculated via Eq. (8.1)] by reducing the overall system modulus while maintaining the high desired T_g (for thermal stability), thus improving product reliability. Figure 8.10 shows an example of one such commercial application via detailing the effect of including a second-phase silicone rubber modifier in the matrix (shown in arbitrarily increasing amounts) of a commer-

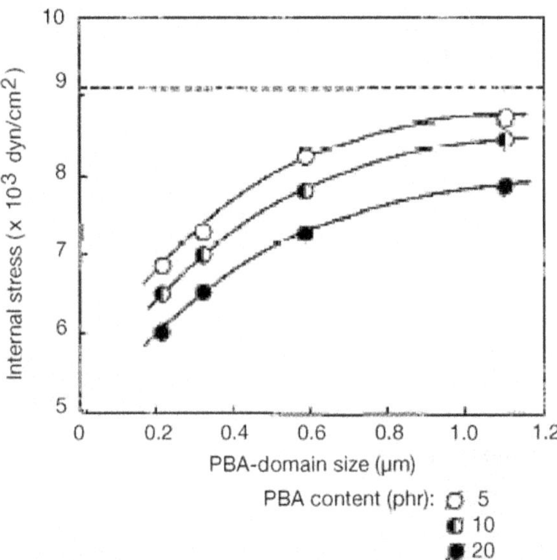

Figure 8.9 Effect on thermoset internal stress [as calculated by Eq. (8.1)] as a function of unreacted second-phase polybutylacrylate content and domain size. *(After Ref. 9.)*

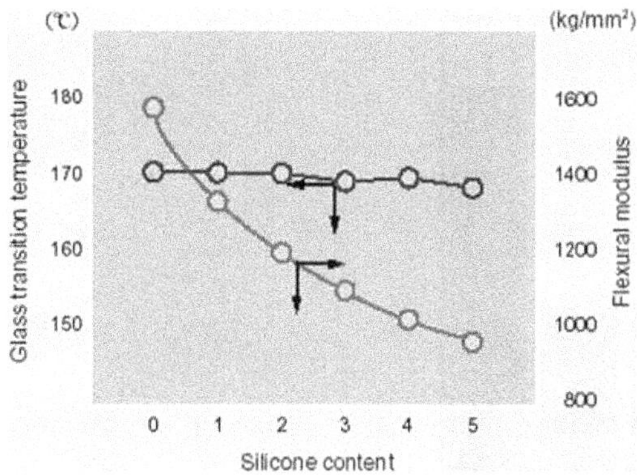

Figure 8.10 Effect of the T_g and flexural modulus of a commercially available molding compound as a function of silicone rubber content. *(Courtesy of Shin Etsu Corporation, KMC product data sheet.)*

cially available molding compound. It is clear that dramatic changes to the molding compound physical properties can be imparted with careful attention to second-phase modifier inclusion.

Tougheners can be included in the same fashion as stress reducers. An example of how toughness can be imparted on a thermoset matrix by the inclusion of a second-phase material is given in Ref. 8. Here, the authors modified an anhydride-cured bisphenol A/bisphenol F mixture (1:1 ratio) with a fixed amount (5 pph) of various reactive siloxane (silicone prepolymer, $T_g = -120°C$, E = 0.0008 GPa) modifiers. Table 8.6 details the various toughener formulations used in this study. Figure 8.11 shows the results from this work, detailed in terms of toughness as measured by K_{Ic} values obtained through a single

Table 8.6 Formulations from Ref. 8 for Toughening Epoxy Matrices

Formulation table	Siloxane	Reactive group	Equivalent weight to epoxy or amine	Average siloxane domain size, μm
S00	Nne	N/A	N/A	N/A
S01	X01	Amine	1800	~0.5
S02	X02	Epoxy	9000	~0.5
S03	X03	Epoxy	3000	>10
S04	X04	Epoxy	4000	>10
S05	X05	Amine	600	0.1–0.2

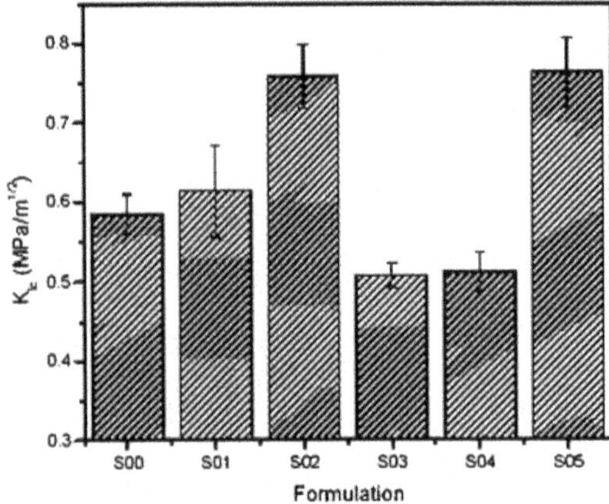

Figure 8.11 K_{Ic} values for the various siloxane-toughened Bis A/Bis F resins cured with an anhydride curing agent (formulations from Table 8.7) compared to pure anhydride-cured resin. *(After Ref. 8.)*

edge notch testing method. All materials were mixed in the epoxy and cured, at which time the siloxane regions phase separated and formed discrete second phases of sizes that depended on the miscibility of the formulation in the material. For amine-terminated formulations, a very fine second-phase particle size was obtained that was in some sense interfacially compatible with the bulk resin but had poor enough miscibility to form small domains with ample mixing. From the results summarized in Table 8.6 and Fig. 8.9, it is clear that including a second-phase rubbery material, of appropriate size and level of interaction, will act to toughen an epoxy matrix and can be used to some degree to improve the performance of molding compounds for aggressive applications. Figure 8.12 shows that, while toughness improvements can be realized with the inclusion a small amounts of a second phase material, this does not have to happen at the expense of T_g and CTE, which should remain unchanged to retain the performance of the starting formulation. Figure 8.12 shows that toughness can indeed be obtained with only very small changes to these two important physical properties.

In both cases where second-phase materials are included to modify the fundamental material properties of the cured system, care must be taken so that, in gaining performance in one important area, performance in another region does not diminish. Including low-modulus materials to reduce stress and increase toughness is a very attractive means to realize performance goals. However, these materials can possess much higher moisture uptake percentages or moisture diffusivity,

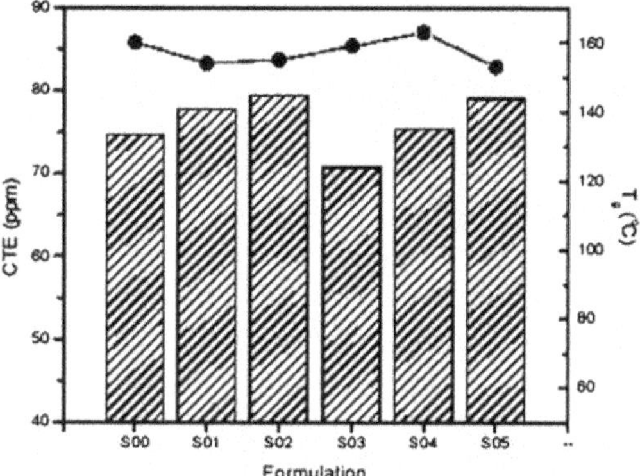

Figure 8.12 CTE and T_g values for the various siloxane-toughened Bis A/Bis F resins cured with an anhydride curing agent (formulations from Table 8.7) compared to pure anhydride-cured resin. *(After Ref. 8.)*

thus allowing more moisture to reach the sensitive silicon surface. Subsequent high-temperature reflow events can create delaminations, even in tougher, lower-stress materials, due simply to the increase in moisture content at the interface. This phenomenon is, of course, application dependent and may not apply to all low-modulus modifiers. Care must be exercised in any case. Additionally, the second-phase regions may exclude silica (or other) filler materials, leading to the inability to realize the moisture uptake, reducing and mechanical property benefits of the filler addition to the bulk. They may also increase the thermal resistance due to the exclusion phenomenon in the second-phase regions.

8.2.2.5 Backbone chemistry modifications. In the same manner that low-modulus tougheners were included in the matrix as discrete phases to modify the properties of the bulk material, modifications to the fundamental chemistry of the constituent materials used in the formation of the bulk can also be realized. In this case, realizing lower-stress, tougher, or otherwise different materials can be achieved via the modification of the resin or curing agent backbone chemistry. By this it is meant that changes to the nature of the raw materials used in the formulation are accomplished by introducing specific changes to the monomer (the basic unit in a polymer structure) to achieve the desired results. For example, to realize stress reduction, a flexible unit on the backbone structure of the polymer can be introduced. If increased stiffness or higher thermal stability is the goal, the appropriate substitutions can be made to the chemical structure to some degree. Imparting changes at the molecular level rather than at the formulation level (macroscopic) has many advantages. The properties will be more homogeneous and will not suffer from the drawbacks presented by filler exclusion and nonuniform physical properties due to the discrete nature of a second phase modifier. However, while it is possible to lower the stress and toughen a bulk material using both backbone changes and second-phase modifiers, it is very it is difficult to impart these changes with a backbone modification without altering other mechanical properties that should remain fixed. In the previous section, it was demonstrated that one could toughen and reduce stress in a material without changing important properties, such as CTE and T_g with the inclusion of a second-phase modifier. When altering the basic chemical structure of the constituent materials in a molding compound, it is difficult to retain all of the desired properties of the original resin. To combat this effect, often mixtures of modified resins and altered filler amounts and materials are used.

8.2.2.6 Adhesion promoters and coupling agents. In recent years, the use of functionalized surfaces to improve polymer adhesion to, and wetting of, inorganic surfaces has drastically changed the role of the interface in polymer composites. Polymer/inorganic interfaces have gone from a highly sensitive, volatile associations, especially in the presence of water, to integral sources of toughness. The microelectronics industry has benefited from the advances made in controllable surface chemistry in several areas, including photoresist wetting on silicon, underfill compatibility with substrates, and chip passivation as well as interface resistance to moisture damage. This is especially true in the case of molding compounds filled with organic materials. The most common way to modify a surface containing an appreciable density of surface hydroxyls (OH groups), as in the case of minerals and oxidized metals (fillers), is to use silane coupling agents. The diversity of silane chemistry is such that many different surface properties can result from the organofunctionality of a molecule that covalently reacts with the target surface (substrate).

Figure 8.13 General structure of an organic silane.

Silanes have the general structure shown in Fig. 8.13, where R represents an organofunctional group, and X is generally alkoxy. The adhesion "promotion" is realized through the bridging of an organic species (an adhesive or molding compound) to an inorganic or metal surface through the chemical reaction (or compatibility) of the silane with the both the adhesive and the surface to be adhered to. Organosilanes can be terminated with many different functional groups that are either miscible (soluble or compatible) in the organic adhesive phase or reactive. Careful choice should be made of this end functionality. In many cases, this bridge is required to achieve any appreciable adhesion at all, or the bridge is more robust in the presence of higher moisture, temperature, or stress levels.

The most common way that surface treatment occurs to create the adhesive bridge is to chemisorb the silane from an aqueous or water-alcohol solution. Silanes usually require a critical amount of water in the solution to facilitate hydrolysis of the alkoxy silanes to silane triols. In some cases, alcohol is needed to promote solubility of a hydrophobic organofunctional group as well as wetting on the target surface during application. The reaction pathway for hydrolysis is given in Fig. 8.14 where, in the case of R = CH_3 (R′ here is the organofunctional group) for a trimethoxysilane, the by-product of hydrolysis is methanol. The kinetics of hydrolysis in the presence of at least a stoichiometric amount of water was studied, and it was determined that, to form the silanols to form, water must indeed be present. Once the formation of silanols occurs, these highly reactive groups are available to condense to siloxanes either with another silanol, known as *polysiloxane formation,* or with a surface hydroxyl. This reaction proceeds according to the depiction in Fig. 8.15, where R can represent another silanol molecule, or it can represent a mineral or metal oxide surface. R′ represents an alternate functionality that can readily condense with the silane triol.

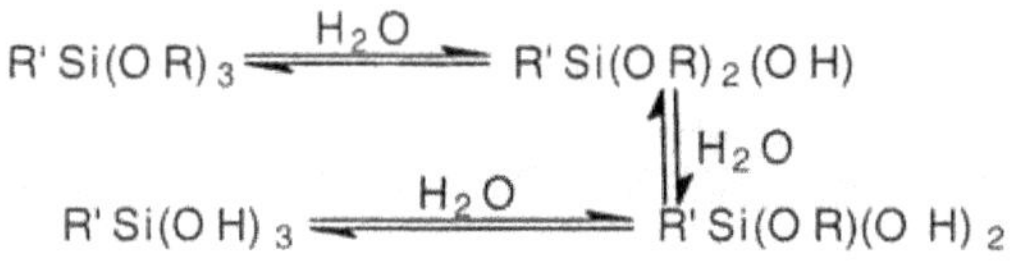

$$R'Si(OR)_3 \xrightleftharpoons{H_2O} R'Si(OR)_2(OH)$$

$$R'Si(OR)_2(OH) \xrightleftharpoons{H_2O} R'Si(OR)(OH)_2$$

$$R'Si(OH)_3 \xrightleftharpoons{H_2O} R'Si(OR)(OH)_2$$

Figure 8.14 Reaction pathway of the hydrolysis of an organic silane compound.

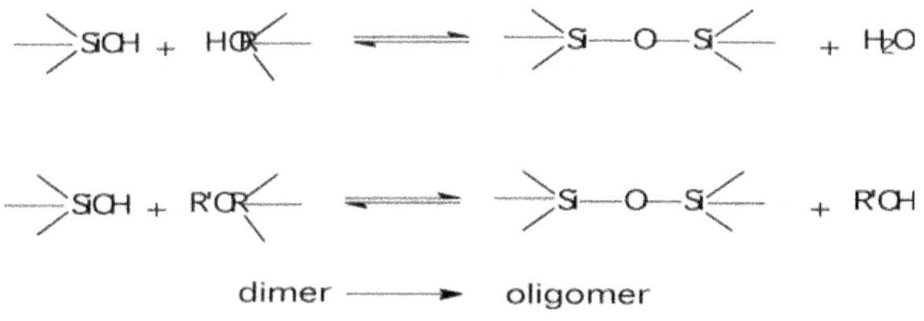

$$\equiv SiOH + HOR\equiv \;\rightleftharpoons\; \equiv Si{-}O{-}Si\equiv + H_2O$$

$$\equiv SiOH + R'OR\equiv \;\rightleftharpoons\; \equiv Si{-}O{-}Si\equiv + R'OH$$

dimer ⟶ oligomer

Figure 8.15 Condensation reaction of a hydrolyzed silane with the surface hydroxyls of an organic or mineral surface (filler in the case of a molding compound).

In many cases, it is impractical to complete a surface preparation in the real manufacture of IC packages. Therefore, coupling agents are commonly included in the molding compound mixtures in proprietary amounts to create coupling. Coupling or adhesion improvements to improve performance can be achieved in other manners as well. Surfaces of substrates, leadframes, and other interfaces (even within the substrate) can be treated to improve adhesion. In some cases, a combination chemical and mechanical approach is used to perform this function. In BGA substrates, copper conductors and polymer dielectric surfaces are often etched with acids to improve adhesion during lamination and plating steps.[12] This is mainly because the commonly held theory of bonding between polymers and plated metals relies strongly on the presence of mechanical interlocking.[13] Chemical etching will only improve this situation and will improve the adhesion of laminated layers in a similar fashion.

Copper has inherently weak adhesion to molding compounds and solder masks such that, beyond surface treatments to improve plated copper adhesion in BGA construction, copper leadframes and top metal layers in BGA substrates are excellent targets for adhesion promotion to improve performance. This is typically held to be true because of the formation of weak oxides on the surface of the exposed copper during production. One means of improving the adhesion of molding compound to copper (especially under high-temperature/high moisture conditions consistent with lead-free reflow) is to deposit a thin layer of Zn-Cr (commercially known as "A2") on the surface of the copper after fabrication (prior to significant oxide formation). The presence of the Zn-Cr layer acts to prevent further oxidation and provides a high-quality surface area for bonding, a result of the unique surface morphology of the Zn-Cr coating.

Another method to improve the copper/polymer interfaces in the package is to simply utilize aggressive etches to remove the oxide and subsequently roughen the surface to provide more surface area for bonding. Others range from utilizing thin organic surface treatments to prevent the formation of the oxide on the copper surfaces to depositing plated copper (electroplated or electroless) on the copper leadframe or runners to increase the surface area for adhesion.

8.2.3 Lead-Free-Compatible Die Attach

Failures in lead-free packages are not limited to the molding compound/package interface (or within the molding compound). In fact, in many cases, the die attach interface more susceptible to failures as the molding compound interface. While many of the advances to improve the adhesion and mechanical properties of the molding compound will apply directly to the die attach, many considerations must be made that are unique to this package assembly material, based entirely on its specialized role.

The role of the die attach in a package (plastic or otherwise) is to mechanically seat the silicon to the carrier (lead frame, substrate, or other) so that interconnect can occur (wirebonding and other lead-attach technologies) and so that the chip will stay positioned during molding. The die attach is also an important element in the heat dissipation path and acts to ground the backside of the die such that most die attach materials (if a polymeric adhesive or a low-melting-point glass—if the material is a metal, no filler is required) will incorporate a metal conductor (heat and electrical) in the matrix. Silver flakes have been used at an overwhelming rate due to the highly thermally and electrically conductive nature of silver as well as its relatively low corrosion and cost attributes.

As die sizes have increased over the years, the role of the die attach has not changed. However, the properties of the materials needed to change so as to provide reliable solutions. Materials with very high processing temperatures and moduli (e.g., Au/Si eutectics, silver glasses, and solders), which can result in extremely high stresses in the package assembly (especially for large die) needed to be replaced with reliable, lower-stress materials. In the early days of hermetic packaging, the relatively high amounts of outgassing of polymer-based adhesives (which compromised hermeticity) prevented their widespread use. With the advent of plastic packaging and more advanced materials for die attach (including extremely low-outgassing alternatives), polymer die attach is now the norm. This discussion will be limited to this technology.

8.2.3.1 Polymer die attach structure and properties. Typically, polymer-based die attaches are formulated from similar starting structures (as in the case of molding compounds (for epoxies). This is shown in Table 8.3 for epoxy molding compounds. However, due to the number of formulations available in the industry, and in the interest of space, the specific structures of common die attaches are not listed. Table 8.7 does, however, list the common types and important physical property attributes for polymer-based die-attach adhesives. For the most part, all modern die attaches consist of low-ionic, high-adhesion-strength materials that must flow well under applied pressure (usually from a syringe) and must retain their dispensed shape when the pressure is removed (thixotropic). The resin must also cure void free, with minimal shrinkage and no bleed-out. To facilitate subsequent wirebonding steps, the material must not outgas to any significant degree. Mois-

Table 8.7 Common Polymeric Die Attach Types

Material	Cure temp. (°C)	Typical T_g (°C)	Typical modulus (GPa) (@ 25°C)	Comments
Silver-filled polyimide	>300	>250	3–5	High-temperature compatible, fairly rigid, high moisture uptake, moderate adhesion
Ag/Au-filled epoxies	25–175	20–160	0.2–9	Wide variety of properties based on modifiers and fundamental resin chemistry, low–moderate moisture uptake, range of solids content, good processability
Ag/Au-filled cyanate esters	80–300	140–240	8–10	Good adhesion, excellent thermal stability, 100% solids, low outgassing, fast cure, high stress (unless modified or mixed), moderate moisture uptake
Ag/Au-filled bismaleimide	<100	–90	<2.0	100% solids, fast cure, hydrophobic (low moisture uptake), low stress, thermally stable
Ag/Au-filled modified cyclo-olefin thermoset	100–175	~75	0, 1–1	Moderate adhesion, very low moisture uptake, low stress, low outgassing, 100% solids
Ag/Au-filled siloxane	~150	<90	<1.0	Low moisture uptake, thermally stable, low stress, good adhesion
Ag/Au-filled thermoplastic	N/A	Variable	<5	Reworkable, lower adhesion, moderate moisture uptake

ture uptake and diffusivity must be kept to a minimum, in the same manner as molding compounds. The important physical property features that render each material unique to the application for which it is chosen for will lie in the CTE, T_g, and modulus. Most materials for large-die applications (and any low-stress application) will tend to have a low room-temperature modulus that increases only moderately at low temperatures. The T_g for large-die materials will tend to be in the 30 to 80°C range, also to keep stresses low. In fact, much attention has been paid to determining the correct material properties for new low-stress die attaches.[14] This analysis will prove particularly useful when evaluating materials for high-temperature applications. In Ref. 14, a trilayer structure, shown in Fig. 8.16, was analyzed to reveal critical properties to monitor (called *figures of merit* or *FOMs*) when choosing die-attach materials. This analysis was based on a bending

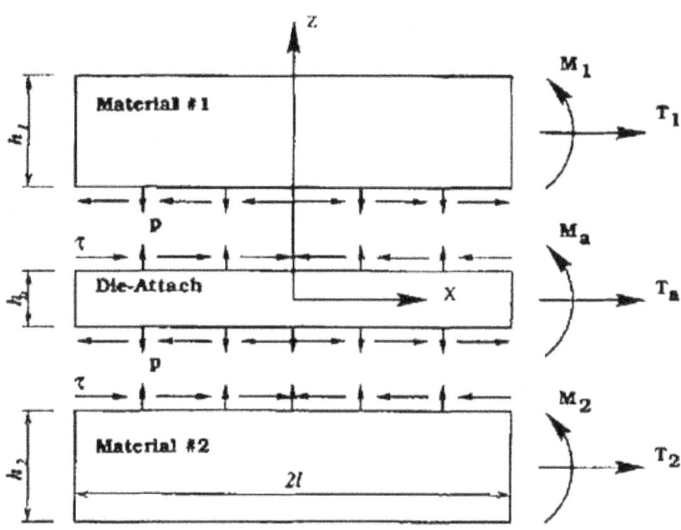

Figure 8.16 Free body diagram of a tri-material structure used to evaluate stresses in a silicon/die attach/substrate (leadframe) structure. *(After Ref. 14.)*

beam approach that aimed to evaluate the role of various materials on the performance of the material in an IC package during stressing. These authors showed that, for both axial and shear stresses arising from thermal excursions (such as in the case of reflowing a microelectronic device), the FOMs (which are rooted in the maximum stresses possible in a particular structure under thermal loading [i.e., stress at the corner of a chip]) depended to a large degree on only the shear and tensile modulus and the adhesive strength (empirically gathered). These relationships are given in Eqs. (8.2) and (8.3), respectively.

$$FOM_{shear} = \frac{G_a^{1/2}}{\tau_{adh}} \tag{8.2}$$

$$FOM_{axial} = \frac{E_a^{1/2}}{\sigma_{adh}} \tag{8.3}$$

where G_a and E_a are the shear and tensile moduli (for the adhesive) respectively, and τ_{adh} and σ_{adh} are the shear and axial adhesive strengths for the adhesive/substrate (lead frame)/die system, respectively. Understanding that a smaller FOM is better for assembly performance, it is clear that low moduli and high adhesion values are desired for best performance. Since the FOMs are rooted in maximum stress values attained during thermal loading, although not explicit in Eqs. (8.2) and (8.3), ΔT is also a strong factor for determining FOMs, in the same manner that internal stress was calculated in Eq. (8.1). Through ΔT considerations, material T_g values and temperature-dependent material properties will come in to play (i.e., the temperature range used for loading a structure to determine the FOM will change the FOM in accordance with the changing properties of the elements of Fig. 8.1).

8.2.3.2 Structure and properties for lead-free processing conditions. Using the FOMs, one can postulate that materials that reduce stress, either through lower T_g values or lower moduli (and possibly through lower curing temperatures to take advantage of the ΔT effect) would be preferred such that, in the presence of a high-stress environment (e.g., a high-temperature reflow scenario), the stress is minimized. In fact, lead-free materials do tend to favor the lower-stress materials outlined in Table 8.7. However, simply lowering the stress associated with the increase in the thermal loading range may not be enough. Recalling that the FOMs from the previous section are strongly dependent on the adhesive strength of the material/interface of interest, elements that will control this factor become extremely important. It is very important to consider the effect that the environment will have on the adhesive strength for the particular material system. High temperatures and the presence of moisture or increased strain (resulting from the expansion of the structure at the lead-free reflow temperature, as some materials are very weak at this temperature) must be considered in determining the correct material to use. Clearly, low-stress materials with good high-temperature adhesion, high-temperature stability and strength, and low moisture uptake would be preferred. However, few pure materials display all of these attributes. In fact, newer materials specifically designed for lead-free applications are increasingly becoming hybrid combinations (mixed formulations or even modified materials at the backbone level as described for molding compounds in Section 8.2.2.2) of materials found in Table 8.7 to satisfy the aggressive needs of this extreme application.

8.3 Concluding Remarks

Clearly, IC packaging for lead-free applications will need to adapt to the challenging conditions that will be found during the assembly of these devices. The adaptations that will be required are extremely broad and cannot be captured completely within the scope of this text. However, through the examples of changes required within two key IC package assembly materials for lead-free applications, the extension to other components of the device should be made easier. Materials and interfaces will continue to evolve and adapt to the conditions presented to them. This fact is a key element in developing the next generation of failure-free devices, no matter what the application or the intrinsic weakness of a particular package technology. Hopefully, this topical treatment of how molding compound and die attach material properties are selected and/or altered for lead-free applications will lead to an understanding of how important materials performance can be to the integrity of the device. Creating failure-free IC packages is certainly rooted in good practices to identify a defect and formulate a corrective action. However, adopting a proactive approach in designing failure-free IC packages should begin with properly selecting and tuning the materials used so as to prevent the failures in the first place.

8.4 References

1. Directive 2002/95/EC of the European Parliament and of the Council on the restriction of the use of certain hazardous substances in electrical and electronic equipment, *The official Journal of the European Union,* 27 January 2003.
2. Fujii, A., Andoh, M., Yamamoto, I., Ibuki, H., Uchida, K. and Yoshizumi, A., "New Epoxy Molding Compounds for SMT with Pre-Plated Lead Frame System," *Proc. IEEE/CPMT International Electronics Manufacturing Technology Symposium,* 1998, p. 478.
3. Mogi, N., and Yasuda, H., "Development of High-Reliability Epoxy Molding Compounds for Surface Mount Devices," *Proc. 42nd Electronic Components and Technology Conference,* 1992, p. 1023.
4. Terashima, K., and Toyoda, T., "Investigation of the Popcorn Phenomenon in Overmolded Plastic Pad Array Carriers," *IEMT/IMC Proc.,* 1998, p. 195.
5. Konarski, M., "Effects of T_g and CTE on Semiconductor Encapsulants," Loctite Corporation Technical Paper.
6. Proctor, P., "Mold Compound—High Performance Requirements," *Advanced Packaging,* October, 2003.
7. Ko, M., Kim, M., Shin, D., Lim, I., Moon, M. and Park, Y., "The Effect of Filler on Molding Compounds and Their Moldability," *Proc. 47th Electronic Components and Technology Conference,* 1997, p. 108.
8. Moon, K., Fan, L. and Wong, C.P., "Study on the Effect of Toughening of No-Flow Underfill on Fillet Cracking," *Proc. 51st Electronic Components and Technology Conference,* 2001, p. 1023.
9. Nakamura, Y., Uenishi, S., Miki, K., Tabata, H., Kuwada, K., Suzuki, H., and Matsumoto, T., "New Profile of Ultra Low Stress Resin Encapsulants for Large Chip Semiconductor Devices," IEEE *Trans. Components, Hybrids and Manufacturing Technology,* Vol. CHMT-12, No. 4, 1987, p. 502.
10. Kinloch, A. J., and Young, R.J., "Fracture Behaviour of Polymers," *Applied Science,* 1983, p. 421.
11. Lee, J., and Yee, A. F., "Inorganic Particle Toughening II: Toughening Mechanisms of Glass Bead Filled Epoxies," *Polymer,* 2001, **42**, p. 589.
12. Chen, L., Crnic, M., Lai, Z., and Liu, J., "Process Development and Adhesion Behavior of Electroless Copper on Liquid Crystal Polymer (LCP) for Electronic Packaging Application," *IEEE Trans. Electronic Packaging Manufacturing,* Vol. 25, No. 4, 2002, p. 273.

13. Suzuki, M., Kawamoto, M., and Takahashi, A., *Polymer, Eng. Sci.*, Vol. 39, No. 2, 1999, pp. 321–326.
14. Gektin, V. and Bar-Cohen, A., "Mechanistic Figures of Merit for Die-Attach Materials," *Proc. Intersociety Conference on Thermal Phenomenon,* 1996, p. 306.

Index

G

H

Q

R

S

T

About the Editors

Charles Cohn is a distinguished member of the technical staff at Agere Systems, Allentown, Pennsylvania, where he is the lead resource on PCB technology, supporting the development of advanced organic PBGA substrates for wire bonded and flip chip IC interconnections. He has authored chapters in several McGraw-Hill electronic packaging handbooks and presented numerous papers on electronic packaging. He has been awarded 11 U.S. patents on IC packaging. Mr. Cohn holds B.S., M.S., and M.E. degrees in mechanical engineering from Columbia University.

Charles A. Harper is president of Technology Seminars, Inc., Lutherville, Maryland, an organization dedicated to the presentation of educational seminars on electronic packaging and materials. He has authored over a dozen well-known books in the field and is among the founders and past presidents of the International Microelectronics and Packaging Society. He is also series editor for the McGraw-Hill Electronic Packaging and Interconnection Series. Mr. Harper is a graduate of the Johns Hopkins University School of Engineering, where he also served as adjunct professor.

www.ingramcontent.com/pod-product-compliance
Lightning Source LLC
Chambersburg PA
CBHW050453190326
41458CB00005B/1270

* 9 7 8 0 0 7 1 4 3 4 8 4 3 *